International Library of Technology
361

Electric Railway Engineering

287 Illustrations

Prepared Under Supervision of
FRANCIS H. DOANE, A. M. B.
DIRECTOR, ELECTRICAL SCHOOLS, INTERNATIONAL CORRESPONDENCE SCHOOLS
IN COLLABORATION WITH
E. C. PARHAM, M. E.
ELECTRICAL ENGINEER, BROOKLYN RAPID TRANSIT SYSTEM, BROOKLYN, N. Y.

ELECTRICAL RAILWAY SYSTEMS
ELECTRIC RAILWAY LINE CONSTRUCTION
TRACK CONSTRUCTION
ELECTRIC RAILWAY CALCULATIONS
RAILWAY MOTORS
ELECTRIC-CAR EQUIPMENT
SPEED CONTROL

Published by
INTERNATIONAL TEXTBOOK COMPANY
SCRANTON, PA.
1927

Printed in U. S. A.

ISBN #978-1-935327-99-8 1-935327-99-2
WWW.PERISCOPEFILM.COM

PREFACE

The volumes of the International Library of Technology are made up of Instruction Papers, or Sections, comprising the various courses of instruction for students of the International Correspondence Schools. The original manuscripts are prepared by persons thoroughly qualified both technically and by experience to write with authority, and in many cases they are regularly employed elsewhere in practical work as experts. The manuscripts are then carefully edited to make them suitable for correspondence instruction. The Instruction Papers are written clearly and in the simplest language possible, so as to make them readily understood by all students. Necessary technical expressions are clearly explained when introduced.

The great majority of our students wish to prepare themselves for advancement in their vocations or to qualify for more congenial occupations. Usually they are employed and able to devote only a few hours a day to study. Therefore every effort must be made to give them practical and accurate information in clear and concise form and to make this information include all of the essentials but none of the non-essentials. To make the text clear, illustrations are used freely. These illustrations are especially made by our own Illustrating Department in order to adapt them fully to the requirements of the text.

In the table of contents that immediately follows are given the titles of the Sections included in this volume, and under each title are listed the main topics discussed.

INTERNATIONAL TEXTBOOK COMPANY

B

CONTENTS

NOTE.—This volume is made up of a number of separate Sections, the page numbers of which usually begin with 1. To enable the reader to distinguish between the different Sections, each one is designated by a number preceded by a Section mark (§), which appears at the top of each page, opposite the page number. In this list of contents, the Section number is given following the title of the Section, and under each title appears a full synopsis of the subjects treated. This table of contents will enable the reader to find readily any topic covered.

ELECTRIC-RAILWAY SYSTEMS, § 17

ELECTRIC-RAILWAY LINE CONSTRUCTION, § 18

TRACK CONSTRUCTION, § 19

ELECTRIC-RAILWAY CALCULATIONS, § 20

ELECTRIC RAILWAY CALCULATIONS
(*Continued*)

RAILWAY MOTORS, § 21

ELECTRIC-CAR EQUIPMENT, § 22

ELECTRIC-CAR EQUIPMENT—(*Continued*)

SPEED CONTROL, § 23

ELECTRIC RAILWAY SYSTEMS

INTRODUCTION

1. A number of railway systems have been developed in which electric motors installed on the cars are utilized to drive the cars and trains. Electric energy for the operation of these motors is generated at one or more main stations and, for the larger systems, is distributed through substations to conductors leading to the moving cars.

The larger number of electric cars are propelled by direct-current, or continuous-current, energy. In large systems alternating-current energy is usually supplied by the main stations to the substations, where it is converted by rotary converters or motor-generators into direct-current energy for the car motors.

Alternating-current motors are also used to propel the cars in some railway systems; in such cases, the rotary-converter substation is not required. Alternating-current energy is generated in the main station and utilized on the moving cars.

2. Electric railway systems are broadly classified according to the kind of electric energy supplied to the motors on the cars, as *direct-current* or *alternating-current systems*. Further classification is based on the method of current collection or supply used on the moving cars, as *trolley*, *third-rail*, *slot*, *storage-battery car*, and *gasoline-electric car systems*. The most generally used is the the direct-current, trolley system. The other systems mentioned have been adopted where other methods of current distribution and collection must be employed to meet operating conditions or comply with city laws.

DIRECT-CURRENT RAILWAY SYSTEMS

CURRENT-COLLECTION SYSTEMS

TROLLEY SYSTEM

3. The methods of supplying current to the cars usually depend on the local conditions affecting operation of the system. In most of the city and suburban roads, the trolley system is employed because of its reasonable cost of installation and the small liability of people or vehicles coming in contact with the *trolley wire*, which is a bare conductor sus-

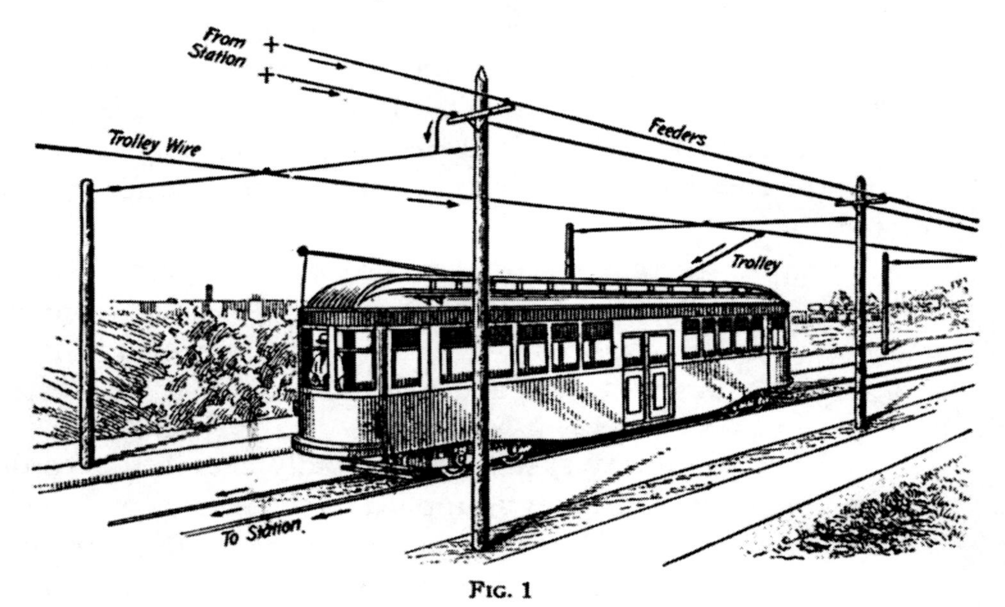

Fig. 1

pended over the center of the tracks. This wire is connected at intervals with large conductors, called *feeders*, which are often mounted on the poles supporting the trolley wire. The feeders are connected to the positive terminals of the main generators or of the rotary converters. Each car is provided with a device called a *trolley*. This consists of a pole mounted on the roof of the car and having at its upper end a wheel that runs in contact with the lower surface of the trolley wire. Current passes through the wheel and pole to the motors on the car and thence to the rails. The rails are connected through the earth or through copper conductors to the negative terminals of the generators or rotary converters. A complete electric circuit is thus formed.

4. Fig. 1 shows the more important current-collection features of a trolley system. All cars are in parallel between the trolley wire and the ground; therefore, the operation of any car is independent of all others that are in normal condition. An accident to one car cannot prevent the others from operating unless the line is short-circuited by the accident.

5. A trolley system using a single trolley wire is called a *ground-return system*, because the ground forms at least a part of the negative side of the circuit, the rails being in contact with the earth. In some early systems, two trolley wires placed side by side were used. One wire was connected to the positive bus-bars and the other wire to the negative bus-bars at the station. Such a system is a *metallic-return system*, because both sides of the circuit are metallic conductors; the rails and earth do not form a part of the circuit. Two trolleys on each car must be used and the overhead wiring in the streets is complicated.

THIRD-RAIL SYSTEM

6. The **third-rail system** is electrically the same as the trolley system. The trolley wire is replaced by a third rail, which is usually mounted on insulators to one side of the track and a little above the level of the track rails. The current passes to the car circuits through shoes that slide on the contact rail and returns to the station through the track rails. In a few cases, both positive and negative conductor rails have been used, the track rails in such cases not being used as a return circuit.

Fig. 2 shows the collecting-shoe arrangement for third-rail equipment. The third rail *a* is of standard **T** section and is supported on insulators *b* resting on every fifth tie, which is extended for this purpose. The link-suspended cast-iron shoe *c* has a limited vertical movement and the whole collecting device is mounted on a wooden beam *d* supported by the truck. A

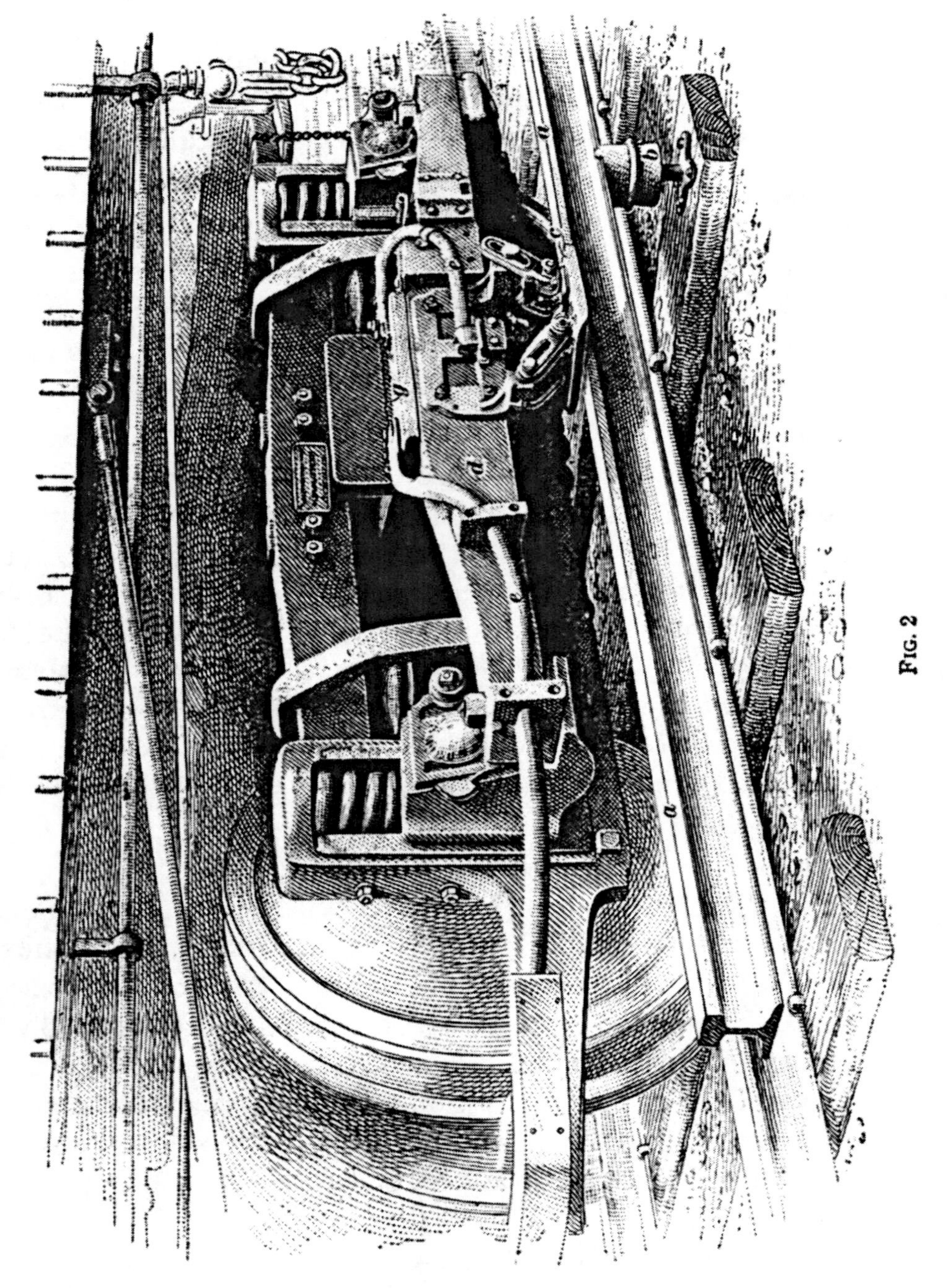

FIG. 2

cable *e* leads to the controlling devices on the car and connection is made to the shoe through a bare, flexible, copper cable *f* which is called the *shoe shunt* because it shunts the current around the link-pin connections, thereby preventing their becoming blistered by the current crossing their comparatively poor contacts. In some cases a *shoe fuse* is connected in the circuit at *g* to cut off the current in case of a short circuit; the fuse may be of either the enclosed or open type. The third-rail construction is much used for heavy-traffic, high-speed service over private right of way where the live rail is not a menace to persons and cattle. As the cross-section of the contact rail has a large copper equivalent, long stretches of track can be supplied with current without using feeders.

SLOT SYSTEM

7. The *slot system*, or *open-conduit system*, Figs. 3 and 4, is used only in large cities where heavy traffic warrants the great

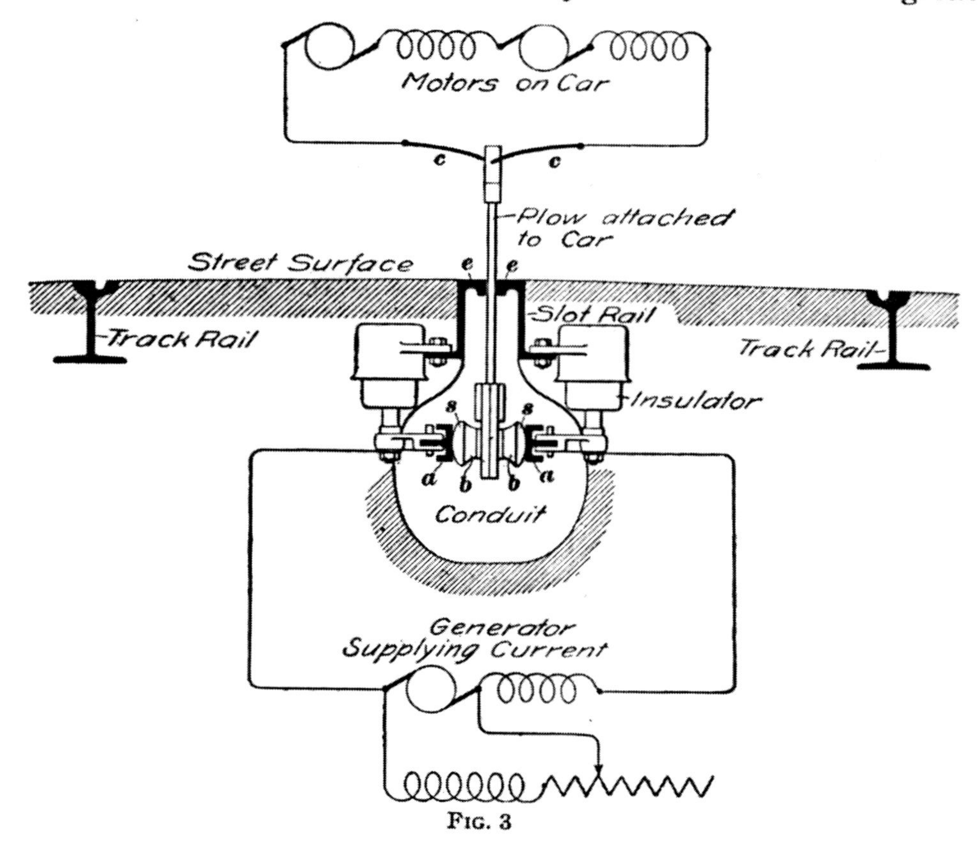

Fig. 3

expense of installation and city ordinances prohibit overhead trolley wires and feeders. Two conductor rails *a*, one positive and the other negative, are mounted in the conduit and are connected through feeders run in adjacent ducts to the positive and negative terminals of the generators at the station, thus forming a metallic-return system. At the top of the conduit is a $\frac{5}{8}$-inch slot between rails *e*, through which passes a plow, suspended from the car truck. Flat steel springs *b* press two cast-iron or soft-steel shoes *s*, mounted on the lower part of the plow, against the conductor rails. Flexible cables *c* extend through the body of the plow and connect through fuses, or shoe shunts, *d*, Fig. 4, to the plow shoes. The upper ends of cables *c*, Fig. 3, connect to the car circuit.

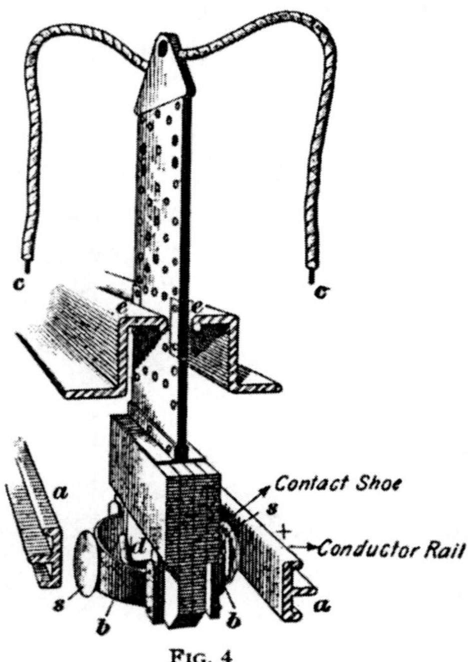

Fig. 4

ENERGY DISTRIBUTION SYSTEMS FOR DIRECT-CURRENT OPERATION

8. The term **energy distribution systems** as here used applies to the arrangement of conductors by which electric energy is transmitted from the main station, either directly or through substations to the trolley wire or third rail, from which the current for the car is taken. Systems using direct-current motors on the cars are here considered; of these there are two main divisions: (1) Direct-current generation and supply to the cars; (2) alternating-current generation and direct-current supply to the cars.

DIRECT-CURRENT GENERATION AND SUPPLY

9. 550-Volt System.—The simplest method of supplying energy to cars is by direct current transmitted from the generators in the main station to the trolley wire without any intervening transforming devices. The voltage between the positive and negative bus-bars is generally from 500 to 600 volts;

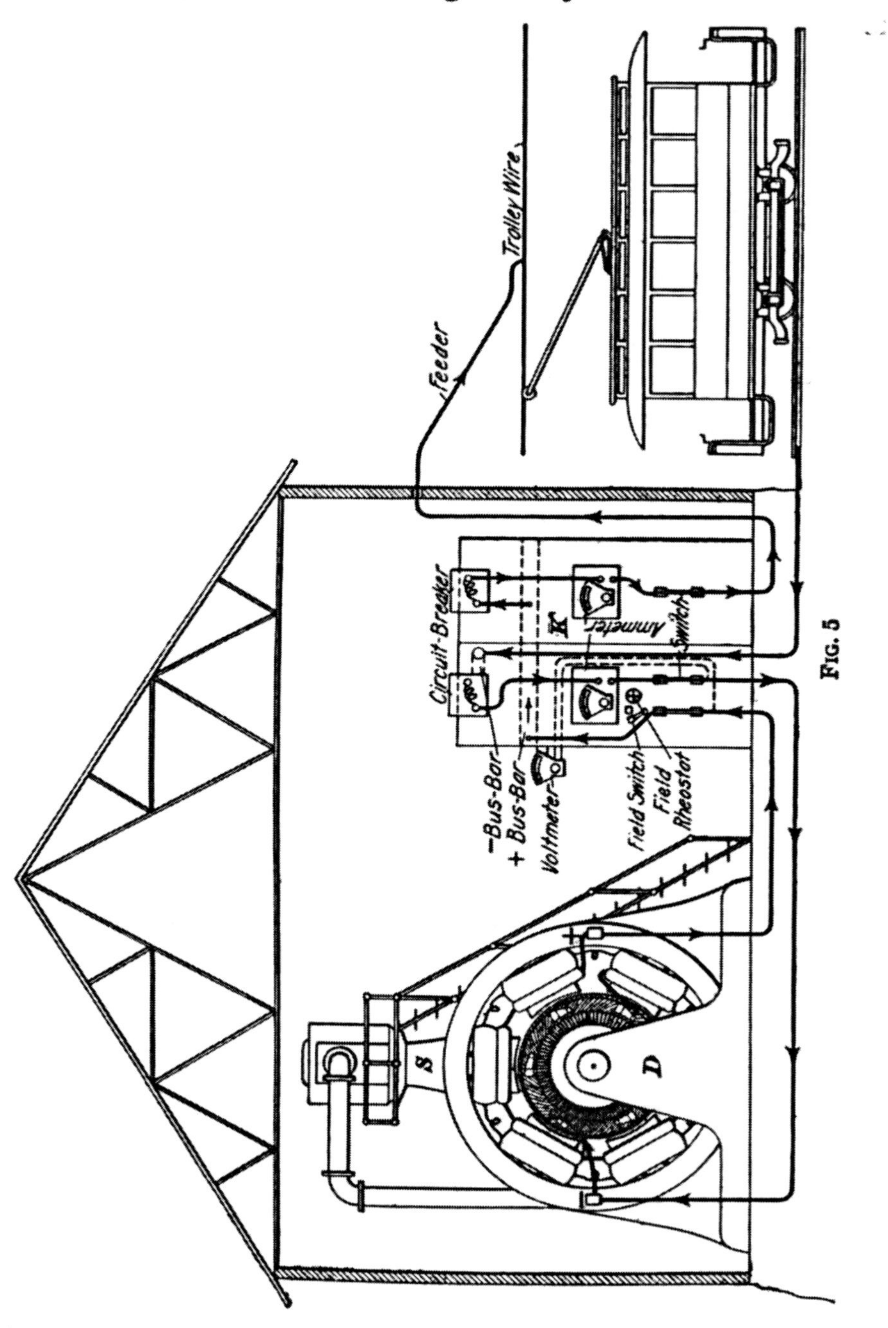

Fig. 5

the latter voltage is the later development. This system is adapted to operation in sections of dense population where the distances are not too great; it is not adapted to economical transmission of large amounts of energy over long distances, because at low voltage the current required is very large and the line losses excessive. Efforts to decrease these losses by greatly increasing the copper result unprofitably.

10. Fig. 5 indicates the more important connections of the parts of a railway system of this simple type. An engine *S* drives a direct-current generator *D*, which is connected with the switchboard *K*; this is connected with the trolley and rails forming a current path as indicated by the arrowheads.

11. Each car requires a current proportional to the power necessary to operate it. The sum of the currents taken by the cars forms the load on the station. The total current is indicated by the ammeter connected to the negative bus-bar, as in Fig. 5.

In a very small system, the trolley wire, if only in one section, may be connected directly to the positive bus-bar without the use of feeder cables. In large systems, the trolley wire is usually in sections and is too small to carry all the current; feeder cables are then provided to connect the positive bus-bar directly with the more distant trolley wire sections.

12. Return cables connecting points on the track to the negative bus-bars are installed in some systems. The conductivity of the track-return portion of the circuit is thus improved and the liability of damage to iron pipes, due to the pipes forming part of the current-return system, is lessened.

13. The maximum distance it is economical to operate cars with direct-current energy generation at a voltage of from 500 to 600 depends on the character of the traffic and on the frequency of the service; the limiting distance has been often stated in round numbers, as 7 miles. Many roads operate over greater distances, but it is probable that operation would be improved if the voltage were raised. By using boosters in the station or storage batteries on distant parts of the line, the permissible limit of economical operation is appreciably increased. Most medium-sized roads are operated over a radius of 6 to 8 miles without using alternating-current energy.

14. High-Voltage Bus-Bars.—Usually the heavier traffic is confined to a comparatively small area, the traffic on lines running to remote points being light; by raising the voltage of the feeders running to these distant sections, normal voltage at the cars can be maintained. Assuming voltages of 550 and 650 to be available at the station, feeders to the near sections could be connected to the 550-volt bus-bars and the long feeders of distant sections to the 650-volt bus-bars, thereby providing a permissible extra voltage drop of 100 volts. The radius of operation can thus be extended 3 or 4 miles.

15. Fig. 6 shows the connections for both high- and low-voltage bus-bars. The negative terminals of compound-wound generators *1*, *2*, and *3* are connected to the ground bus, while each positive terminal is connected through a main switch, the shunt of ammeter *A*, and the circuit-breaker, to the middle of a single-pole double-throw switch. By means of this switch, the positive side of the generator can be connected to either the upper or the lower bus-bar.

Generators *1* and *2* are shown connected to the lower bus-bar and operating in parallel, their equalizing switches being closed. Generator *3* is shown connected to the upper bus-bar and its equalizer switch is open. Most standard railway generators will generate 650 volts on full series field; generator *3* is assumed to generate 650 volts and generators *1* and *2*, 550 volts. Any machine, however, can be connected to either bus-bar. The feeders have single-pole double-throw switches for connection to either bus-bar. If feeders *A* and *B* are supplying near sections or if their loads are light, they can be connected to the 550-volt bus-bar; feeder *C* on a distant or heavily loaded section can be thrown on to the 650-volt bus-bar. By this arrangement, a fairly uniform voltage can be maintained at the cars under widely varying load and transmission conditions. In Fig. 6, the generators are shown equalized on the negative side; this connection is the prevailing practice in railway plants. The main and equalizer switches are mounted side by side near the machine and the negative leads are carried directly to the rail bus-bar.

16. Installation of Boosters.—On sections distant from the station, overcompounding of the generators may be inadequate to compensate for the feeder drop; in such cases *boosters* may be used in connection with the main generators in order to obtain high voltage at the distant section of the road. A **booster** is a generator connected in series with the feeder or

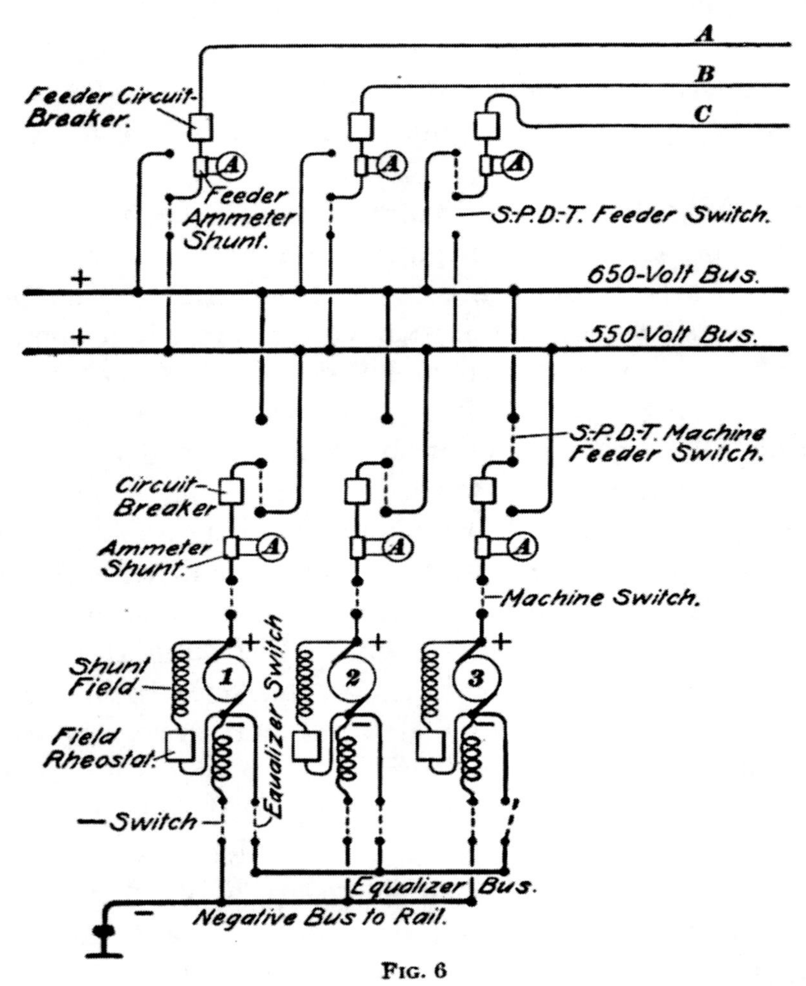

FIG. 6

feeders, the voltage of which is to be raised, so that the booster voltage is added to that of the main generators.

In Fig. 7, *a* represents the armature of a station generator, and *b*, the armature of a booster. Short feeders connect the positive bus-bar of the generator with near-by sections of the trolley wire; the circuits of long feeders include the booster, which in this case is assumed to generate 200 volts.

Railway boosters are generally of the *series type;* that is, the field winding is in series with the armature so that the voltage generated increases in proportion to the booster current. When the feeder load is light, the booster voltage is low because its field is weak; a heavy load increases the booster voltage and compensates for the drop in the boosted feeders. This automatic voltage regulation is indispensable, for were the booster to generate constant voltage regardless of the value of the feeder current, then at light loads with negligible line drop, the voltage applied to the cars would be excessive; this would be a tax on the motors and would cause continuous trouble with lights, heaters, and other auxiliary devices on the cars.

17. Street-railway boosters are driven at constant speed, generally by shunt-wound motors, though other constant-speed

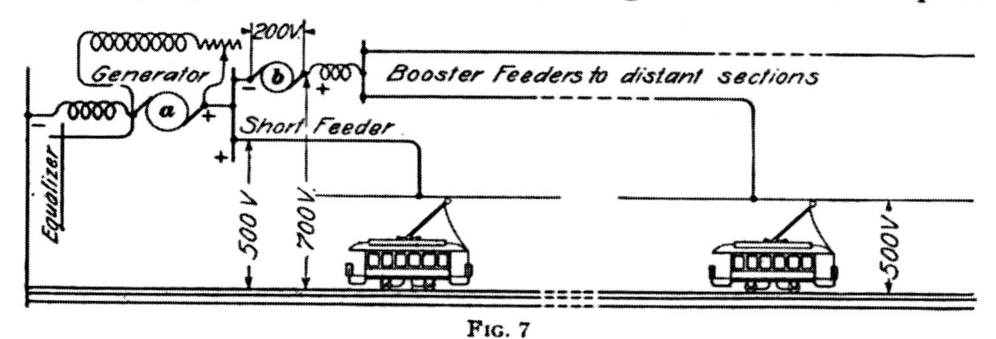

Fig. 7

drive can be used. Fig. 8 shows a booster set consisting of a shunt-wound motor *M* coupled to a series-wound booster *B*. Boosters do not differ in their general construction from standard generators except that their commutators may be larger owing to their comparatively large current rating. Recent types of booster are equipped with commutating poles to give perfect commutation without brush shifting.

18. Booster Rating.—Booster output in watts is the product of the booster amperes by the booster volts. If a booster carries 600 amperes and boosts this current 200 volts, the booster output is $600 \times 200 = 120,000$ watts or 120 kilowatts. A decrease of current to 300 amperes would lower the booster voltage to 100 if the voltage varies directly with the current, that is in a straight line, and the output in this case

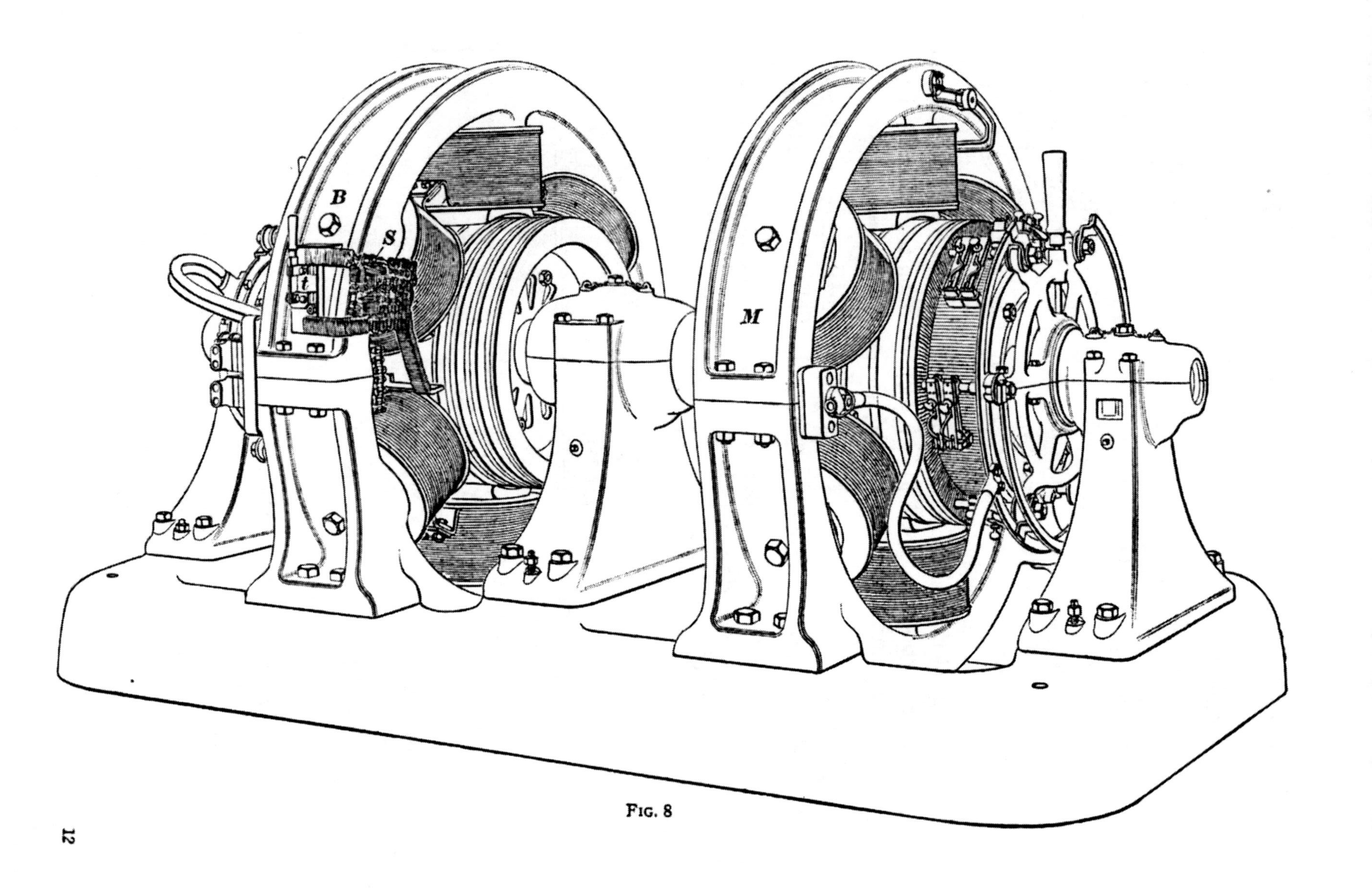

FIG. 8

would be $300 \times 100 = 30,000$ watts or 30 kilowatts. At zero booster current, the booster output would be zero, but its voltage would be 25 or 30 volts due to residual magnetism. Booster voltage rarely varies in an absolutely straight line, and the amount by which the variation departs from the straight-line variation is important, because too much departure will result in trouble with lamps and other devices designed for rated voltage.

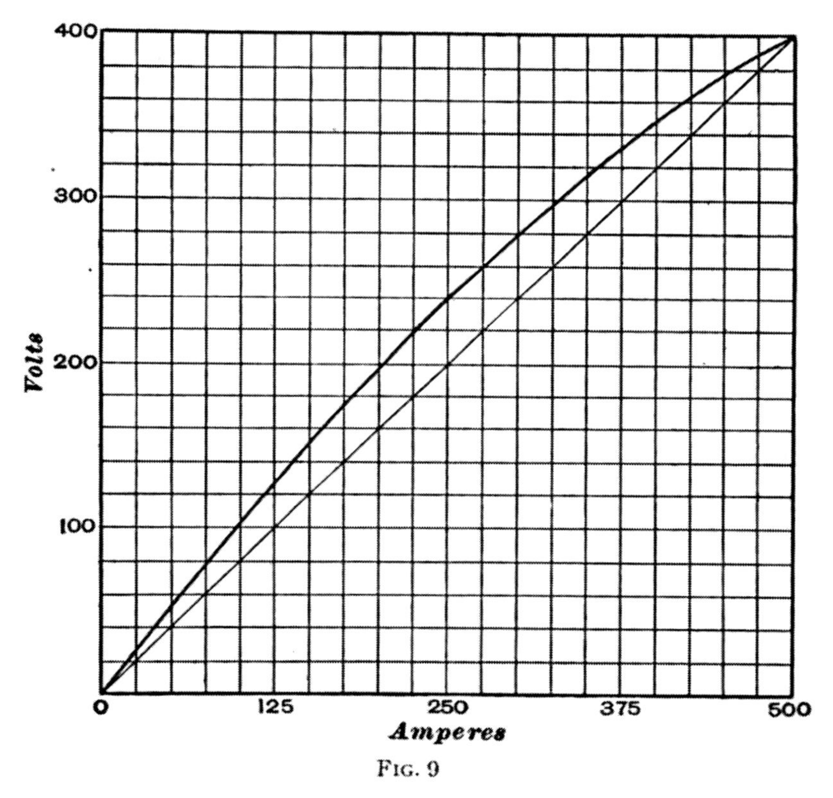

FIG. 9

In Fig. 9 the straight line shows voltage variation of a booster in which the change of voltage per ampere is constant. When the amperes are 500, the booster voltage is 400 and when the amperes are 250, the booster voltage is 200; this proportion is maintained over the current range. A booster with such a perfect characteristic would cost a prohibitive amount of money, and such perfection is unnecessary. The curved line illustrates more nearly the actual manner in which booster voltage varies with the booster current. Here, at 500 amperes the booster voltage is 400 as before, but at 250 amperes the voltage is 240, or 40 volts high, and at no point on the curve does the same proportion between voltage and current exist. This variation is satisfactory, provided the departure is within safe limits, which are given in Table I.

TABLE I

BOOSTER VOLTAGE VARIATIONS

Full-Load Voltages of Boosters	Allowable Maximum Variation of Full-Load Voltage at Partial Load Per Cent.
50 to 100	20
100 to 150	15
150 to 250	$12\frac{1}{2}$
250 to 500	10

19. The rating of a booster required for a given case depends on the drop in the feeders to be boosted, but the drop in the trolley-wire section is not usually considered as there are ordinarily but few cars per section in the outlying districts. The resistance of the feeder and the maximum current to be transmitted are usually known or may be calculated and the size of the booster is determined by an application of Ohm's law.

Suppose that the feeder has a resistance of .3 ohm and carries a maximum current of 600 amperes. The drop is $600 \times .3 = 180$ volts and for the same current the drop will be the same whether the feeder voltage is boosted or not; therefore, the correct rating for the booster in such a case will be 600 amperes at 180 volts. The output is then $600 \times 180 = 108$ kilowatts.

A size found well adapted to average railway service has an output of 120 kilowatts or 600 amperes at 200 volts. For smaller roads an output of 60 kilowatts or 600 amperes at 100 volts is convenient. To adapt boosters to a variety of conditions, it is customary to provide the series-field winding with taps, so that the number of active turns can be changed to suit existing conditions; or to furnish a variable shunt by means of which some of the current can be diverted from the series-field winding. Where a shunt is used, it should be of the inductive type; otherwise, the booster is apt to respond sluggishly to sudden changes in the load. The objection to a non-inductive shunt is that when the current increases suddenly, most of the increase at first will pass through the shunt and but little through the series-field in parallel with it.

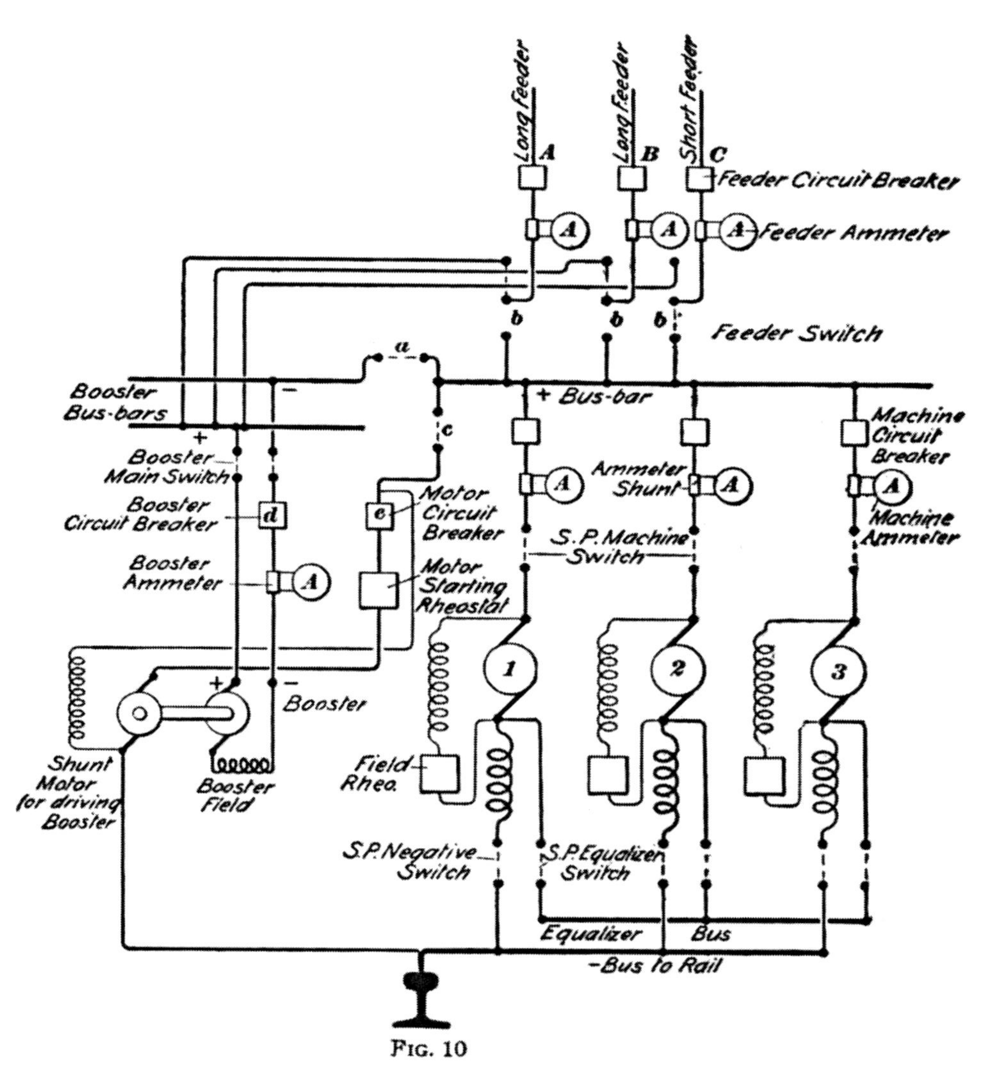

Fig. 10

20. Booster Connections.—The connections of the booster are such that the booster voltage is added to that of the main generators. The booster negative terminal is connected to the positive generator bus-bar, as shown in Fig. 10. The main generators *1*, *2*, and *3* connect to the positive bus-bar, which connects to the booster negative bus-bar through a switch *a*. Each feeder includes a double-throw single-pole switch *b*, with which to connect it either to the main positive bus-bar or to the positive bus-bar of the booster. As indicated by the dotted lines, the long feeders *A* and *B* are shown connected to the booster, while the short feeder *C* is connected directly to the main generator bus-bar. In order that feeder *C* will have the same voltage impressed on it as the feeders *A* and *B*, the right-hand feeder switch *b* should be thrown to its upper position. The booster is driven by a shunt motor operated from the main generators and equipped with a circuit-breaker and starting rheostat. The booster is connected to the booster bus-bars through an ammeter *A*, circuit-breaker *d*, and a double-pole main switch. The connections of Fig. 10 are equivalent to those of Fig. 7. All current supplied to feeders *A* and *B*, Fig. 10, passes through the booster and the feeder voltage is increased by an amount proportional to the current. The motor circuit-breaker should be so arranged that on opening it trips the booster circuit-breaker. A simple way to do this is to mount the two circuit-breakers *d* and *e* side by side and interlock them. They should also be so arranged that the booster circuit cannot be closed until after the motor has been started. If current were cut off the motor but not off the booster, the latter would be driven as a series motor reversed in direction of rotation, because series generators and motors run in opposite directions for given connections.

21. At times, it is desirable to use a small 550-volt generator as a booster. This can be done by connecting the armature and series coil of the generator in series with the feeder to be boosted and adjusting the excitation of the shunt coils to that necessary to give the desired electromotive force. This adjustment is sometimes made by connecting the shunt-field coils in parallel and then connecting the group of coils in parallel with a length of feeder in which the drop is sufficient to give the desired exciting electromotive force. This makes the shunt-field excitation of the booster proportional to the voltage drop in the feeder.

22. Location of Booster.—The location of the booster is preferably in the power station, where no extra attendant is required; if a motor-driven booster is located out on the line, its operating current must be transmitted from the station and the line loss is therefore increased. Boosters driven from some other source of power, such as a steam or gas engine, are free from this objection but the cost of attendance remains.

23. Economy of Boosters.—Booster practice represents an excess of electric energy generated in order that a sufficient amount may reach the locality of useful consumption, and all that does not reach that locality is wasted; but the real difference in the economies of boosters and of high-tension alternating-current transmission is not so great as the apparent difference. Boosters cause heavy losses only when the load is heavy, and this is a comparatively small part of the total time. With alternating-current transmission, certain transforming losses are constant irrespective of the useful load, and the annual cost of substation attendance may cause the total expense of losses and attendance to exceed those incident to booster operation. Boosters permit line extensions without expensive changes in station equipment or excessive outlay for copper. These advantages are more fully realized if storage batteries are installed out on the line. The battery will charge from the booster when the load is light and discharge when it is heavy, thereby keeping a fairly uniform load on the feeder and working it to best advantage.

The annual cost of operation, under certain conditions, with boosters may be less than that with alternating-current transmission. The system to adopt can be decided only by comparing their costs of operation. There are roads on which cars are operated over a radius exceeding 20 miles with the aid of boosters and storage batteries; these roads give satisfactory service, are as economical in operation as similar roads for which alternating-current transmission is used, and are less liable to interruptions from breakdowns that characterize transforming appliances necessary with alternating-current transmission and direct-current distribution.

24. Storage-Battery Auxiliaries.—On systems dependent entirely on direct current, the storage-battery auxiliary may be in the form of a portable substation, which is a motor-driven car loaded with batteries and which can be used where most needed; or it may be in the form of a permanently located battery that *floats* on the line. In the first case, the batteries are charged from the line during the periods of light load and then dicharged into the line when help is required on heavy loads. The disadvantage of this system is that it requires an attendant during the charging period, unless the charging is done at the station. The advantage of the system is that the car can be placed at the point where it is most needed and thereby meet overload shifts due to attractions at different terminals.

25. A floating battery is connected permanently across trolley and ground on some distant line section and, by properly proportioning the number of cells connected in series, charge and discharge currents may be made approximately the same. The automatic operation of the set depends on the drop in the line. A heavy load on the line causes sufficient line drop to allow the battery to discharge and help the station; when the line load is light, the line drop is small and the line voltage at the battery is sufficient to charge the battery.

ALTERNATING-CURRENT GENERATION AND DIRECT-CURRENT SUPPLY

26. Electric energy can be economically transmitted at greater distance by alternating current than by direct current, and is more economically generated in one large station than in several small ones. Therefore, where many cars are to be operated by direct current over a large area, it is the practice to concentrate generation in one or two large stations equipped with polyphase alternators and transformers and to transmit the energy to centrally located substations equipped with transformers and rotary converters or motor-generators, which supply direct-current energy of the desired voltage to the cars.

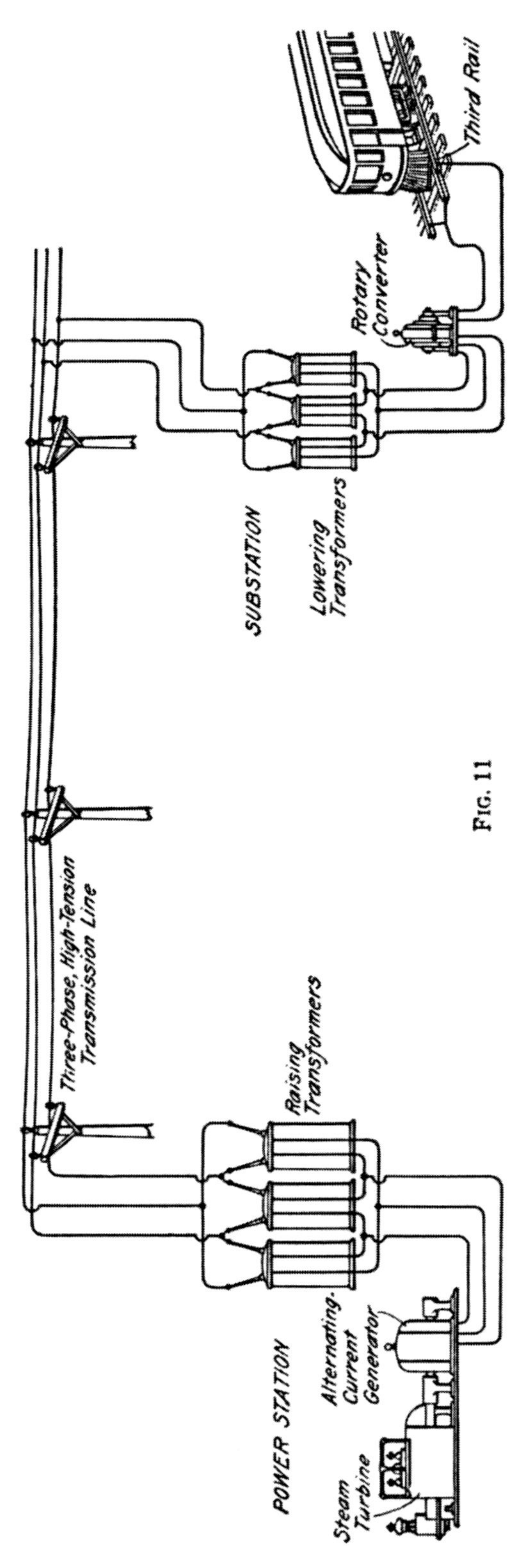

Fig. 11

In the main station, the alternators are of the rotary-field type, because this construction places the high-voltage conductors on the stator of the machines which affords more room for insulation and effective ventilation. In the substations, six-phase converters are often used because their capacity is nearly twice that of an ordinary generator of the same weight. Where considerable direct-current energy is required near the main station, a substation is frequently installed in the same building.

27. 600-Volt Supply. Fig. 11 shows the general features of the electric-energy distribution system of a large railway system. Alternating-current energy is generated in a main station, the voltage raised by transformers for the transmission lines, and the energy transmitted to substations where transformers and rotary converters supply direct-current energy to the cars at approximately 600 volts.

28. 1,200- to 1,500-Volt Supply.—In order to transmit direct-current energy economically over very considerable distances, the voltage of the direct-current supply lines in some systems is 1,200 volts or more. The development of the direct-current commutating-pole motor, which allows a high electromotive force to be impressed safely on its terminals, made possible the successful development of the 1,200-volt system. This high-tension system is especially adapted for extensions to 600-volt systems and for interurban roads.

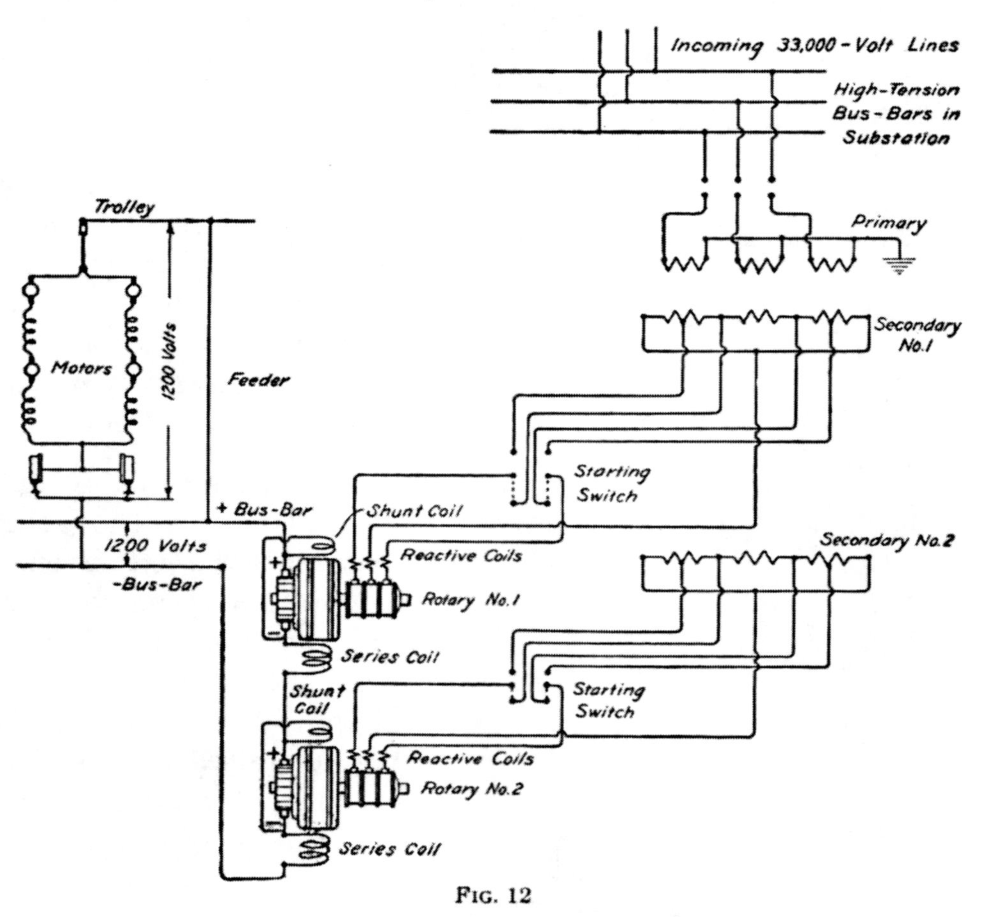

FIG. 12

29. Fig. 12 shows the more important connections of a substation designed to supply direct-current energy at 1,200 volts. The two rotary converters are supplied with alternating-current energy from two separate secondaries of the transformer, in order to prevent circulation of current between the alternating-current ends, which might occur if both rotaries were connected to one secondary. The direct-current ends of the two rotaries are connected in series to produce the 1,200 volts required.

In some cases, one or more 1,200-volt rotaries are used instead of sets consisting of two 600-volt rotaries in series. Motor-generator sets are also used in some systems. The motor drives either two 600-volt direct-current generators connected in series or one 1,200-volt generator. If there are four motors on a car, two groups are formed, each of two motors connected permanently in series. For low speeds the groups are connected in series, and for high speeds the groups are connected in parallel by means of the control apparatus on the car. A maximum electromotive force of about 600 volts is impressed across the terminals of each motor. Motors have been developed intended for an impressed electromotive force of 1,200 volts on their terminals.

Current collection by trolley wire is usually employed in these high-tension systems, but the third-rail method has also been used successfully.

30. 2,400-Volt Supply.—In the 2,400-volt, direct-current, supply system, motor-generator sets in the substation are used. An alternating-current motor drives two 1,200-volt, direct-current generators, which are connected in series.

The locomotives are equipped with four 1,200-volt motors. For low speed the four motors are in series, and for high speed the motors are arranged in two pair so that only two motors are in series across the 2,400-volt circuit in each current path.

31. Portable Substation.—On large railway systems, events are likely to occur that attract unusual travel in certain directions at irregular intervals. It is not often advisable to install direct-current feeders to meet these temporary conditions, but the application of a portable substation consisting of a rotary converter and its auxiliary apparatus, a motor-generator, or a storage battery will often solve the problem.

The necessary apparatus is mounted on a car, moved to a central position on the loaded section, and connected to the alternating-current and direct-current feeders if a rotary or motor-generator is used, or if a storage battery is used, it is connected across the direct-current feeder or trolley circuit. In some cases, the transformer and high-tension switches are mounted on a steel tower erected near the car. By this means, the temporary load may be carried satisfactorily with little if any change in the regular energy distribution system. It may be necessary to run an alternating-current feeder circuit to the portable converter substation, but the circuit wires would be comparatively small.

INDEPENDENT ENERGY SYSTEMS

32. Storage-Battery Cars.—Improvements in both the lead storage cell and the Edison storage cell and in the construction of the cars on which a battery is installed for motive purposes have adapted storage-battery cars for special classes of railway service, among which are: (1) Steam-road branch lines in suburban service and in local service where the population is sparse. (2) At steam-road terminals in the larger cities and in tunnels to avoid the smoke nuisance. (3) For construction and emergency cars on lines on which breakdowns may be such as to interrupt the energy transmission lines. (4) As "trippers" to be charged during hours of comparatively light station load and to be operated during morning and evening rush hours. (5) As "night hawks" to be operated at night, thereby permitting the power station to be shut down. (6) On new extensions of all kinds to develop traffic. (7) In small towns of 7,500 inhabitants and more. (8) As independent railways of light construction to serve agricultural interests as feeders to standard steam trunk lines.

33. The storage battery that furnishes energy for propulsion can be charged while in position on the car or after removal; in the latter case, another battery is substituted if the car is to continue in service. The car is usually able to travel from 80 to 100 miles on a single charge. The voltage of the battery depends on the kind and number of cells connected in series; in some cars 88 lead cells in series are used and the discharge is at an average voltage of 173. As the car carries its own source of electric energy, it can be operated over any tracks of suitable gauge independent of feeder or trolley installation. The control apparatus on the car serves to vary the portion of the battery electromotive force that is applied to the motors in such manner as to start and to regulate the car speed, as explained in another Section.

34. Gasoline-Electric Cars.—A gasoline-electric car is equipped with a gasoline engine mechanically connected to a direct-current generator, which in turn is electrically connected through speed-control apparatus to motors used for propelling the car. This type of car was developed to provide frequent single-car service on short railroad lines and to replace trains operated by steam locomotives at infrequent intervals. The operation of the speed control for these cars is explained in another Section.

ALTERNATING-CURRENT RAILWAY SYSTEMS

ENERGY DISTRIBUTION SYSTEMS FOR ALTERNATING-CURRENT OPERATION

SINGLE-PHASE GENERATION AND SUPPLY

35. The development of a successful single-phase alternating-current, railway motor, lead to the construction of many electric-railway systems in which alternating-current is generated in the stations and utilized to propel the cars. In the United States, the development has been mostly on the single-phase system using a series-wound commutator motor.

Single-phase generators may be installed at the power station, but three-phase generators are often employed; only one or two of the windings are however used to supply energy for car propulsion. All three phases of these generators are, in some cases, used to supply energy to the railway shops.

In some systems, transformer substations are installed, as indicated in Fig. 13, to reduce the transmission line voltage to that desired for the trolley system. In other systems, the voltage of the high-tension generators is applied through feeders

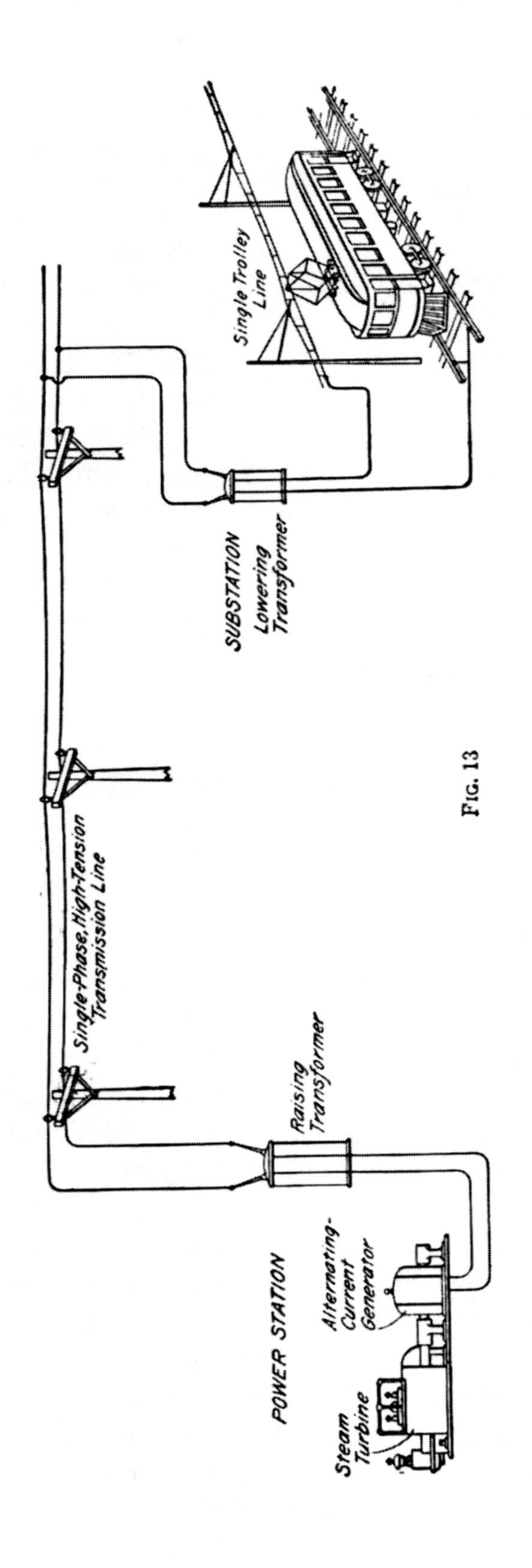

POWER STATION
Steam Turbine
Alternating-Current Generator
Raising Transformer
Single-Phase, High-Tension Transmission Line
SUBSTATION
Lowering Transformer
Single Trolley Line

FIG. 13

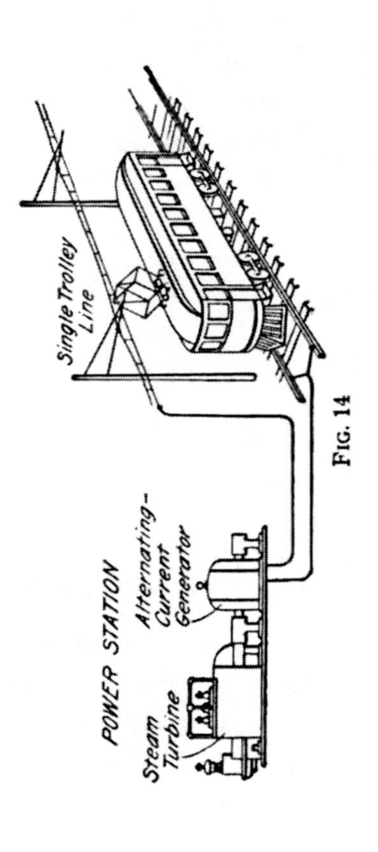

POWER STATION
Steam Turbine
Alternating-Current Generator
Single Trolley Line

FIG. 14

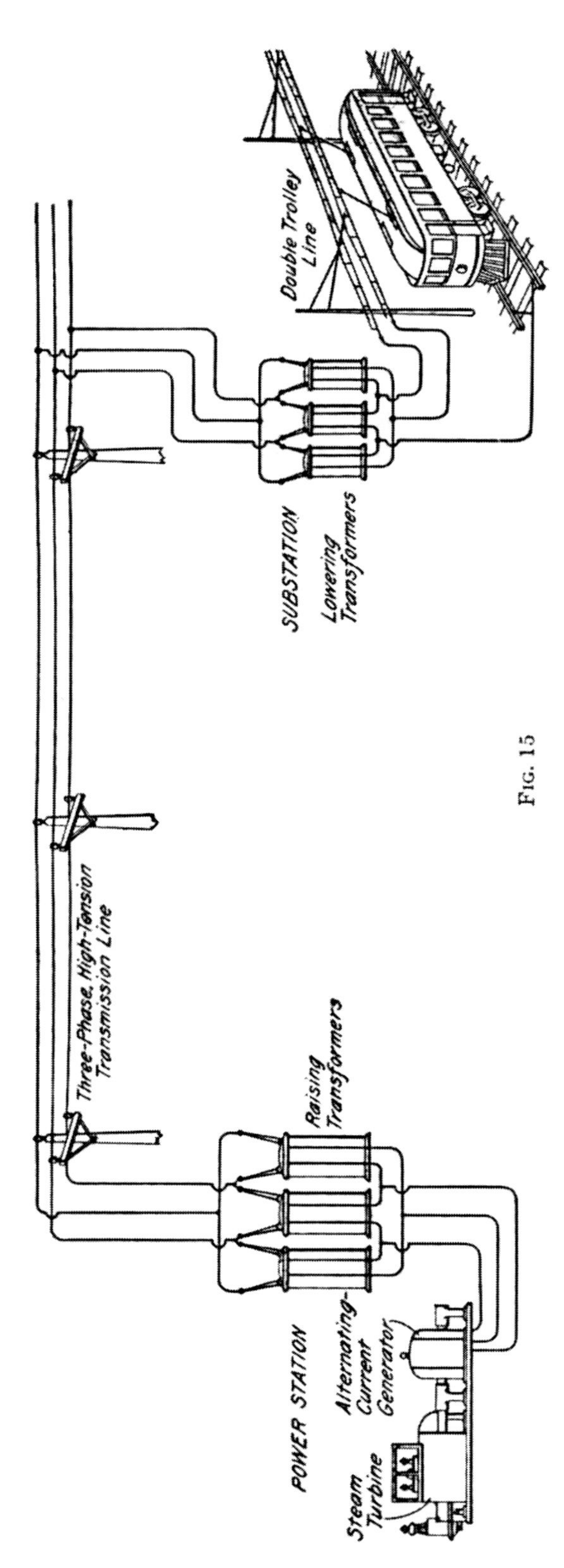

Fig. 15

to the trolley and rail without transformation of any kind, as indicated in Fig. 14.

Various voltages between trolley and rail have been employed on alternating-current roads. Among these values are 11,000 volts and 6,600 volts. A transformer on the car reduces the voltage to that suitable for the motors. The trolley system of current collection is usually employed. The trolley wire is installed with particular attention to its insulation and the high-speed service to which it is subjected.

THREE-PHASE GENERATION AND SUPPLY

36. The use of three-phase railway motors is rare in the United States. It is customary in systems of this kind to install two trolley wires over each track and to use the track rails for the third conductor of the three-phase energy-supply system, as indicated in Fig. 15.

In one installation of this kind, three-phase energy is generated at the station and the voltage stepped up by transformers for transmission to the substation. The transformers there installed reduce the voltage of the energy for the trolley and rail supply to 6,600 volts. Three-phase transformers on the locomotive further reduce the voltage to a value suitable for the three-phase induction motors, approximately 500 volts. Taps at different points on the secondaries of the transformers and adjustable resistances in the rotor circuits of the motors serve to start the motors without taking excessive current from the supply system. The motors run, however, at approximately constant speed for all loads and grades.

Motors of this kind are better fitted for railroad work in which infrequent stops are made and the speed desired is nearly constant, as they are essentially constant-speed motors.

Two or more speeds may be obtained from three-phase induction motors by changing the connections of the coils on the stators so as to produce a different number of poles, and by the cascade system of connecting two motors. In the cascade system, the rotors of the two motors are provided with windings connected to slip rings. The stator of No. 1 motor is connected to the three-phase line wires; the rotor of this motor is connected to the stator of the No. 2 motor, the rotor of which is connected to **Y**-connected resistors. The combination will run at such speed that the frequency of the current in the No. 1 rotor is of the proper value to drive the rotor of the No. 2 motor at the combination speed. Both motors must rotate at the same speed if they are connected to the same size of driving wheels. If the motors have the same number of poles, the combination speed is about half the speed of each motor when connected separately to the line wires.

ENERGY CALCULATIONS

GENERAL CONSIDERATION

37. The station-generator capacity required to support a given schedule on a given road depends on the number of cars and their schedule speed, the weight per car, topography of the road, character of traffic, and manner of handling equipment, and the condition of the line and of the track return. In new developments it is fair to assume that the condition of line and of track return is good. On large systems, it is usual to assume that the saving of power due to cars coasting down grades offsets the extra power demand of cars ascending grades. On small systems, however, where the average and maximum demands on the station are so widely different, the acceleration of one loaded car on a grade may easily treble the existing current demand on the station.

The maximum demand, hence maximum capacity, of a station can be greatly reduced by the installation of current-limiting devices upon the cars and by the enforcement of a rule requiring cars to be operated up grades on the series-notches of the controllers.

On a small road the energy required for lighting cars is inconsiderable, but on all roads the energy consumed by electric heaters may exceed 20 per cent. of the total energy required, because the heaters draw current continuously while the motors do not. The probability of snow equipment being in operation at the same time as schedule cars must also be considered.

A relation that works out satisfactorily on large and average-sized roads is that, assuming 25 per cent. of the station output to be consumed in transmission and conversion, the total kilowatt capacity of the station must equal the total horsepower capacity of all car motors likely to be in operation at the same time; to this, assuming that the summer and winter schedules are the same, must be added the extra capacity required by electric heaters.

Car speeds are largely fixed by the character of the service. City cars may not average more than 10 miles per hour, while interurban cars may average 40 to 50 miles per hour, according to the number of stops. The number of cars required depend on the length of line, frequency of service and schedule speed. To determine the best schedule speed and frequency of service for any proposed road requires a close study of dependent conditions, such as probable traffic and returns therefrom, competition, etc. In all cases, traffic estimates should consider traffic growth.

PASSENGER FACTOR AND LENGTH OF TRACK

38. The length of the proposed road depends largely on the population to be served and its distribution. The curves in Fig. 16 show what may generally be expected; they are plotted from data obtained under average conditions in cities and towns. The curve marked *passenger factor* p shows approximately the number of rides to be expected each year per inhabitant of the territory to be served by the system. This includes the center of population and the suburban sections reached by the road. With a population of 1,500,000, for example, there may be expected a passenger factor of 240. With a population of 250,000, the passenger factor expected is 190.

The curve marked *track factor* t shows the number of miles of single track to each 1,000 inhabitants. With a population of 375,000, the miles of single track for each 1,000 inhabitants is .61 mile, approximately. With a population of 1,500,000, the track factor is .49, approximately.

Let l = length of single track, in miles;
n = population to be served, say 10 years from time of building road;
t = track factor.

Then,

$$l = \frac{n\,t}{1{,}000}$$

For example, if the population of a territory in 10 years will be 1,000,000 and the track factor t, Fig. 16, is .51, the length of

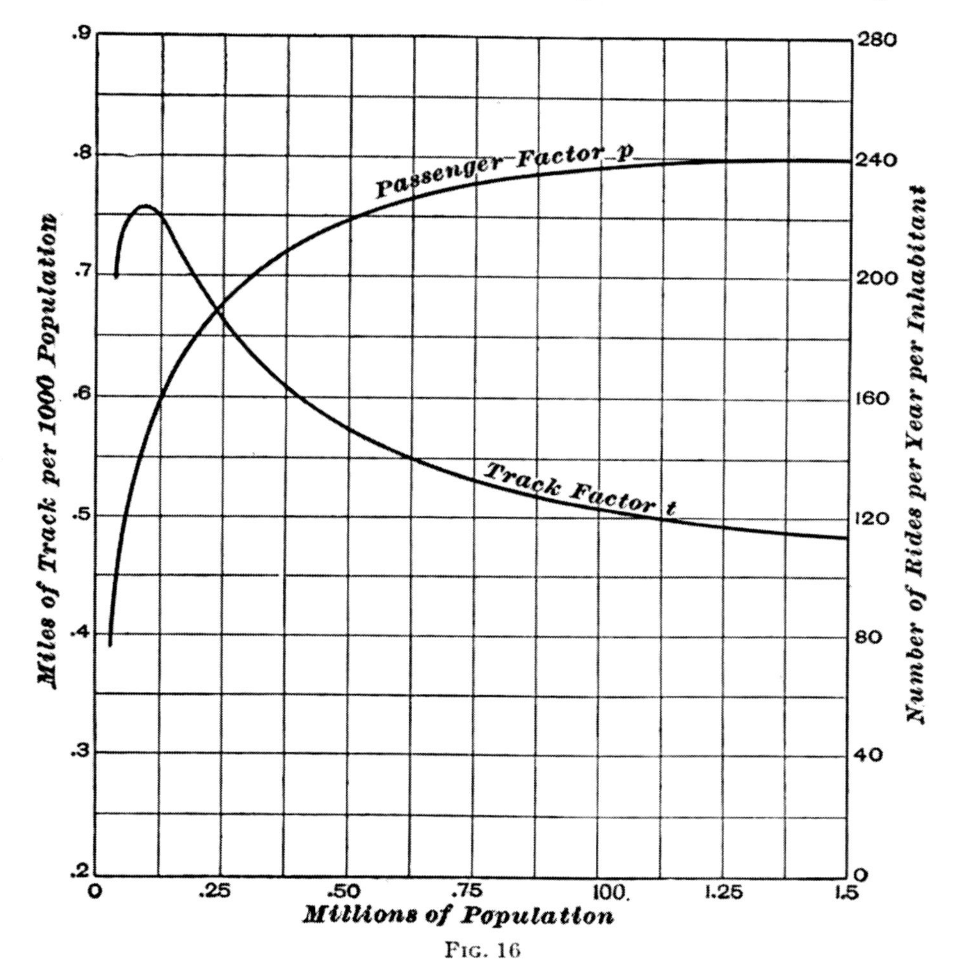

FIG. 16

single track required to serve this population is $l = \frac{1,000,000 \times .51}{1,000}$ $= 510$ miles.

NUMBER OF CARS

39. Knowing the number of contributing inhabitants, the fare per ride, and the probable number of rides each year per inhabitant, Fig. 16, the annual income in dollars can be computed. The average income for each car per mile of run, or as it is expressed per car-mile, is 21.56 cents on seventeen representative New York State roads; and the income on sixteen roads in various portions of the country is 27.31 cents per car-mile. Call the average for the thirty-three roads 24 cents. If the estimated annual income of the proposed road is divided by an assumed income per car-mile, the result will be the number of car-miles that must be made in a year to realize the estimated income; this, divided by the total number of hours that cars must run each year, gives the necessary number of car-miles each hour; and this divided by the schedule speed, in miles per hour, gives the number of cars required.

The annual income is equal to the product of the contributing population, n; the passenger factor p, Fig. 16; and the single fare, f, expressed in dollars. The number of cars c required is equal to the total annual income $n\,p\,f$ divided by the product of 365 (the days in a year); the operating hours per day, h; the schedule car speed in miles per hour, v; and the income per car-mile, r, expressed in dollars.

$$c = \frac{n\,p\,f}{365\,h\,v\,r}$$

EXAMPLE.—The following estimates are made for a proposed railway system to serve a population of 250,000: The single fare is to be 5 cents; the cars are to operate, 19 hours a day at an average speed of 10 miles per hour; and the income per car-mile is estimated as 24 cents. How many cars are required?

SOLUTION.—The passenger factor p, Fig. 16, for the population n of 250,000 is 190; f is \$.05; h is 19 hr.; v is 10 mi. per hr.; and r is \$.24. The number of cars

$$c = \frac{250{,}000 \times 190 \times .05}{365 \times 19 \times 10 \times .24} = 143 \text{ cars. Ans.}$$

SIZE AND WEIGHT OF CARS

SIZE

40. The size of the cars to be used on any one continuous line of a railway system depends on the number of passengers to be transported each day on that line and the number of cars available for the service. An approximate estimate of the passenger capacity of any one of these cars may be made as follows: The estimated number of passengers per day divided by the number of cars on the line gives the passengers per car per day. The schedule speed in miles per hour multiplied by the number of hours' operation of the car each day gives the total distance that the car travels during the day; and this divided by the length of the line gives the number of single trips made each day by the car. The average number of passengers per trip is equal to the number per day divided by the number of trips made by the car. The average ride of a passenger in city work is estimated as 3 miles; therefore, the capacity of the car may be considered as equal to the passengers per trip divided by one-third of the number of miles length of the line.

CAR WEIGHTS

41. The weight of cars depends on the nature of the traffic in which they are engaged. Interurban cars are heavier than city cars and frequently approach in size and weight those used on steam roads. Many complete double-truck trolley cars of older design but now in service weigh about 1,000 pounds per seated passenger and single-truck trolley cars about 800 pounds. Some of the more recent interurban cars weigh as low as 900 pounds, but most of them average about 1,250 pounds per seated passenger. The weight of ordinary trolley cars has been reduced 25 per cent.

These figures are, of course, only approximately correct, because the weight of trucks, car bodies, motors, and auxiliary devices vary greatly.

To the dead weight of the car should be added the live weight of the passengers estimated as 150 pounds per passenger. The live weight seldom exceeds 10 or 15 per cent. of the dead weight.

TRACTIVE-EFFORT FORMULAS

42. The following formulas for calculating the tractive force required by electric cars are only of approximate accuracy because of the many factors that modify the running conditions, among which are the conditions of the running gear, the road-bed, the wind, and the weather.

43. Tractive effort, or **tractive force,** refers to the force applied to a car to keep it in motion. On motor-driven cars this force is applied at the rim of the wheel. The resistance to motion, which the tractive effort must overcome, is due to four causes: *Train resistance,* which includes the resistances due to bearing friction, rolling friction, flange friction, and wind pressure; *grades,* which require that the car or train be lifted gradually at a rate depending on the steepness of the grade; *curves,* which increase the flange friction; *acceleration,* because all bodies at rest or in a certain condition of motion resist any effort to change that condition, therefore force must be used to overcome this resistance.

44. Train Resistance.—Were it not for train resistance, a car once in motion at a certain speed on a straight, level, rigid track would stay in motion without decrease of speed and without application of any force. All train-resistance formulas assume good condition of track and bearings and freedom from wind pressure, except that due to the train motion.

Let T_r = train resistance, or its equivalent, the tractive effort required to overcome it, in pounds;
w = weight of car, in tons;
v = car speed, in miles per hour;
s = number of square feet of head-end surface area of car.

Then,

$$T_r = 50\sqrt{w} + \frac{w\,v}{25} + \frac{s\,v^2}{400}$$

The first two elements of the right-hand side of the formula relate to the mechanical friction; the last element relates to the air friction.

EXAMPLE.—What is the tractive effort required to overcome the train resistance of a 16-ton car operating at 16 miles per hour on straight level track, the head end area of the car being 80 square feet?

SOLUTION.—Here $w = 16$ tons; $v = 16$ mi. per hr.; and $s = 80$ sq. ft. Substituting these values in the formula, gives

$$T_r = 50 \times \sqrt{16} + \frac{16 \times 16}{25} + \frac{80 \times 16 \times 16}{400} = 261 \text{ lb.} \quad \text{Ans.}$$

45. Effect of Grades.—When a car is traveling a 1-per-cent., a 2-per-cent., or a 3-per-cent. grade, the car is lifted vertically 1, 2, or 3 feet, respectively, for every 100 feet of distance traveled along the track. For the comparatively small grades usually met in railway practice, it is sufficiently accurate to consider the 100-foot section as measured along the rail, because the horizontal distance moved over, on which the percentage of grade is really based, is practically the same as the rail distance.

Any body on a grade has a tendency to move down the grade and, neglecting friction, a force must be applied to the body in a direction parallel to the rail to overcome this tendency. This force depends on the weight of the body and the grades. For small grades, the force is one one-hundredth of the weight of the body for every per cent. grade.

Assuming a weight of 1 ton, or 2,000 pounds, and a grade of 1 per cent., the force acting parallel to the rail is $2{,}000 \times \frac{1}{100} = 20$ pounds. The work done by a force of 20 pounds acting through a distance of 100 feet is equivalent to the force of 2,000 pounds acting through a vertical distance of 1 foot, which is the distance that the body rises for every 100 feet of rail at 1-per-cent. grade.

The extra tractive effort T_g, required on account of the grade of the tracks, is equal to 20 pounds for each ton of weight and for each per cent. grade.

Let w = weight of car, in tons;
g = per cent. grade, expressed as a whole number.

Then,
$$T_g = 20\, w\, g$$

EXAMPLE.—What is the extra tractive effort required to force a 16-ton car up a 3-per-cent. grade?

SOLUTION.—In the formula $w = 16$ and $g = 3$.
$$T_g = 20 \times 16 \times 3 = 960 \text{ lb.} \quad \text{Ans.}$$

46. Effect of Curves.—When a car is passing around a curve, the flanges of the wheels rub against the side of the rail head, thus causing friction and thereby increasing the necessary tractive effort over that required to overcome train resistance and grade. Curves are usually estimated to add to the tractive effort $\frac{1}{2}$ pound per ton of train weight for each degree of curvature. The degree of curvature of a railway curve is the angle of an arc of that curve between two points 100 feet apart.

If c = number of degrees curvature;
w = weight of train, in tons;
T_c = extra tractive effort required to overcome resistance of curve, in pounds.

Then, $$T_c = \frac{w\,c}{2}$$

EXAMPLE.—What is the extra tractive effort required to force a 16-ton car around a 3° curve?

SOLUTION.—In the formula $w=16$ and $c=3$; therefore,

$$T_c = \frac{16 \times 3}{2} = 24 \text{ lb.} \quad \text{Ans.}$$

47. Acceleration.—So far, uniform car motion has been assumed. It takes more energy to start a car than it takes to keep it moving at the speed acquired. Car-motor capacity based on assumptions of uniform motion will be insufficient unless the station stops are very infrequent. To start a car from rest and accelerate it to a given speed, energy must be expended in excess of that required to balance train resistance and the resistances of grades and curves; this extra energy is stored in the car by virtue of its weight and motion and it must be dissipated as heat when the car is stopped with the brakes. The extra tractive effort required to accelerate a car depends on the weight of the car and on the rate at which it is accelerated. In car-operation problems, the rate of speed increase (acceleration) or speed decrease (retardation or deceleration) is expressed in miles per hour per second. For example, an acceleration of $1\frac{1}{4}$ miles per hour per second, means that during each second, the car speed is increased $1\frac{1}{4}$ miles per hour; a car started from rest will, at the end of the first second, be moving at the rate of $1\frac{1}{4}$ miles per hour; at the end of the second second, $2\frac{1}{2}$ miles per hour, and so on. Again, if it takes a car 16 seconds to accelerate from rest to a speed of 16 miles per hour, its acceleration will be 1 mile per hour per second.

In general, if the difference in the two speeds, expressed in miles per hour, is divided by the seconds taken to accelerate from one speed to the other, the quotient is the average acceleration in miles per hour per second.

If a = acceleration, in miles per hour per second;
w = weight of the train, in tons;
T_a = extra tractive effort required for acceleration, in pounds.

Then,
$$T_a = 100\, w\, a$$

EXAMPLE.—What is the extra tractive effort required to accelerate a 16-ton car from rest to a speed of 16 miles per hour at the rate of 1 mile per hour per second?

SOLUTION.—In the formula $w=16$ and $a=1$; therefore, $T_a = 100 \times 16 \times 1 = 1{,}600$ lb. Ans.

48. Total Tractive Effort.—The total tractive effort T, taking into consideration the resistances offered by train resistance, grades, curves, and acceleration, may be expressed by a formula combining the right-hand members of the formulas in Arts. **44** to **47**.

$$T = 50\sqrt{w} + \frac{w\, v}{25} + \frac{s\, v^2}{400} + 20\, w\, g + \frac{w\, c}{2} + 100\, w\, a$$

substituting the values of T_r, T_g, T_c, and T_a in the examples under the preceding formulas, $T = 261 + 960 + 24 + 1{,}600 = 2{,}845$ pounds.

This is the total tractive effort required to accelerate a 16-ton car at the rate of 1 mile per hour per second on a 3-per-cent. grade and in a 3° curve. This is a possible condition. With a straight level track $20\, w\, g + \frac{w\, c}{2}$ will become zero and the total tractive effort required to accelerate the car against train resistance and inertia will be $T = 261 + 1{,}600 = 1{,}861$ pounds.

49. Tractive Effort Per Ton.—If the tractive effort per ton is desired, the right-hand members of the formulas for T_r, T_g, T_c, T_a, and T should be divided by w, the weight in tons of the car. Thus, in the example in Art. **44** the tractive effort per ton t_r required to overcome train resistance is

$$t_r = \frac{50}{\sqrt{w}} + \frac{v}{25} + \frac{s\,v^2}{400\,w} = 12\tfrac{1}{2} + \tfrac{16}{25} + \tfrac{16}{5}$$

$$= 16 \text{ pounds, approximately} \qquad (1)$$

The combined tractive effort per ton, Art. **48,** is

$$t = \frac{50}{\sqrt{w}} + \frac{v}{25} + \frac{s\,v^2}{400\,w} + 20\,g + \frac{c}{2} + 100\,a \qquad (2)$$

Substituting the values given in the examples, $t = 12\tfrac{1}{2} + \tfrac{16}{25} + \tfrac{16}{5} + 60 + \tfrac{3}{2} + 100 = 178$ pounds.

50. Limit of Adhesion.—The pressure of the brake shoes against the car wheels tends to stop their rotating, but the pressure of the car wheels on the rails tends to keep the wheels rolling; the most effective pressure will prevail. This means, that, when braking, if the braking pressure is excessive, the car wheels will stop rotating, but the car will still move; also, when accelerating, if the tractive effort tending to rotate the wheels exceeds the tendency of the wheel-rail friction, called *adhesion*, to prevent wheel slippage, the wheels will rotate, but the car may not move. The rate at which a car can be accelerated, or the steepness of the grade that it can ascend, is limited by the rail-wheel adhesion, which depends on the weight on the drivers and on the coefficient of friction between wheel and rail. The latter is lower for street railways, where the rails are liable to be dirty and slippery, than it is for elevated, subway, or interurban roads, where the rails are clean.

Table II gives *coefficients of adhesion* for rails under different conditions and assumes uniform tractive effort applied to the wheels. The **coefficient of adhesion** expresses the ratio between the tractive effort that will just slip the wheels and the weight on the rails directly under the driving wheels. The application of sand increases adhesion.

The coefficient of adhesion of 30 per cent. is the maximum value given in the table, but tests with electric locomotives have recorded as high a coefficient as 35 to 40 per cent. under favorable conditions. Except in infrequent-stop, high-speed service, where lower rates of acceleration prevail, it is customary to provide sufficient motive power to slip the wheels on dry rail; in high-speed service, this practice is not observed because high-speed equipments are not intended for high rates of acceleration and efforts to qualify them as such result in enormous currents and unnecessarily high cost of equipment.

51. On interurban or elevated roads, the adhesive force may be safely taken as 15 per cent. of the weight on the rails under the drivers; on city street railways, a safe value is 12 per cent. of the weight on the rails under the drivers; this gives $2,000 \times .15 = 300$ pounds per ton for elevated and interurban service and $2,000 \times .12 = 240$ pounds per ton in city trolley service.

TABLE II

COEFFICIENTS OF ADHESION

Condition of Rail	Without Sand Per Cent.	With Sand Per Cent.
Clean dry rail	30	
Wet rail	18	22
Rail covered with sleet	15	20
Rail covered with dry snow	10	15

On single-truck, two-motor cars, all wheels are drivers and the weight on the rails under the drivers equals the total weight of the car. This style of equipment is well adapted to hill climbing and to operation under unfavorable rail conditions.

On double-truck, two-motor cars, from 55 to 70 per cent. of the total weight rests on the rails under the drivers, thus limiting the tractive effort to from 165 to 210 pounds per ton weight of car at 15 per cent. adhesive force.

On four-motor, double-truck cars, all weight is on the rails under the drivers; these equipments are desirable for roads operating double-truck cars in hilly localities.

52. Limiting Grades.—The car wheels on separate axles are usually not rigidly connected together, therefore, on grades, where weight is transferred from the forward to the rear drivers, less tractive effort can be applied than is the case on a level, because of the tendency of the forward wheels to slip. The motors are usually started while connected in series and, if the wheels connected to one motor slip, the high counter electromotive force of the spinning motor reduces the current through both motors to an insufficient value for satisfactory operation.

If t_s = tractive effort required to start car on level, in pounds per ton;
g = grade, expressed as a percentage;
w = car weight, in tons;
p = per cent. of total weight on rails under drivers, expressed as a decimal;
r = coefficient of adhesion, expressed as a decimal.

Then,

Weight on rails under drivers, in pounds $= 2{,}000\ w\ p$
Adhesive force, in pounds $= 2{,}000\ w\ p\ r$
Force required to start car on grade g $= w\ t_s + 20\ w\ g$

Each per cent. of grade requires 20 pounds per ton additional tractive effort over that required on a level. When the grade is such that the tractive effort required to start on it is just sufficient to cause wheel slippage; then,

$$g = \frac{2{,}000\ w\ p\ r - w\ t_s}{20\ w} = \frac{2{,}000\ p\ r - t_s}{20}$$

EXAMPLE.—If 65 per cent. of the weight of a car rests on the rails under the drivers and if the coefficient of adhesion is 15 per cent., what is the maximum grade that the car can start on without wheel slippage, assuming that it requires a tractive effort of 70 pounds per ton weight of car to start on the level?

SOLUTION.—In the formula, $p = .65$; $r = .15$; and $t_s = 70$; therefore,

$$g = \frac{2{,}000 \times .65 \times .15 - 70}{20} = 6.25 \text{ per cent. Ans.}$$

The wheels will slip if the car is started on a grade exceeding this percentage, assuming the conditions stated in the example.

53. Acceleration Curve.—Fig. 17 shows typical curves for an electric train equipped for rapid acceleration. Curve *A* shows the relation between speed and time, curve *B* shows tbe total current supplied, and curve *C* shows the voltage. Starting from rest, the speed increases almost uniformly up to 25 miles per hour; the speed curve then bends over, thereby indicating decrease in acceleration; at 37½ miles per hour, the curve has become almost horizontal; the speed is then nearly uniform and the acceleration has become practically zero. After 93 seconds, the current is shut off; the train then coasts by virtue of the energy stored in it and the speed is gradually decreased because this stored energy has to overcome train

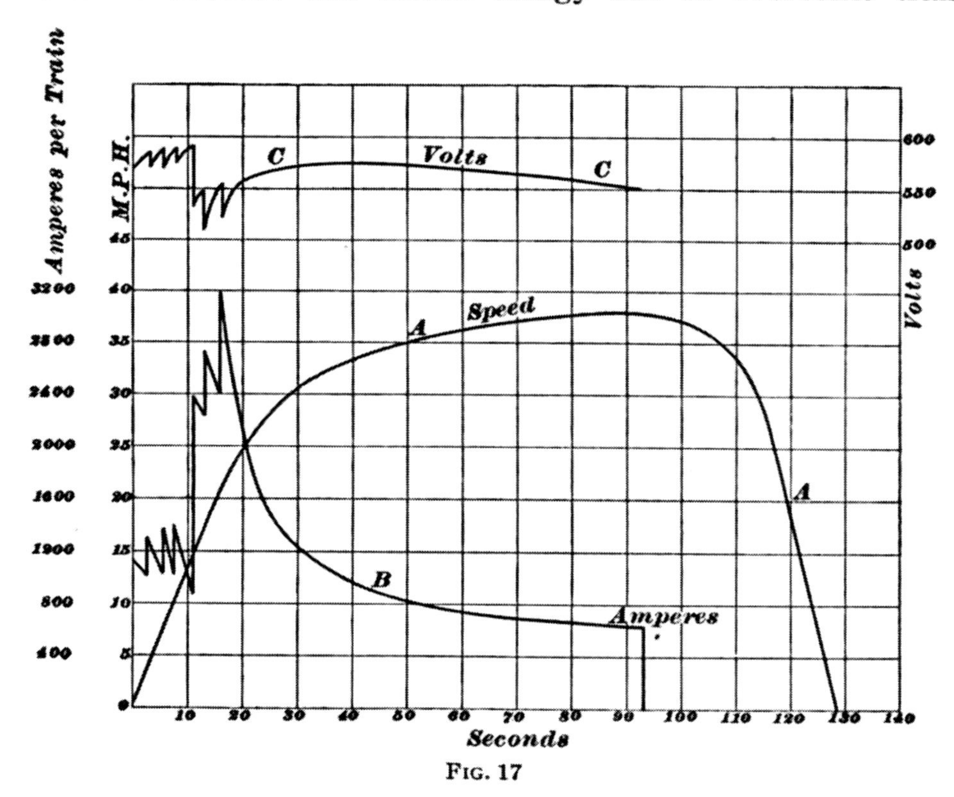

FIG. 17

resistance. Later, the brakes are applied and the train retarded, as indicated by the straight sloping line at the right, and finally brought to a stop. With all starting resistance in series, the total current is about 1,100 amperes; and as the resistance is cut out, the current varies as indicated by the notches in the curve during the first 10 seconds. The motors are then thrown into parallel and the total current rises to about 2,400 amperes, after which it further increases to 3,200 amperes, as the resistance on the parallel notches is cut out. Up to this point the current through each motor and the tractive effort have remained practically constant, since the motors of each car are first in series and then in parallel. As the train resistance remains almost constant for moderate speeds, the speed is accelerated during the first 20 seconds almost uniformly from 0 to 25 miles per hour. The average acceleration during this interval is 1.25 miles per hour per second. Beyond 25 miles per hour, the current and tractive effort diminish, thereby decreasing the rate of acceleration; and when the current has dropped to about 650 amperes, the speed has become nearly uniform. The tractive effort is then wholly utilized in overcoming train resistance; during acceleration, a large part of the total effort is used to increase the speed and thereby store energy in the train, the remainder being used to overcome train resistance.

54. Tractive Effort for Trains.—When trains are composed of two or more cars, the tractive effort per ton of train is less than for single-car operation because, while the head-end air resistance remains the same in the two cases, for given speed, the weight of a long train is greater than that of a single car so that the tractive effort per ton required by head-end air resistance is correspondingly less. Each car added, however, increases the amount of surface exposed to side friction, also adding a certain amount of air friction where the cars couple together. The latter effect can be largely eliminated by solidly vestibuling the coaches in a manner to present a smooth surface where they join. The total air resistance of a train is, therefore, much less than the air resistance of one car multiplied by the number of cars. The effect of head-end air resistance can be much decreased by so shaping the head end of the train as to cleave the air and turn it off to the side.

A corrective factor of $1+\frac{n-1}{10}$, where n is the number of cars in the train, is applied to the third element of the right-hand side of the formula for the total tractive effort when train operation is to be considered. This element relates to air friction. The formula then becomes:

$$T=50\sqrt{w}+\frac{w\,v}{25}+\frac{s\,v^2}{400}\left(1+\frac{n-1}{10}\right)+20\,w\,g+\frac{w\,c}{2}+100\,w\,a$$

EXAMPLE.—A 400-ton, eight-car train operates at 60 miles per hour on a straight level track. The surface area of the head end of the train is 120 square feet. (*a*) What is the total tractive effort? (*b*) What is the tractive effort per ton weight of train?

SOLUTION.—(*a*) In the formula $w=400$; $v=60$; $s=120$; and $n=8$. As the train is operating on a straight level track at constant speed, the last three elements of the formula, which relates to grades, curves, and acceleration are eliminated. The total tractive effort

$$T=50\times\sqrt{400}+\frac{400\times 60}{25}+\frac{120\times 60\times 60}{400}\times\left(1+\frac{8-1}{10}\right)$$
$$=1{,}000+960+1{,}836=3{,}796 \text{ lb. Ans.}$$

(*b*) $3{,}796\div 400=9.49$ lb. per ton. Ans.

55. It should be noted that the tractive effort, 1,836 pounds, required to overcome air resistance almost equals the effort required, 1,960 pounds, to overcome all mechanical resistance. At very low speeds, train resistance decreases as the speed increases, reaching a minimum around 5 miles per hour; the speed corresponding to minimum train resistance, however, varies and depends on the condition of the cars and track. As the speed further increases the train resistance increases rapidly, due to the rate at which air resistance increases. Head-on high winds can easily double the train resistance, while high winds from the rear have the opposite effect. The shorter a train is, the lower is the speed at which the air resistance becomes equal to the mechanical resistance, because the larger portion of the resistance due to air friction is caused by the surface at the head end of the train, and this remains the same whatever the number of cars

TABLE III

RELATION OF WEIGHT OF CAR TO SURFACE AT END

Weight of Car Tons	Area of Surface at One End Square Feet
20	90
30	100
40	110
50	120
60	120

may be, while the weight of the train, and consequently the mechanical resistance, is greatly lessened by decreasing the number of cars.

The more frequent are the stops, the greater is the percentage of total train energy input expended on acceleration; therefore, in low-speed, frequent-stop service, the trains should be light, while in high-speed service, with few stops, light weight is not so important.

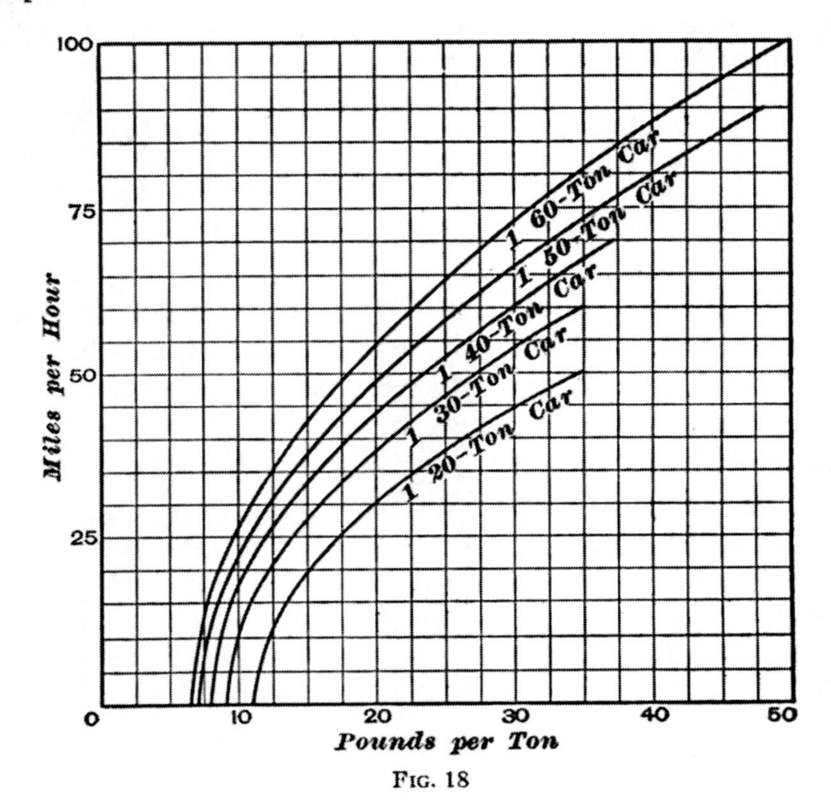

Fig. 18

The approximate relation between the weight of a car and the area of the surface at one end is shown in Table III.

56. Train Resistance Curves for Single Cars and Trains.—Fig. 18 indicates the train resistance per ton for single cars of different weights; Fig. 19 gives train resistance per ton for five-car trains composed of cars of similar weights to those of Fig. 18. For example, the train resistance per ton of a single 20-ton car, Fig. 18, operating at 25 miles per hour is $17\frac{1}{2}$ pounds per ton; that of a train of five 20-ton cars, Fig. 19, at the same speed is about $7\frac{1}{2}$ pounds per ton. The train resistance per ton of a single 60-ton car, Fig. 18, operating at

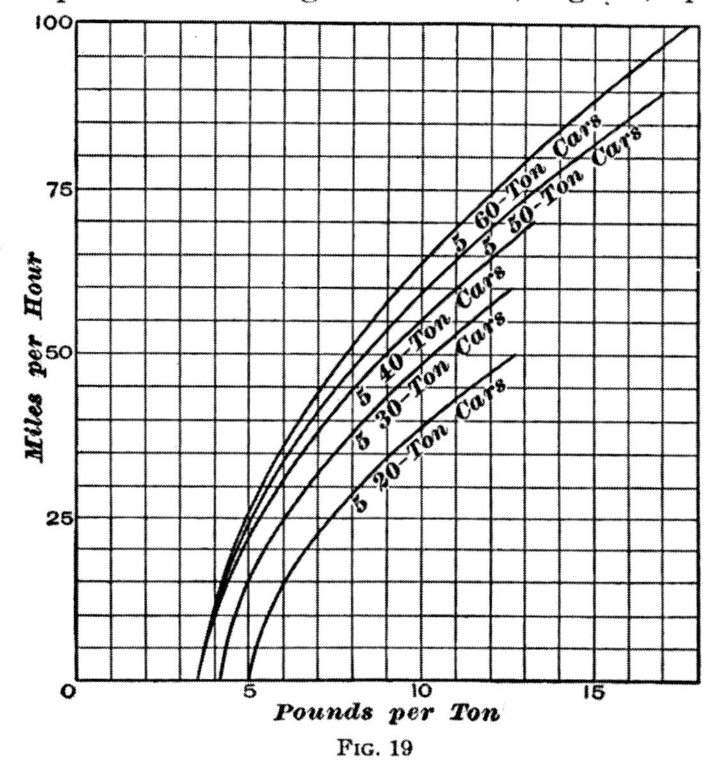

Fig. 19

60 miles per hour, is $22\frac{1}{2}$ pounds per ton; for a train of five such cars, Fig. 19, it is about $9\frac{1}{2}$ pounds per ton.

POWER FORMULAS

57. By **horsepower of operation** is meant the power required to operate a car of given weight and with certain assumptions in regard to train resistance, grades, and acceleration. The effects of curves on power requirements can usually be ignored; because, if the curves are of long radius the power is negligible, and if the curves are of short radius, not only is the speed reduced, but the headway of the car helps it to round the curve. The power may be expressed either as horsepower or kilowatts.

58. Kilowatt Input for Train Resistance.—The kilowatt input to a car operating at uniform speed on level track can be calculated from the following formula:

Let p = kilowatt input;

t_r = train resistance, or tractive effort required to overcome it, in pounds per ton;

w = weight of car, in tons;

v = speed, in miles per hour;

e = full-speed efficiency of motor equipment, expressed as a decimal.

$$p = \frac{2\,t_r\,w\,v}{1{,}000\,e}$$

EXAMPLE.—(*a*) A 16-ton car runs at a uniform speed of 16 miles an hour on level track against a train resistance of 17 pounds per ton. Assuming the efficiency of the motor equipment at full speed to be 70 per cent., what is: (*a*) the kilowatt input? (*b*) the horsepower input? One kilowatt equals 1.34 horsepower.

SOLUTION.—(*a*) In the formula $t_r = 17$, $w = 16$, $v = 16$, and $e = .7$. Substituting these values,

$$p = \frac{2 \times 17 \times 16 \times 16}{1{,}000 \times .7} = 12.4 \text{ K. W.} \quad \text{Ans.}$$

(*b*) $$\text{H. P.} = 12.4 \times 1.34 = 16.62 \text{ H. P.} \quad \text{Ans.}$$

59. Total Kilowatt Input for Train Resistance and Grades.—The combined tractive efforts per ton for train resistance and grade is $t_r + t_g$. The kilowatt input of a car operating at uniform speed on a grade can be calculated by means of the following formula:

$$p_{rg} = \frac{2(t_r + t_g)\,w\,v}{1{,}000\,e}$$

EXAMPLE.—A 16-ton car moves at a uniform speed of 16 miles per hour against a train resistance of 17 pounds per ton up a 3-per-cent. grade. The efficiency of the motor equipment at full speed is assumed to be 70 per cent. (*a*) What is the input in kilowatts? (*b*) What is the input in horsepower?

SOLUTION.—(*a*) In the formula, $t_r = 17$, $t_g = 3 \times 20 = 60$ (Arts. **45** and **49**), $t_r + t_g = 17 + 60 = 77$, $w = 16$, $v = 16$, and $e = .7$. Substituting these values,

$$p_{rg} = \frac{2 \times 77 \times 16 \times 16}{1{,}000 \times .7} = 56.3 \text{ K. W.} \quad \text{Ans.}$$

(*b*) $$56.3 \text{ K. W.} \times 1.34 = 75.4 \text{ H. P.} \quad \text{Ans.}$$

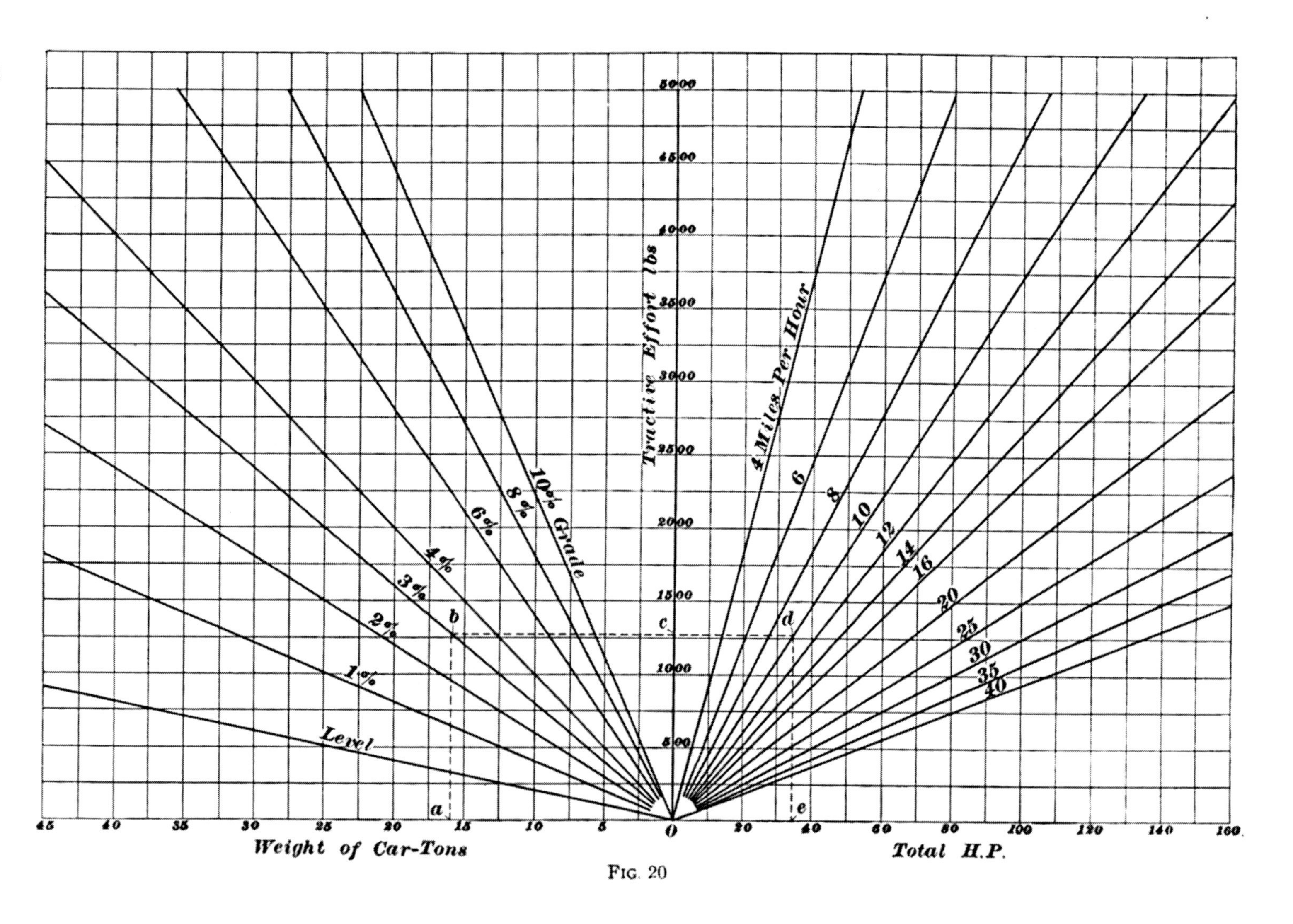

Tractive Effort lbs
5000
4500
4000
3500
3000
2500
2000
1500
1000
500
0
10% Grade
8%
6%
4%
3%
2%
1%
Level
4 Miles Per Hour
6
8
10
12
14
16
20
25
30
35
40
a
b
c
d
e
45
40
35
30
25
20
15
10
5
20
40
60
80
100
120
140
160
Weight of Car-Tons
Total H.P.

FIG. 20

60. Tractive Effort and Horsepower-Output Curves. The curves of Fig. 20 show approximately the tractive effort and the horsepower output of the motors required to operate a car under given conditions. The train resistance on a level track is assumed to be 20 pounds per ton, which is a safe figure for city operation. An example will serve to illustrate the use of the curves.

EXAMPLE.—What is: (*a*) the tractive effort, and (*b*) the output of the motors required to move a 16-ton car up a 3-per-cent. grade at the rate of 10 miles per hour?

SOLUTION.—(*a*) First find the point *a* on the weight data line corresponding to 16 tons; proceed vertically upwards to point *b* on the 3-per-cent. line; then horizontally to the right to point *c* on the tractive effort line; and read 1,280 pounds. Ans.

(*b*) From point *c* continue horizontally to the right to point *d* on the line marked 10 (mi. per hr.) and then vertically downwards to point *e* on the total-horse power line and read 34.1 H. P. Ans.

NOTE.—The movements for this particular solution are indicated in Fig. 20 by dotted lines.

To obtain the horsepower input, the horsepower output derived from the curves must be divided by the efficiency of the motors under running conditions selected.

61. Kilowatt Input for Acceleration.—Calculations relating to the kilowatt input for acceleration are only of approximate accuracy because they involve changing conditions of the tractive efforts during acceleration, the energy expended in the starting resistances or other speed regulating devices, and the efficiency of the motors. If the acceleration is uniform, the average speed may be taken as one-half of the final speed.

If the tractive force T_a in pounds and the average speed v in miles per hour are known and a value of the combined efficiency e of the starting apparatus and motors is assumed, the approximate value of the kilowatt input for acceleration is

$$p_a = \frac{T_a v}{503\, e}, \text{ or approximately } \frac{.002\, T_a v}{e}$$

EXAMPLE.—A 16-ton car is accelerated from rest to a speed of 16 miles per hour in 16 seconds with a tractive force of 1,600 pounds. The combined efficiency of the starting apparatus and motors is assumed to be 60 per cent. What is the kilowatt input required for acceleration?

SOLUTION.—In the formula, the tractive force $T_a = 1,600$ and the average speed $v = 16 \div 2 = 8$; $e = .6$. Substituting these values,

$$p_a = \frac{.002 \times 1,600 \times 8}{.6} = 42.7 \text{ K. W.} \quad \text{Ans.}$$

ENERGY TESTS

INTERURBAN ROADS

62. Tests made on cars in every-day operation afford reliable data for estimating the probable energy required for a given service; such tests include observations of power, speed, time, voltage, current, grades, curves, and everything else likely to affect the energy consumption. The following figures are taken from the record of tests made with 40-foot cars each weighing 63,000 pounds complete and each equipped with two 150-horsepower motors mounted on the forward truck. The energy consumption was measured for both limited and local service so that the effect of stops could be determined.

In local service, the average speed for the whole run of 56.5 miles was 22.6 miles per hour, but part of the run was at reduced rates through cities; outside of the cities, the local-service speed averaged 26.6 miles per hour.

In limited service, the speed averaged 28.3 miles per hour for the whole run and 35.3 miles per hour, exclusive of slow running, in the cities. The speed between stations frequently rose to 40 and 45 miles per hour and on one part of the road reached 60 miles per hour. Most of the grades were less than 2 per cent., but a few short ones were as high as 3 per cent. The weight of the car with passengers was usually from 34 to 34.5 tons. The energy consumption, as indicated by the average of a large number of watt-hour-meter readings, is given in Table IV. From these figures it would be safe in making a preliminary estimate on a similar road, to allow from 70 to 75 watt-hours input per ton-mile for limited service and 85 to

90 watt-hours input per ton-mile for local service. The energy consumption in local service runs higher than in the limited service at moderate speeds because of the greater number of stops. Each stop is followed by a start with heavy accelerating current, which is required to overcome the inertia of the car weight. If the limited service had been at very high continuous speed, the energy consumption in watt-hours per ton-mile would probably exceed that for the local service because of the great effect of air resistance at the higher speeds.

TABLE IV

ENERGY CONSUMPTION OF INTERURBAN CARS

Class of Service	Kilowatt-Hours per Car-Mile	Watt-Hours per Ton-Mile
Local service, outgoing trips	2.24 to 2.78	66.7 to 81
Local service, return trips	2.62 to 3.05	77 to 89.5
Local service, average for six round trips	2.62	76.6
Limited service, outgoing trips	2.10	58.7
Limited service, return trips	2.31	71.6
Limited service, average for round trips	2.20	65.1

63. Tests made with 25-ton cars give results quite consistent with those of Table IV. The average energy consumption for regular trips at speeds varying from 8 to 29 miles per hour was 2.16 kilowatt-hours per car-mile, or 86.4 watt-hours per ton-mile. The average consumption for test runs made with speeds varying from 19 to 27.5 miles per hour, was 1.96 kilowatt-hours per car-mile, or 78.4 watt-hours per ton-mile. The greatest consumption was for a short run of 1.46 miles at a low speed of 8 miles per hour; here the energy consumption was 3.7 kilowatt-hours per car-mile, or 148 watt-hours per ton-mile.

64. A 7-ton, storage-battery car in regular service over grades ranging from 3.8 to 4.5 per cent., at 22 miles per hour schedule speed, including five stops per mile, required 63.4 watt-hours per ton-mile for a daily mileage of 110 miles with an average of seven passengers per car-mile.

65. The application of the preceding data is illustrated in the following example:

EXAMPLE.—An interurban electric road is to operate ten cars, each weighing 30 tons when loaded; six are for local and four for limited service. Local cars are to average 20 miles per hour and limited cars, 32 miles per hour. Estimate the approximate station capacity, assuming the total loss between generators and cars to be 18 per cent. of the delivered energy at the cars.

SOLUTION.—The average energy consumption, in watt-hours per ton-mile, may be taken at 72.5 for the limited cars and 87.5 for the local service. (See Art. 62.) In 1 hr. the total number of watt-hours supplied would be:

For local service, 6 cars × 30 tons × 20 mi. ×87.5 watt-hours....................	=315,000 watt-hours
For limited service, 4 cars × 30 tons ×32 mi.×72.5 watt-hours...........	=278,400 watt-hours
	593,400 watt-hours

As the energy supplied to the cars in 1 hr. is 593,400 watt-hours, the power is 593,400÷1=593,400 watts=593.4 K. W. The loss in lines, working conductor, rotaries, and transformers, is 593.4×.18=106.8 K. W. The average station output must be 593.4+106.8=700.2 K. W. On an interurban system where comparatively few cars are operated, the local fluctuations are great and the maximum load may be twice the average; also, considerable energy is required for lighting and heating. In the present case, the station should be capable of furnishing at least 1,000 K. W. and to insure against shut-downs, two 1,000 K.-W. generators would be advisable; or at least three generators of 500 K. W. each, one, two, or three being operated, as occasion might require; by thus operating generators only when needed and then at or near full load, a high-load factor and greater efficiency is maintained.

CITY ROADS

66. The energy consumption per ton-mile is greater for city service than for interurban, because the cars are lighter, the tractive effort per ton greater, the stops more frequent, and, as a rule, the surface of the track is not in as good condition. In frequent-stop service, much energy is wasted in the starting rheostats. The average energy consumption per ton-mile is seldom less than 90 watt-hours and often is more.

Table V gives data of tests made on cars equipped with motors of the sizes ordinarily used in city and suburban work.

TABLE V
ENERGY CONSUMPTION OF CITY CARS

Weight of Car Tons	Horsepower per Motor	Number of Motors per Car	Kilowatt-Hours per Car-Mile	Watt-Hours per Ton-Mile
19.25	35	2	1.67	87.0
16.64	40	2	1.58	94.9
24.35	38	4	2.78	114.0
21.65	40	4	2.58	119.0
19.12	35	4	2.38	124.0
21.65	38	4	2.83	131.0
19.92	40	4	2.85	143.0
22.80	35	4	3.26	143.0
21.73	40	4	3.12	144.0
21.80	50	4	3.29	152.0
21.04	38	4	3.32	157.0
21.19	40	4	4.12	193.0

EXAMPLES FOR PRACTICE

1. If 750 pounds is required to propel a 30-ton car on a level track, what total force must be applied to propel the car up a 2-per-cent. grade?
Ans. 1,950 lb.

2. (*a*) If a car weighs 25 tons, what force must be applied to produce an acceleration of 1.25 miles per hour per second? (*b*) What must be the total force applied to produce the acceleration and overcome the train resistance as well, assuming that the latter amounts to 20 pounds per ton weight of car?
Ans. { (*a*) 3,125 lb. (*b*) 3,625 lb.

3. A certain car has 60 per cent. of its weight resting on the driving wheels and the adhesive force between track and rail is 15 per cent. of the weight on the drivers. A force of 75 pounds per ton weight of car is necessary to start the car from rest. What is the steepest grade on which the car can be started without slippage of the wheels on the tracks?
Ans. 5.25 per cent.

4. If a force of 20 pounds per ton is required to propel a 25-ton car on a level track at the rate of 15 miles per hour, the motor efficiency at that speed being 70 per cent., what is the input, in kilowatts, for the car?
Ans. 21 K. W.

METHOD OF MAKING ENERGY TEST

67. The simplest method of running an energy test on a car is to connect a watt-hour meter into the motor circuit, note the reading at the beginning of a regular run and the reading at the end of the run and take the difference, which will be the watt-hours consumed during the run. This method automatically averages the products of the volts and amperes and multiplies the result by the time in hours; but it gives no information in regard to maximum, average, and minimum volts and amperes. When such information is wanted, or when a watt-hour meter of sufficient capacity is not available, a test can be made with a voltmeter and an ammeter, the first being connected across trolley and ground and the second in the main motor circuit. Where the car equipment includes heaters, compressor, governor, electric headlights and car lights, and only a record of the energy consumption of the motor circuit is desired, either these other circuits must be kept open, or the ammeter or the current coil of the watt-hour meter must be connected only in the motor circuit.

68. The voltmeter-ammeter test consists in taking a series of simultaneous voltage and ampere readings at a certain number of seconds apart; each voltage reading is afterwards multiplied by its corresponding amperage reading and the product, watts, placed opposite in a third column of the record sheet. The sum of the voltage readings divided by their number, gives the average voltage during the test; the sum of the current readings divided by their number gives the average current during the test; and the sum of the watt readings, divided by their number gives the average watts required during the test. Table VI gives 27 of the 426 readings taken during such a test.

The following data are based on the full record of readings: Maximum voltage, 625; maximum current, 355 amperes; maximum power, 189,750 watts; maximum voltage and current readings did not occur at the same time; minimum voltage, 360; minimum current, 0 amperes; average voltage from all readings, 511; average current from all readings, 87 amperes;

TABLE VI

RESULT OF AMMETER-VOLTMETER TEST

Volts	Amperes	Watts	Volts	Amperes	Watts	Volts	Amperes	Watts
520			525	169	88,725	500	280	140,000
510			505	123	62,115	500		
480	212	101,760	475	109	51,775	540	74	39,960
515			500	98	49,000	475		
480	315	151,200	520			510		
490	175	85,750	460	154	70,840	525		
500	170	85,000	490			500	164	82,000
450	130	58,500	475	206	97,850	500		
500	84	42,000	475	107	50,825	490	150	73,500

average power, 43,112 watts, found by taking the average of all of the watt values in the complete record sheet; average of current readings, not considering the zero readings, 149 amperes; average power, considering only readings when electricity flows, 74,445 watts; duration of test, 1.7 hours; energy to drive car during complete test $= \frac{43,112}{1,000} \times 1.7 = 73.29$ kilowatt-hours; energy per hour is equal to the average kilowatts $= 43.112$ kilowatt-hours; weight of car, 21.19 tons; energy per ton-hour $= 43.112 \div 21.19 = 2.03$ kilowatt-hours; length of run 17.78 miles; average speed, 10.46 miles per hour; energy per car-mile $= 43.112 \div 10.46 = 4.12$ kilowatt-hours; energy per ton-mile $= 2.03 \div 10.46 = .193$ kilowatt-hour, or 193 watt-hours.

RELATION OF MAXIMUM AND AVERAGE CURRENTS

69. The maximum current taken by a car during a run depends on the manner in which the motorman handles the controller; if he moves the handle too quickly, the current may be excessive.

The average current depends on the number of times that the car must be started from rest, on the voltage, the weight of the car, grades, curves, gear ratio, and track conditions. A great number of tests under widely different conditions indicate that the maximum current is from three to five times the average current when the latter is based on all test readings, including the periods of coasting and stops.

When the energy test is made with a watt-hour meter, the result obtained directly will be the watt-hours of energy expended on the car during the test trip. If the watt-hours are divided by the time, in hours, of the trip, the result is the average power in watts. If the average watts is divided by the average voltage (ascertained by voltmeter readings or assumed), the result is the average current.

LOCATION OF POWER HOUSE

GENERAL CONSIDERATIONS

70. Purely electrical considerations would place the power-house as nearly as practicable to the center of the system in order to minimize the transmission losses. The location of water-driven stations, however, is dictated by the location of the dam, and the ideal central location of steam-driven stations is often influenced by other features that may be more important than is transmission loss. For example: The best electric location might fall in a section where office buildings would pay better dividends than many well-managed railroads. Again, the load center might fall where coal would have to be hauled at great expense or where there would be lack of water for boilers and condensers. The final selection of a power station site is a compromise of several sites governed by conflicting conditions, each of which must be considered from the point of view of its advantages and disadvantages expressed in dollars.

CENTER OF SYSTEM

71. By **center of system** is meant the center of load or traffic. A portion of the energy transmitted from the station to the distribution points is lost in the distributing system. If the conductors are too small, the loss will be excessive and the voltage at the end of the line low; the loss depends on the line resistance and on the current transmitted. The load center and geographical center seldom coincide; the location of the latter depends on the number of miles of track and on their disposition; the location of the former depends on how the load is distributed.

In Fig. 21, A B represents 10 miles of straight level track on which 14 similar cars operate at regular intervals. The geographical, or mileage, center, is located at P as there is

5 miles of track on each side of it. Assuming equal power requirements for the cars *1* to *7* on each side of the center line *c l* the load center is also at *P*. An absolutely even distribution of load is improbable, however, due to grades, directions in which the passenger loads travel, and density of population at different points on the line. For example, heavier grades and denser traffic near *A* would bring the load center between *A* and *P*.

72. When locating a station site future extension and traffic growth should be considered. The station should be situated at a point that will be at approximately the load center of the

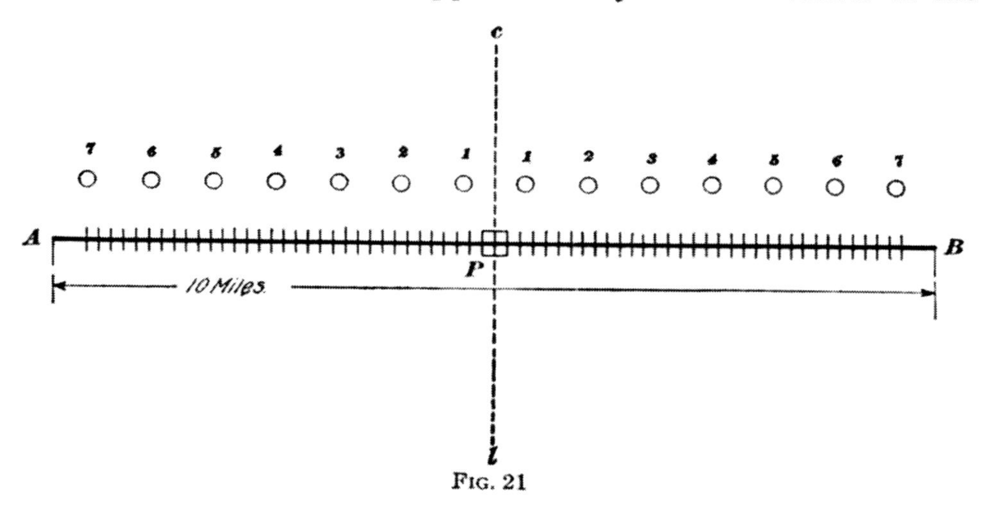

Fig. 21

future enlarged system, even if this location is some distance from the central position under present conditions.

73. The load center of any system can be located with sufficient accuracy for practical purposes by considering the system in sections, each with fairly uniform load distribution, and assuming that the load on each section is concentrated at its geographical center. For example, a system may consist of two lines merging at a point, as represented by Fig. 22, and the load may not be uniform on all parts of the system. The load, being fairly uniform over section *a b*, may be considered as concentrated at its central point *c*; a heavier load distributed over section *b d* is concentrated at *e*; and the uniform load of section *d f*, at *g*.

The concentrated loads should be estimated from the number of cars operating on the section and the energy requirements of each car. The loads can be expressed in horsepower, kilowatts, or in kilowatt-hours, but all must be in the same units. The relation between the loads should be noted. For instance, if load c is taken as 1 and load e is four times as great, it is called 4; and if load g is five times as great as c, it is called 5. The load center for the length of road $a\ d$ is first determined. Suppose that the distance L between c and e is 7 miles, and the distance between the center of load h and e is l miles. The

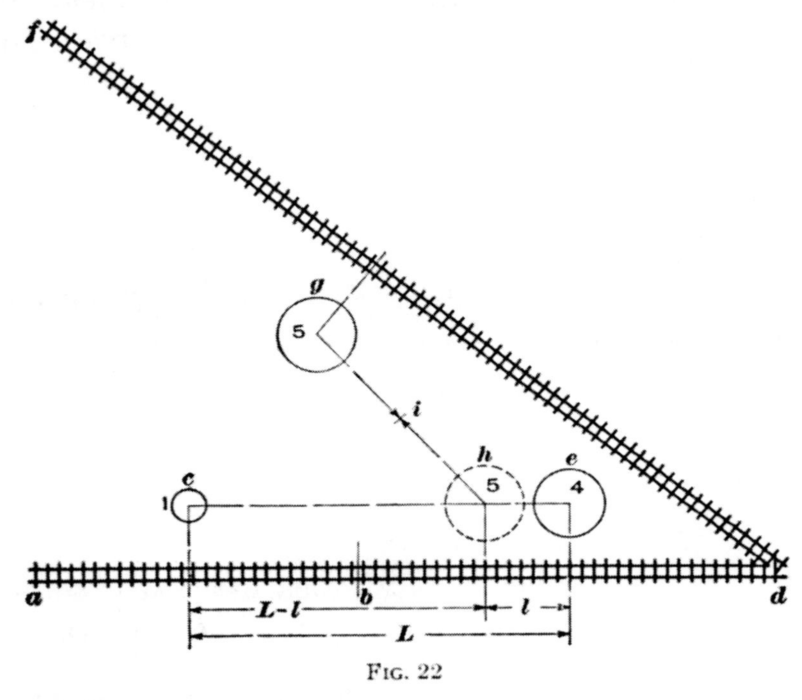

FIG. 22

center is located at such a point that the product of the lesser load 1 and the longer distance $L-l$ is equal to the product of the larger load 4 and the shorter distance l.

$$1\times(L-l)=4\ l, \text{ or } L=5\ l; \text{ as } L=7,\ 5\ l=7 \text{ and } l=1\tfrac{2}{5} \text{ miles}$$

The center of the concentrated load $1+4=5$ for the portion of the road $a\ d$ is at h between e and c and $1\frac{2}{5}$ miles from e.

The problem is now to find the final center of load of the whole system, which will lie between h and g. By consulting a map on which the route of the proposed road is traced, the distance between h and g may be estimated. In this case both loads

g and h have the value 5; therefore, the center of load will be at i midway on a line between h and g. If loads h and g are not equal the center of load is determined in the manner explained for determining h.

Considerations of the price of real estate, coal haulage, and water supply may cause the station to be located at other than the theoretical load center.

ENERGY COSTS

74. The cost of generating electric energy varies, because it includes items subject to wide fluctuations; in the same station the cost will be higher during some months than during others, because the load will vary both in amount and in the manner in which it is distributed throughout the 24 hours of a day. The ideal load condition would be the full station capacity, 24 hours per day 365 days in the year, and without material fluctuation. This is, of course, impracticable because of the comparatively limited demand for lights in the daytime and for power at night; conditions are considerably helped by carrying a mixed light and power load, because their maximums do not coincide. A station must be prepared to meet the maximum demand upon it. The maximum demand, however, may be active only a few hours of the twenty-four; during the time that the demand is below the station capacity, expensive machinery is absorbing interest on idle investment, and the costs of attendance and fixed charges are going on.

The *load factor*, or ratio of the average load to the maximum load, materially affects the cost of power, as shown by Table VII, giving cost per kilowatt-hour. The table is based on the following assumptions: Steam power plant of 10,000 kilowatts capacity; cost per kilowatt installed complete, $100; coal containing 12,000 heat units per pound of combustible, $2 per ton of 2,000 pounds; fixed charges for interest, depreciation, and taxes, 12 per cent. per year. This table shows that the cost per kilowatt-hour, including all charges except that of real estate, varies from $2\frac{1}{4}$ cents when the station operates with a load factor of 10 per cent. to .6 cent when the load factor is 100 per cent. It will be noted that in all cases the cost of coal is a large percentage of the total cost.

75. The cost per kilowatt-hour, not including interest and depreciation, will lie between .65 cent and 1 cent for many steam stations; in others the total cost, including interest, etc., will lie between 1 and 2 cents per kilowatt-hour. In the largest plants the cost, including interest, etc., may be considerably below 1 cent per kilowatt-hour.

When energy is sold to one railroad company by another, a common charge is 3 cents per kilowatt-hour. Assuming that

TABLE VII

COSTS PER KILOWATT-HOUR

Load Factor Per Cent.	Cost of Coal Cent	Cost of Labor Cent	Other Items Cent	Operating Charges Cent	Fixed Charges Cents	Total Cost per Kilowatt-Hour Cents
10	.60	.13	.12	.85	1.41	2.25
25	.45	.07	.08	.60	.56	1.16
50	.40	.05	.07	.52	.28	.80
75	.38	.04	.07	.49	.21	.70
100	.36	.03	.07	.46	.14	.60

a station is equipped with the highest grade generating and economizing devices, the cost of power is strongly affected by the distance of the station from the coal mine and by labor charges.

76. In order that the total cost of generating power in a station may be analyzed and money leaks checked before they become serious, a complete record of the various costs and of the station output must be kept. To this end, every station switchboard should be equipped with at least one recording watt-hour meter for registering the station output, and it is well to provide two such meters so that one can operate while the other is being calibrated. In case only one instrument is available, it should be checked at frequent intervals to insure correct records of energy consumption. By dividing the total cost of operation per day by the total output in kilowatt-hours, as gotten from the recording instruments, the operating cost per kilowatt-hour is obtained.

77. The fixed charges entering into the cost of generating energy consists of depreciation, interest on investment, insurance, and taxes. The depreciation in value of electric machinery depends on many conditions such as length of service, care, obsolescence, which is the quality of becoming out of date, change in load conditions making the capacity of a machine

TABLE VIII

COST OF STEAM POWER PLANTS AND EQUIPMENT

Capacity of Plant Kilowatts	Cost per Kilowatt	Capacity of Plant Kilowatts	Cost per Kilowatt
100,000	$ 60	10,000	$ 90
40,000	70	5,000	100
20,000	80	2,500	140

too great or too small, etc. Many ways of calculating depreciation are in use, the most common being to make a yearly allowance of a fixed per cent. of the first cost. The rate of depreciation is not the same for all parts of a power plant, but for practical purposes in determining the cost of energy an allowance of 5 per cent. for depreciation, 5 per cent. for interest, and 2 per cent. for taxes and insurance will generally give results approximately correct.

In estimating the first cost of a plant the figures in Table VIII will serve as a guide. These costs are fair averages for complete installations exclusive of land and buildings, for which from $10 to $20 per kilowatt must be added.

ELECTRIC-RAILWAY LINE CONSTRUCTION

OVERHEAD SYSTEMS

LINE POLES

WOODEN POLES

1. The term **line construction** as used here relates to the system of conductors and their supports installed on electric railroads for the purpose of transmitting electric energy from the substations or the low-voltage main stations to the cars. On the larger number of electric roads this system consists of some form of trolley wire supported by poles on one or both sides of the street or right of way.

2. Selection.—Wooden poles are used extensively for supporting the trolley wire and feeders on suburban, interurban, and some city roads. Poles with tops less than 24 inches in circumference should not be used except for very light work. The best poles are cut when the sap is down; to secure these, some managers send inspectors to the woods to mark consignments. Redwood is much used in the West; elsewhere cedar and chestnut, preferably second growth, are used in their natural condition. Octagonal poles are generally of hard pine; sawed poles have a better appearance than round ones in natural condition, but their life is shorter. Frequent paintings improve appearances and prolong life by keeping water out of checks and climber holes. Decay is further retarded by coating the ground end of the poles with coal tar or other preservative. The bases of the poles are left untreated, in order that they may absorb moisture and prevent dry rot. Creosote is commonly used as a preservative; but whatever is used, the treatment should extend above the ground line where the tendency to decay is most active.

Poles, when received, should be sorted and the best used on curves and as anchors. Table I gives data on weights and limiting dimensions of wooden poles.

3. Installation.—Wooden poles need not be set in concrete except when the soil is yielding. For poles from 25 to 50 feet long, the holes should be from 5 to 7 feet deep depending on the nature of the ground. In construction where a span wire stretched between poles on opposite sides of the street supports the trolley wire, two or three stones jammed against the pole butt on the side away from the track and two or three more near the mouth of the hole on the side next to the track, will keep the span wires from pulling the tops of the poles together. A piece of timber 8 inches by 8 inches by 3 feet may be used instead of the stones on the track side. The filler should be well tamped around the pole

TABLE I

WOODEN POLES

Length Feet	Net Weight Pounds	Minimum Circumference at Top Inches	Minimum Circumference 6 Feet From Butt Inches
25	325	24	38
30	460	24	38
35	600	24	42
40	710	24	45
45	1,075	24	48
50	1,375	24	51

during the filling, and the span wire should not be attached until the pole is firmly secured in position. On straight track and where side brackets on the poles support the trolley wire, poles should have a backward rake of 2 or 3 inches. With span-wire construction in yielding soil, the rake should be from 8 to 12 inches, according to the kind of pole foundation. To get equal rakes on all poles, a plumb or level must be used. In city work, poles are placed 100 to 125 feet apart.

METAL POLES

4. Construction.—The metal poles used in city work are usually made of three sizes of pipe telescoped together and welded, although poles made of seamless steel tube, pressed-steel parts riveted together, and latticework built up of structural steel are also used. Fig. 1 shows a tubular steel pole adapted to the three types of construction. That shown in (*a*) is suitable for side-bracket work, that in (*b*) for center-pole, and that in (*c*) for span-wire construction. The method of attaching the span wire to the pole is shown in the detail

TABLE II

METAL POLES

Style of Pipe	Diameter of Sections			Length Feet	Weight Pounds
	Bottom Inches	Middle Inches	Top Inches		
Standard......	5	4	3	27	350
Extra heavy...	5	4	3	27	500
Standard......	6	5	4	28	475
Extra heavy...	6	5	4	28	700
Standard......	7	6	5	30	600
Extra heavy...	7	6	5	30	1,000
Standard......	8	7	6	30	825
Extra heavy...	8	7	6	30	1,300

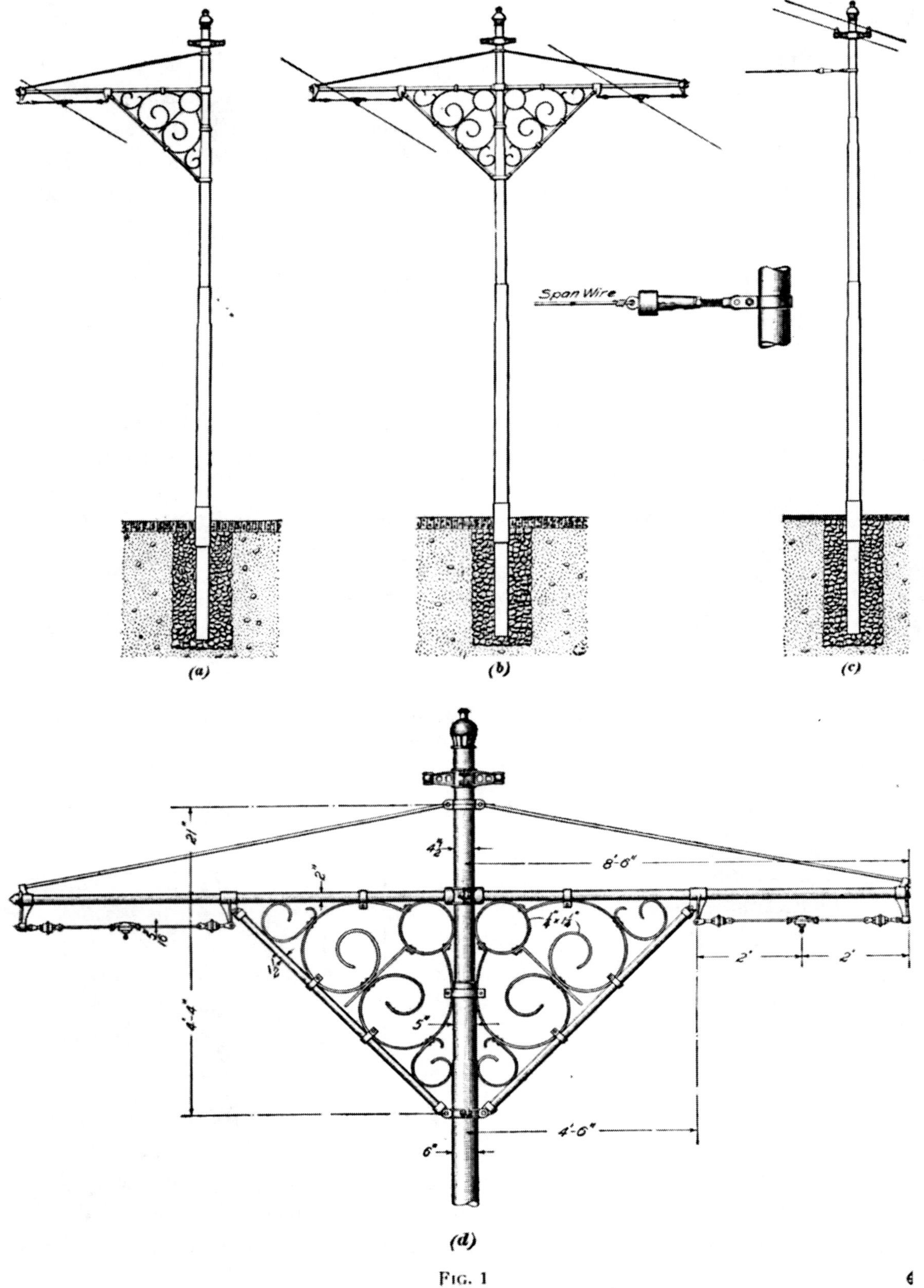

Fig. 1

sketch. Feeders are carried on iron cross-arms bolted to the pole, as in (*c*). An enlarged view of the top of the center pole is shown in (*d*). The trolley-wire hanger is flexibly supported from a short span wire stretched between brackets. Span wires are usually $\frac{5}{16}$-inch stranded steel cable. Table II gives data relating to metal poles.

5. Installation.—Metal poles are always set in concrete, for which the following is a suitable composition: Portland cement, 1 part; clean sharp sand, 2 parts; clean broken stone, 3 parts. The concrete lagging should be brought above the ground line, rounded off to shed water, and its contact surface with the pole well tarred.

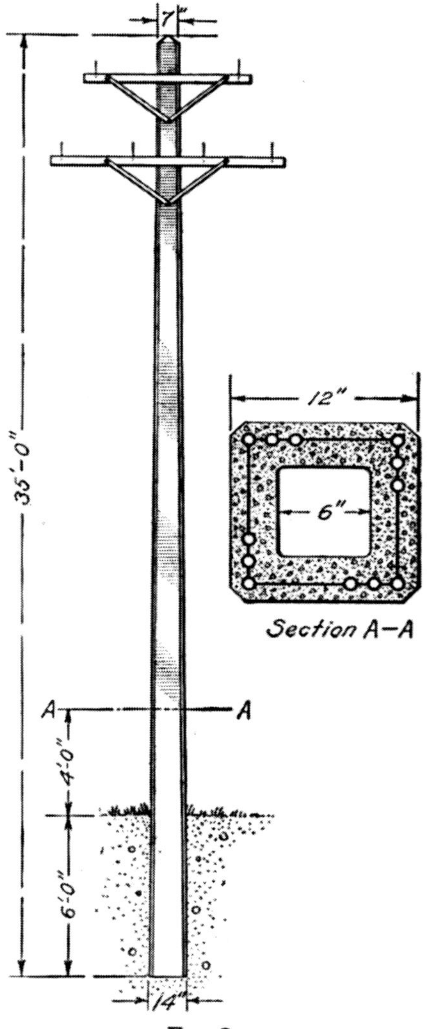

FIG. 2

CONCRETE POLES

6. Construction and Installation.—Poles made of concrete reinforced by metal rods are used on some roads. They are formed by placing in a mold long iron bars separated by spacing pieces and then pouring in the concrete. These iron rods greatly strengthen the pole. Some poles are made with the mold in a horizontal position and are thoroughly dried before being placed in the hole. Others are made with the mold in a vertical position over the hole; some form of elevating device is then necessary to lift the concrete to the top of the mold. Concrete poles have a long life and withstand weather conditions better than wooden poles.

Fig. 2 shows a typical concrete pole of hollow cross-section; it is 7 inches square at the top, 14 inches square at the bottom, and is 35 feet in length, 6 feet of which is in the ground. In the sectional view, the long reinforcing rods are indicated by circles and the spacing pieces are shown connecting them. These pieces are placed at intervals along the pole. The weight without fixtures is approximately 2,700 pounds. The concrete mixture is cement, 1 part; sand, 2 parts; crushed stone (not too large to pass through a $\frac{3}{4}$-inch screen), 3 or 4 parts.

REPAIRS AND RENEWALS

7. Repairs.—The life of tubular steel poles is not definitely known and their repair is practically limited to the results of collisions. They are exceedingly tough and difficult to bend and if bent, they are difficult to straighten. If badly bent the best procedure is to replace them. Cracked wooden poles may be sufficiently strengthened by binding with telegraph wire, the turns of which are held at intervals with staples. Another method of repairing wooden poles consists in forming around the affected length a concrete sleeve reinforced with steel rods.

8. Renewals.—Both metal and wooden poles are replaced without interruption to circuits or service by first installing, alongside the one to be replaced, another pole with cross-arms and insulators to which the wires can be conveniently transferred; the old fixtures and pole can then be removed.

FEEDERS

MATERIAL

9. Feeders serve to transmit electric energy to the different sections of the trolley wire, which is sometimes called the *working conductor*. Sizes 0000 B. S. and smaller feeders are usually of solid wire, but the larger feeders are of stranded wire. Either copper or aluminum cables bare, or with weatherproof braided insulation, are used. The bare aluminum wire has the property of shedding sleet. Feeders, when installed in ducts, are usually in the form of stranded cables, insulated and provided with an outside covering of lead.

TYPICAL FEEDER LAYOUTS

10. The simplest feeder layout is a single wire serving both as trolley wire and feeder. With this arrangement and a heavy load, the voltage at the end of the line will probably be low. On some roads, a cable is run parallel to the trolley wire and tapped to it at intervals. Both trolley wire and feeder then serve as outgoing conductors.

When the trolley wire is divided into sections insulated from one another, an independent feeder for each section is

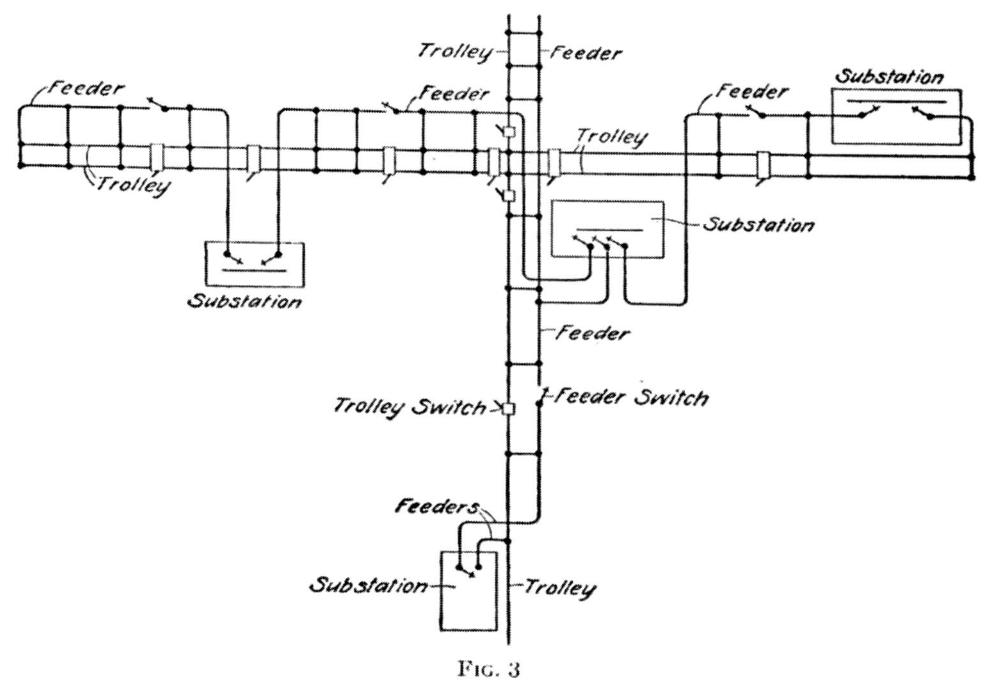

Fig. 3

sometimes run from the station; in other cases, several feeders are used and taps from each feeder are tied into one or more of the trolley-wire sections. An insulator directly connected to the ends of the two trolley wires serves to separate the section. The shape of this insulator is such that the trolley wheel may pass smoothly under it from one trolley wire to the other. In such cases, if trouble occurs on a trolley section or on a feeder, only a portion of the system need be cut out of service while repairs are being made.

Switches in the feeder circuits and switches at the ends of the trolley-wire sections allow considerable flexibility in routing the current from one section of the road to another section. In case of trouble at one substation, the feeders and trolleys normally fed by that substation may be supplied with energy from the feeders and trolleys of the adjacent section of the road, which are fed by another substation.

11. In Fig. 3 is shown the general layout of substations, feeders, and trolleys for a 1,200-volt, direct-current, interurban road. The portion of the system represented by the horizontal lines of feeders and trolleys is a two-track road with two trolley wires. The portion represented by the vertical lines of trolley and feeder is a single-track road with one trolley wire. By means of the trolley switches and the feeder switches at the ends of the sections, the sections may be disconnected from each other or tied together as desired. With this arrangement, if one substation is disabled, the trolley may be supplied with current from another substation.

FEEDER SECTION SWITCHES

12. Feeder Tap Switch.—Fig. 4 shows a section switch of a type that is sometimes used to connect a feeder with a trolley section. The hinge clip of the switch is connected to the trolley wire. Opening the switch cuts the trolley section out of circuit. A fuse is sometimes installed in the switch box as shown. The box is usually mounted on a pole near the tap point of the feeder.

13. Feeder Automatic Sectionalizing Switch.—Independent feeder construction has the advantage that sections are isolated in time of trouble. Construction in which all of the feeders are connected together out on the line has the advantage that the full current-carrying capacity of all feeders is utilized so that less feeder conductor is required and overloaded feeders are helped by the lighter loaded ones. By means of *automatic sectionalizing switches* connected to adjacent ends of trolley sections the full advantages of both systems are realized.

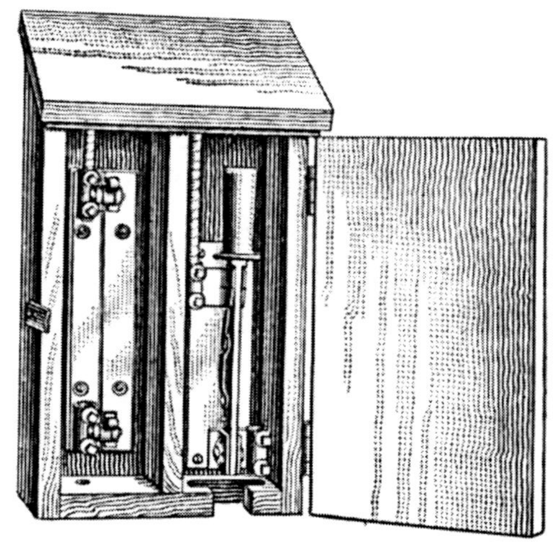

Fig. 4

Fig. 5 shows a layout for three feeders, in which one automatic switch is shown developed and another outlined. Trolley sections *a*, *b*, and *c* are fed from feeders controlled by switches and circuit-breakers *d*, *e*, and *f* in the power house. When circuit-breaker *f* and its switch are closed, the operating coil *g* of the relay is energized through the path including the right-hand switch blade *h* and the ground; the relay then operates and connects together the two contacts below it. When circuit-breaker *e* and its switch are closed, the operating coil *i* of the switch that bridges the line insulator is energized; the switch contacts shown below coil *i* then close and a conducting path is formed around the

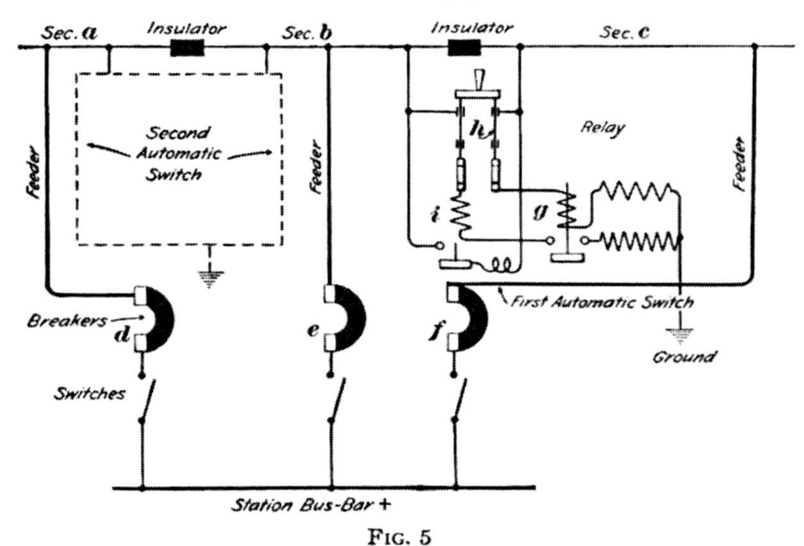

Fig. 5

insulator between the ends of the trolley sections. Current may then pass from one trolley section to the next.

The current path of coil *i* includes the contacts under coil *g*; therefore, if either coil *i* or *g* is deenergized, the switch under *i* is operated and the trolley sections are separated. If a short circuit occurs on any of these sections, all of the circuit-breakers will open because the sections are tied together, and all the automatic switches immediately open and separate the trolley sections. If the trouble is in section *a*, circuit-breaker *d* will not stay closed, but the other circuit-breakers, such as *e* and *f*, may be closed. With *e* and *f* closed, the automatic switch shown on the right becomes active and joins section *b* and *c*, but the automatic switch indicated at the left is not active, because the left-hand operating coil corresponding to coil *i* of that switch does not receive current when circuit-breaker *d* is open. The automatic switch to the left of section *a* (not shown) is also inactive because the right-hand operating coil corresponding to coil *g* of that switch is deenergized; therefore, trolley section *a* is cut out of circuit.

When repairs are made and circuit-breaker *d* is closed, the automatic switches to the right and left of section *a* are closed and all of the trolley sections are again tied together.

SPLICING FEEDERS

14. Copper Wires.—Small feeders of solid wire may be joined by some form of Western Union joint as shown in Fig. 6. The wires are twisted together, soldered, and insulated with tape. A solution of resin in alcohol makes a good soldering flux for this purpose.

Large stranded copper cables may be joined by weaving the strands together as indicated in Fig. 7 (*a*) and (*b*). The copper joint should be soldered by pouring melted solder over and through it.

Fig. 6

In some cases the joints are made by soldering the ends of the cables into a copper sleeve. All splices should be taped to give insulation equivalent to that of the regular insulation on the wire. The tape covering should be painted with water-proof paint.

15. Aluminum Wires.—Stranded aluminum feeders may also be joined by the method indicated in Fig. 7, except

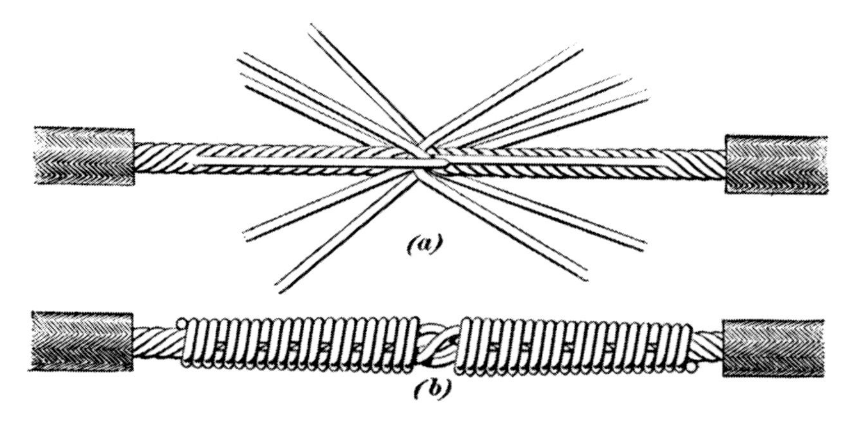

FIG. 7

soldering is not required. The joint is made as follows: Remove the insulation from 2 feet of the ends to be spliced; open out the wires; clean and straighten them; cut out a few of the inner strands so that the final joint will not be too bulky; place the cables together as shown in Fig. 7 (*a*); then wrap all of the strands one at a time around the wires forming the core, the wires of the two cables being wrapped in opposite directions; tape and paint the joint.

16. The McIntyre joint shown in Fig. 8 is much used on aluminum wires or cables not exceeding No. 0000 B. S. in size. The ends of the wires to be joined are inserted side by side into an aluminum tube and gripped by a special tool having a groove of the same shape as the tube; the tube and contained wires are then given from $2\frac{1}{2}$ to 4 complete twists.

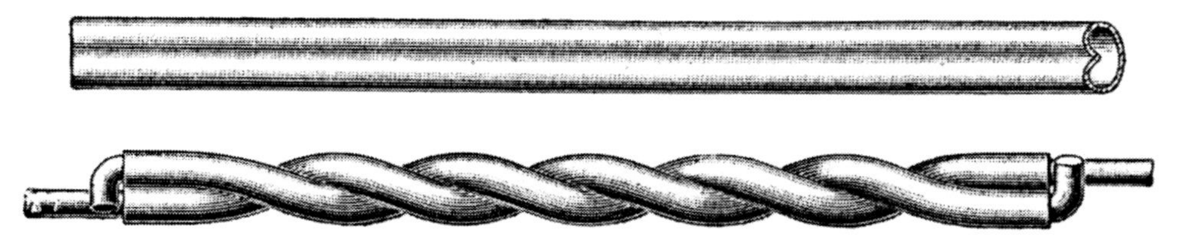

FIG. 8

The joint is easily and quickly made and its mechanical strength and electric conductivity are good.

17. On account of the stiffness of large aluminum cables, joints of the compression type, one form of which is shown in Fig. 9, are often used. The ends of the cable are inserted

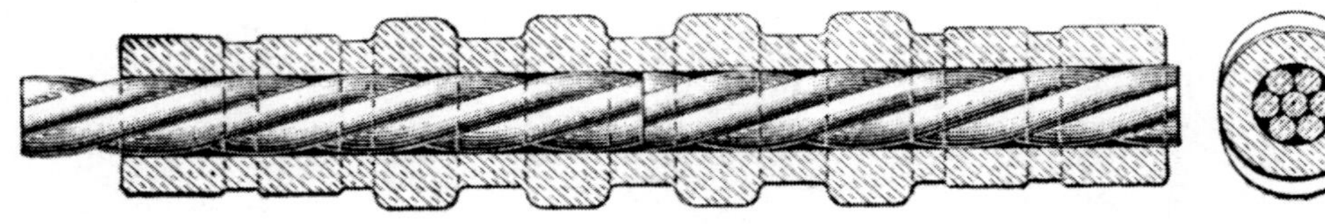

Fig. 9

into a cast aluminum sleeve of proper size, the ends butting together in the center of the sleeve. The sleeve with its cables is then inserted between dies in a portable hydraulic jack, and sufficient pressure is applied to the dies to cause the metal of the sleeve and the metal of the cables to flow together into a solid, homogeneous mass. When tested, the cable will be pulled in two instead of the ends of the cables pulling out of the sleeve.

A modified form of the joint just described is shown in Fig. 10. Terminal pieces of aluminum are compressed on

Fig. 10

the ends of the cables at the factory. One terminal has a right-hand thread and the other terminal a left-hand thread. The ends of two cables are joined by screwing into the two terminals a right-and-left-hand threaded stud *a*, which draws the terminals into firm contact with the stud.

FEEDER INSULATORS

18. Glass insulators may be used for feeders of ordinary size. Fig. 11 shows one that is adapted for either top or side tying. The top groove is usually employed for straight-line construction and the side groove for curves and corners where the stresses that tend to break the insulator are greater.

For large feeders, insulators of the type shown in Fig. 12 are sometimes used. These are provided with metal clips that are bent over the wire, thus holding it in position. The insulator is made of molded compound insulation, the clip

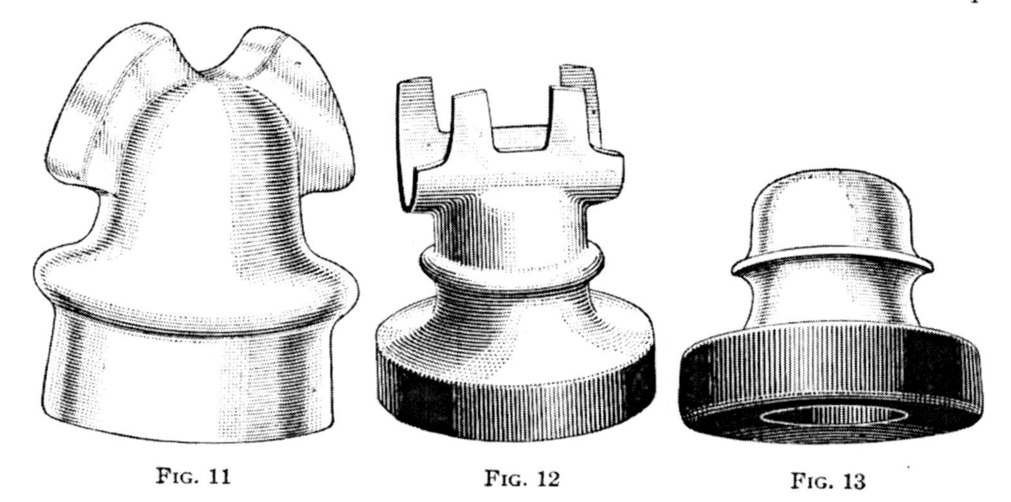

FIG. 11 FIG. 12 FIG. 13

forming part of a metal shell mounted on the top portion.

Fig. 13 shows a type of insulator used for corners only. The top part is covered by a metal shell and the cable lies in the groove at the lower portion of this cap.

TROLLEY WIRES

MATERIAL

19. Trolley wire is usually of hard-drawn copper, but tough composition wire is sometimes used on curves where the wear is heavy. The size is seldom less than No. 0 B. & S.; a few roads use No. 000 or 0000 B. & S., but No. 00 is the most common size.

Hard-drawn copper has greater tensile strength, better wearing qualities, and slightly higher resistance than soft copper; with a good feeder system the trolley carries current such short distances that the effect of higher resistance is negligible.

At curves and at other places where there is heavy wear and stresses, phono-electric wire is frequently used. This is an alloy wire; its tensile strength is 40 per cent. greater than that of hard-drawn copper wire; and its conductivity is 50 per cent. of that of pure copper.

CROSS-SECTIONS OF TROLLEY WIRES

20. Trolley wire for moderate car speeds is usually of round cross-section, Fig. 14 (*a*). An ear, forming part of a hanger used to support the wire, is either clamped or soldered to the trolley wire. With round wires there is some possibility of the trolley wheel striking the edges of the ear and temporarily separating the wheel and wire, thus causing sparking and final destruction of the hanger. By careful installation, however, the obstruction to the passage of the wheel is comparatively slight.

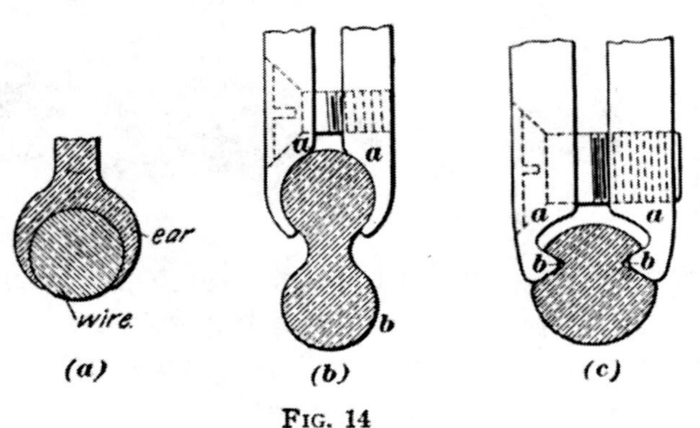

Fig. 14

For high-speed service, the under portion of the trolley wire should offer a perfectly smooth surface for the wheel to run on. Fig. 14 (*b*) shows a cross-section of a wire sometimes used. The upper part of the wire is gripped by the clamp ears *a*, leaving the lower part *b* unobstructed.

Fig. 14 (*c*) shows a standard form of grooved trolley wire for high-speed work. It is supported by clamp ears *a* that fit into grooves *b*. The under portion of the wire is unobstructed and the trolley wheel passes smoothly under the hanger.

SPLICING TROLLEY WIRES

21. Trolley-wire splices must be strong enough to stand heavy mechanical stresses and must offer minimum obstruction to the passage of the trolley wheels. The most common

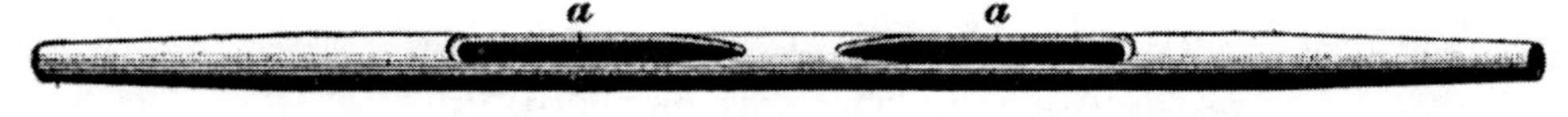

Fig. 15

form of splice, Fig. 15, is made with a tapered tinned brass sleeve; the wires enter at each end and are bent up through the openings *a*; the remaining space is then sweated full of solder and the ends of the wires trimmed.

Fig. 16 shows a form of mechanical splicing sleeve especially adapted to temporary or emergency work. The wires are held by the steel wedges *a*. This sleeve has requisite strength and conductivity, but offers more obstruction to the under running

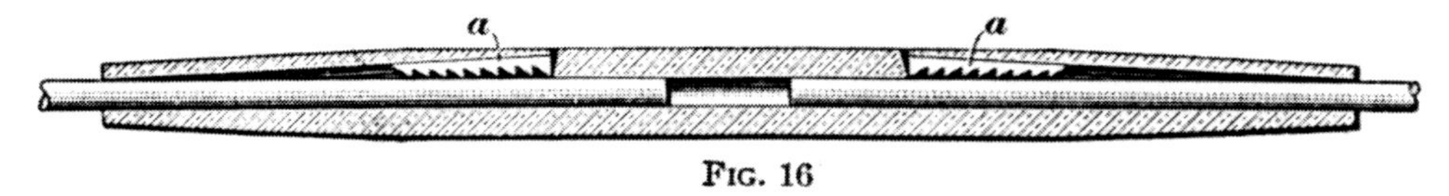

FIG. 16

trolley wheel than does the soldered splice. The sleeve splice may be made anywhere on the wire circuit.

Some forms of hanger ears that are used to support the wire also serve to connect the ends of two adjacent pieces of trolley wire as explained later. This form of splicing device is used only at the point of support of the wire.

TYPES OF TROLLEY-WIRE SUSPENSION

22. Span-Wire Construction.—In Fig. 17 is shown a form of suspension known as the **span-wire construction.**

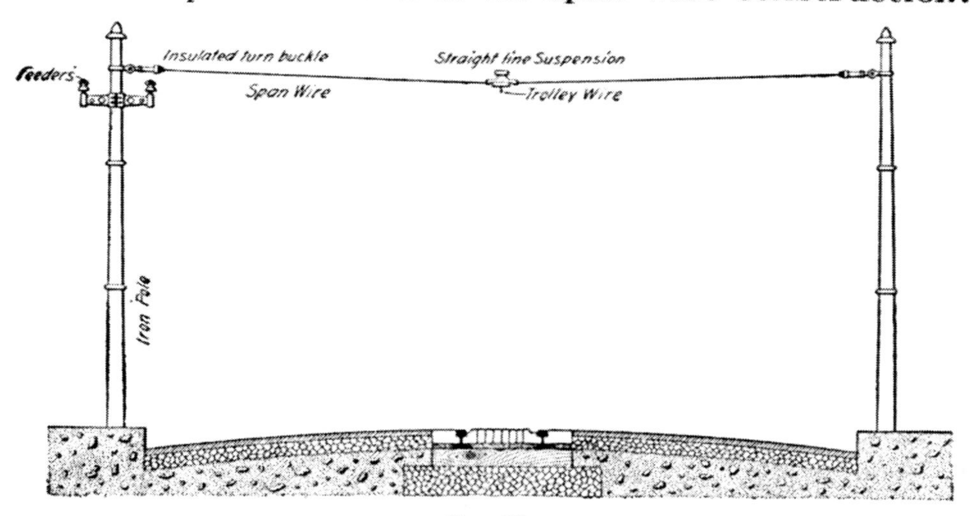

FIG. 17

Steel span wires are supported from poles located on opposite sides of the roadbed, or street, and the trolley wire is suspended from the span wires by means of insulated hangers. In case of double track, two hangers on the same span wire are used; each hanger supports a wire over the center of its track. Feeder cables are often supported on cross-arms placed on the poles as indicated in Fig. 17.

23. Center-Pole Construction.—Center-pole construction is adapted to wide streets where the poles will not interfere

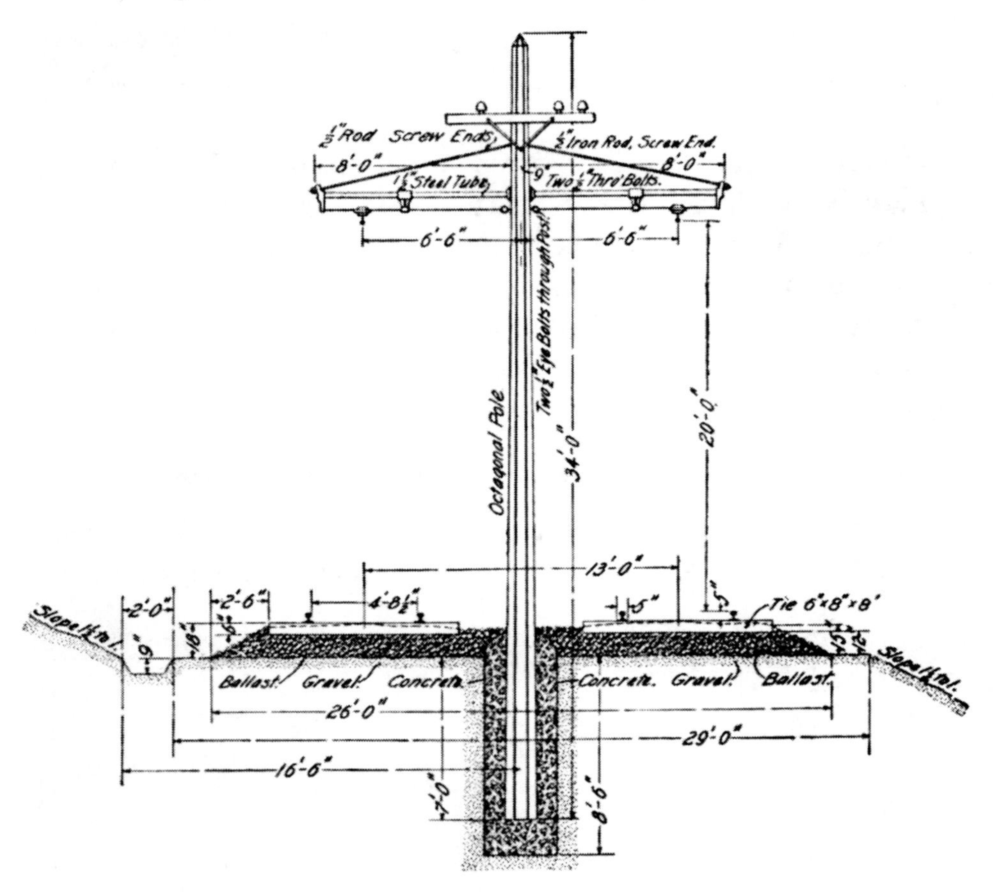

Fig. 18

with traffic; it is much used on interurban roads. Fig. 18 shows one type of center-pole construction, in which yellow-pine octagonal poles are set in concrete and a No. 000 trolley wire is suspended 20 feet above the center of each track. The hangers are attached to small, stranded-steel, span wires, which make the suspensions flexible and receive the blows of the trolley wheel as it passes under the hangers; the life of the overhead rigging is thus extended.

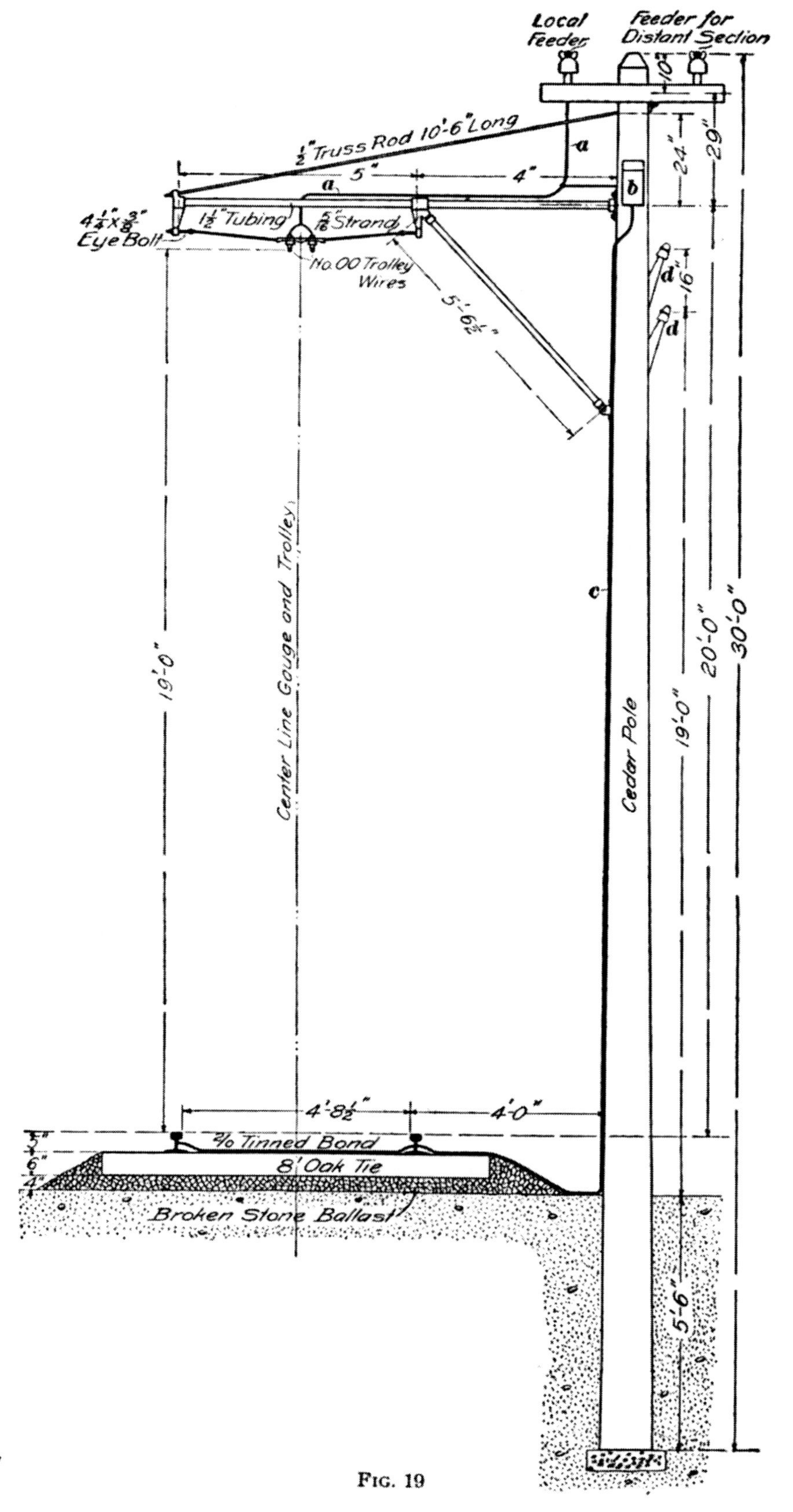
Local Feeder
Feeder for Distant Section
10"
1/2" Truss Rod 10'-6" Long
a
24"
29"
5"
4"
a
b
4 1/4" x 3/8" Eye Bolt
1 1/2" Tubing
5/16" Strand
No. 00 Trolley Wires
5'-9 1/2"
d
16"
d
Center Line Gouge and Trolley
c
19'-0"
Cedar Pole
19'-0"
20'-0"
30'-0"
4'-8 1/2"
4'-0"
5"
6"
4"
2/0 Tinned Bond
8' Oak Tie
Broken Stone Ballast
5'-6"

Fig. 19

24. Side-Bracket Construction.—Side-bracket construction is often used when the track is on one side of the street. Fig. 19 shows one type of side-bracket construction. In this case two trolley wires are used. Cars going in one direction use one wire and in the opposite direction, the other wire. At a *turnout*, which is a short length of double-track construction, one of the two wires is led to a position over one track and the other wire to a corresponding position over the other track. Feeders are carried on cross-arms and the local feeder is tapped to the trolley wires by the conductor *a*. A lightning arrester *b* is provided at intervals and is connected between the feeder tap and the ground wire *c*. Telephone wires are installed on brackets *d*.

TROLLEY-WIRE ERECTION

25. When a dead trolley wire is to be erected, it is generally unreeled under the span wires in the middle of the track and tied with temporary tie wires to the span wire. The wire is then put under tension and permanently fastened to the hangers.

When a live trolley wire is to be erected, the reel is often mounted on a flat car pushed ahead of the construction car. The wire, which is put under tension by means of the forward

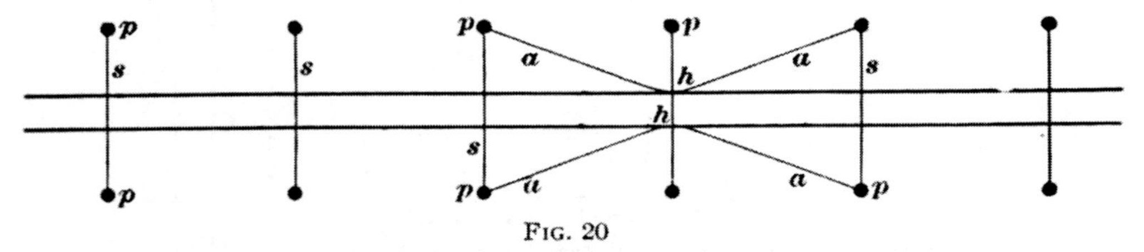

Fig. 20

movement and by brakes on the reel, is immediately fastened permanently to the suspension insulators on the span wire.

26. Typical Tangent Construction.—Fig. 20 shows a common arrangement of span wires and trolley wires for a double-track road. The poles *p* are spaced about 125 feet apart along the road and between opposite poles are stretched the span wires *s*. Anchor wires *a* are provided at intervals of about 500 feet on straight track and at the approach of curves, to hold the trolley wire in place and to prevent tearing it from the hangers if it breaks. The trouble due to the break

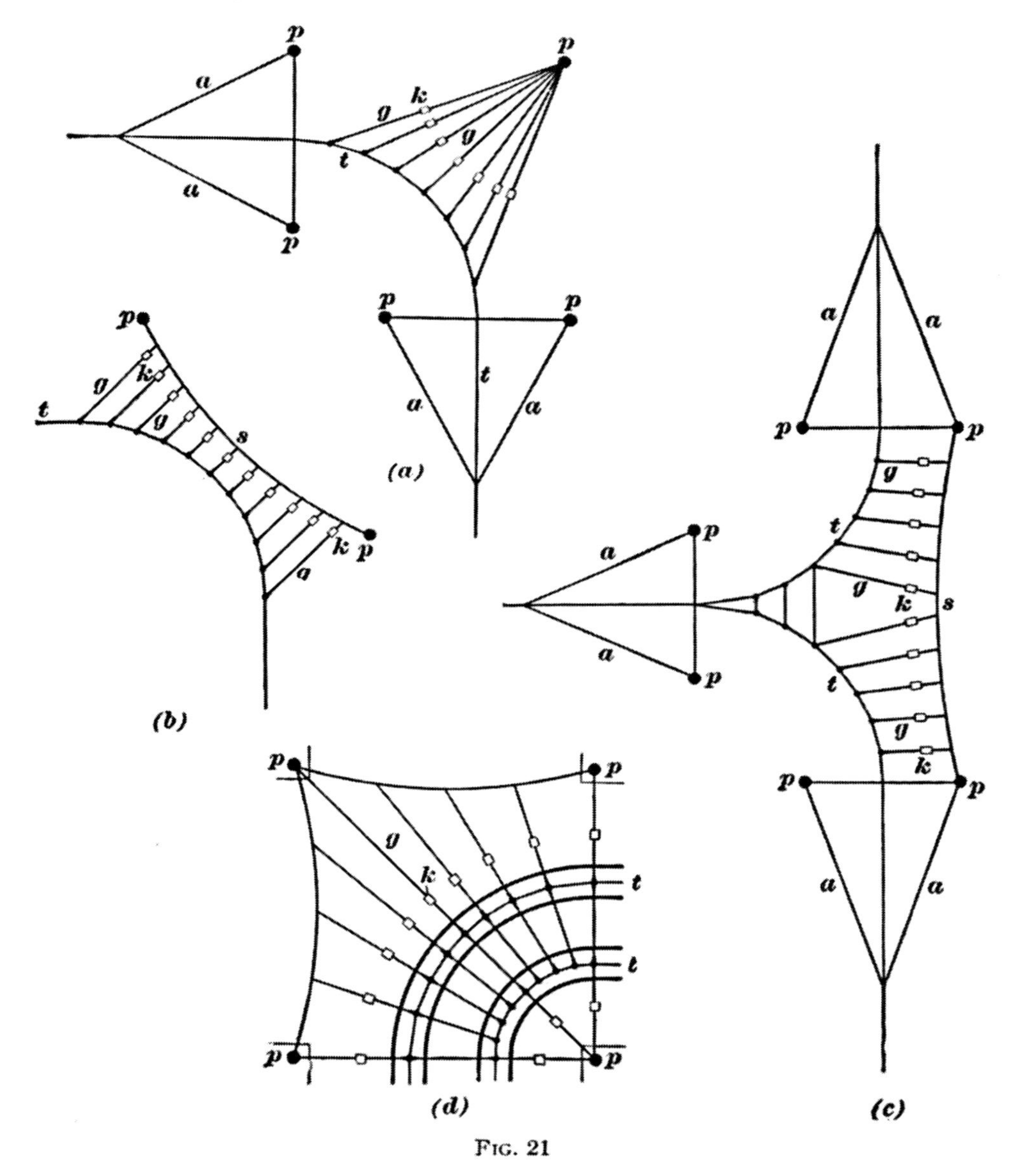

FIG. 21

will be localized to the length of trolley wire between the two adjacent anchor points, as the trolley wires to the other sides of these points remain firmly supported.

27. Typical Curve Construction.—Fig. 21 shows methods of securing the trolley wire at curves. In (*a*) is shown an arrangement of guy wires g, called *pull-over* wires, from a single pole p to the trolley wire t. Strain insulators are shown at k. The straight portion of the wire at the ends of the curve is held in place by anchor wires a.

In (b) is shown a flexible method of suspension. A large span wire s supports the guy wires; this construction tends to equalize the stresses on the span wires and it is quite generally adopted.

A method of suspension for a double curve is shown in (c). In (d) is shown a method of suspension used for the two trolley wires of a two-track construction. The pull-over wires are connected to two heavy span wires and the pulls are made nearly at right angles to the trolley wires.

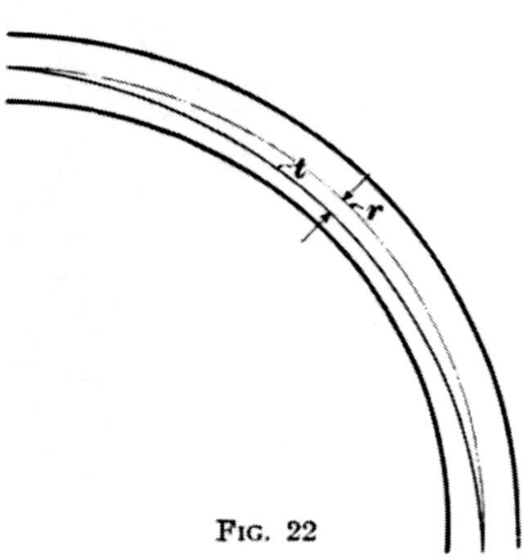

FIG. 22

28. Offset in Trolley Wire at Curves.—In rounding a curve, the trolley wire does not follow the track center line but is shifted toward the inside rail of the curve; the distance depends on the radius of the curve. The offset from the track center line is indicated in Fig. 22, in which the curve r is the track center line and t the path of the trolley wire. A set of values for the offset, indicated in Fig. 22 by the distance between the arrows, and measured at the middle of a curve of 90° arc, is given in Table III. These values may be modified to suit operating conditions.

The offset allows the trolley wheel to follow better the curve of the trolley wire than it would be if the wire followed the track center line. This feature decreases the amount of wear on the trolley wheel and the wire at curves and the tendency of the wheel to climb off the wire.

29. Trolley-Wire Tension.—Wire strung in hot weather must be allowed more sag than wire strung in cold weather, otherwise, contraction may break the wire or strain the whole overhead construction. A range of 80° F. between summer and winter temperatures corresponds to a variation of nearly

4 feet a mile in the length of the trolley wire. For a 125-foot span of No. 0 wire installed with a tension of 2,000 pounds, the sag at the center of the span should be 3.8 inches; for a

TABLE III

OFFSET IN TROLLEY WIRE

Radius of Curve Feet	Offset Inches	Radius of Curve Feet	Offset Inches
40	16	100	6
50	13	120	5
60	12	150	4
80	8	200	3

tension of 1,500 pounds, 5 inches; for 750 pounds, 9.5 inches; for 500 pounds, 15 inches. It has been recommended that for localities where the temperature does not fall below −20° F., the sag should equal three-fourths of 1 per cent. of the length of the span when the wire is strung at the ordinary temperature of 60° to 65° F. Experience has shown that for a 125-foot span if a sag of $125 \times 12 \times .0075 = 11\frac{1}{4}$ inches is allowed, the sag in the warmest weather will not exceed 15 inches.

30. Span Wires.—Span wire is usually of galvanized iron or of steel; a common size is $\frac{5}{16}$ inch in diameter and composed of seven strands of No. 12 B. W. G. In heavier construction, $\frac{3}{8}$-inch cable composed of seven strands of No. 11

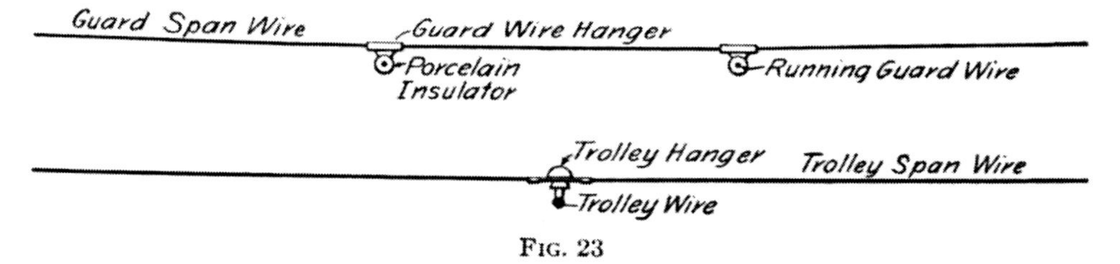

Fig. 23

B. W. G. is used. Solid span wire is not recommended, but if used, should in no case be smaller than No. 1 B. & S. for No. 0 trolley wire. Span wires should be so placed that the trolley wire will be from 19 to 20 feet above the top of the track rails. At steam-railroad crossings it may be necessary to place it higher and under elevated structures, lower than 19 feet.

31. Guard Wires.—When guard wires are used, they are generally in pairs suspended from a separate span wire; each guard wire is located about 18 inches above and to one side of the trolley wire, as shown in Fig. 23, to prevent telephone and other wires from falling on the trolley wire. No. 6 or No. 8 B. W. G. wire with weather-proof insulation is commonly used for this purpose. The installation of guard wires has ceased to be standard practice except in special cases such, for example, as at the crossing of high-tension wires over the trolley wire, because their expense is out of proportion to their utility, except under special conditions.

AUXILIARY TROLLEY-WIRE DEVICES

32. Span-Wire Insulators.—The trolley hangers are so made as to provide insulation between span wire and trolley wire, except when the hanger itself forms part of a tap circuit from feeder to trolley. With wooden poles, the hanger insulation is sufficient; with iron poles, one and sometimes two

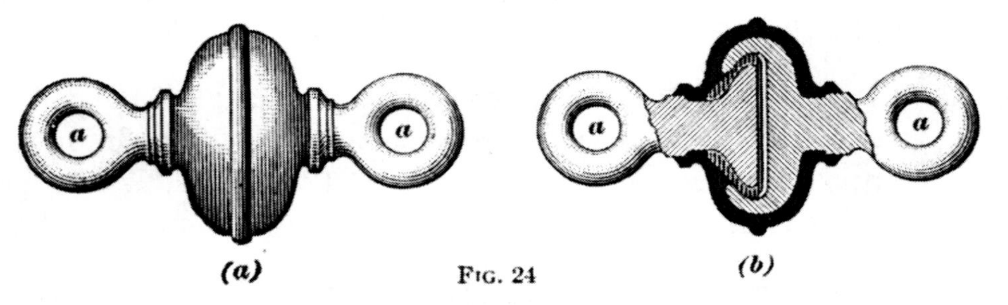

(a) Fig. 24 (b)

span-wire insulators are usually connected to the span wire between each iron pole and the hanger, in order to improve the insulation of the circuit.

Fig. 24 illustrates a **span-wire insulator**, which is commonly known in the trade as a **strain insulator**, view (*a*) showing the exterior and view (*b*) the interior construction. The span wires are attached to metal eyes *a*, and these are insulated from each other by sheet mica and by the molded insulation which covers all of the device except the eyes. The metal parts within the insulation are so interlocked as to withstand the stress to which they are subjected by the span wire without pulling apart.

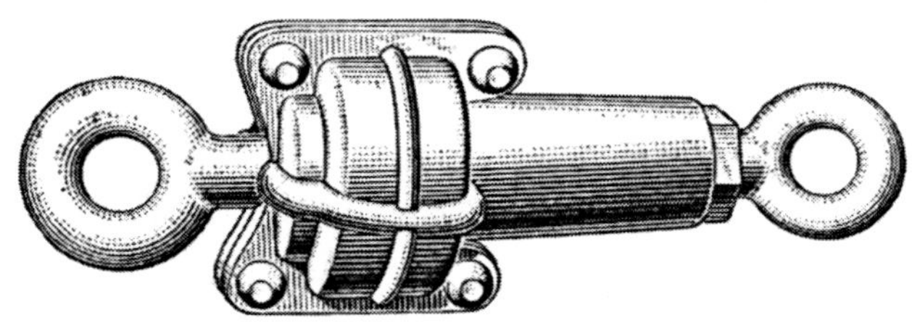
Fig. 25

Fig. 25 shows an insulated turnbuckle that may be used to tighten and to insulate the span wire for construction using iron poles. Fig. 17 shows two of these turnbuckles in place on the span wire.

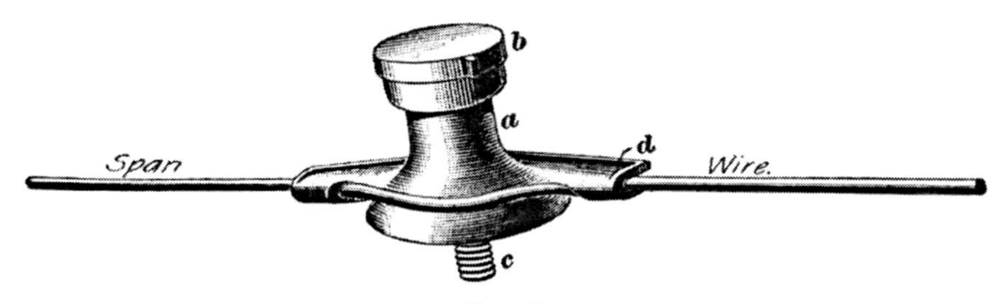

Fig. 26

33. Suspensions.—In general, **suspensions,** or **hangers,** consist of a body casting, which is attached to the span wire, an ear, which grips or is soldered to the trolley wire, and insulation between the ear and the body casting. Fig. 26 shows a common form of suspension with the ear removed. The body casting is shown at *a*, a screw cap that holds the insulated bolt in position at *b*, the insulated bolt that holds the ear at *c*, and the grooved arms that hold the hanger in place on the span wire at *d*.

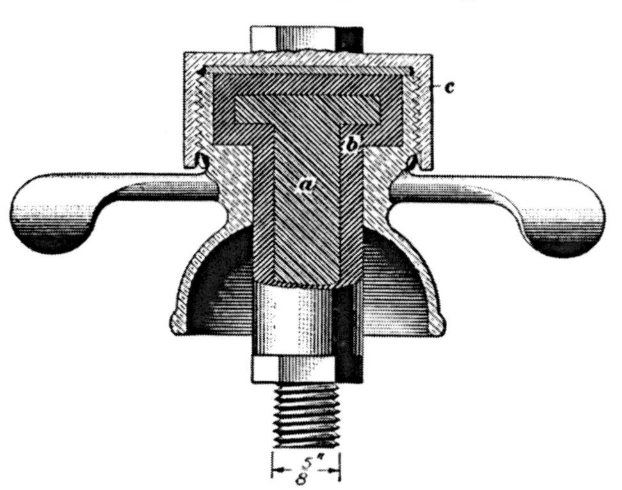

Fig. 27

Fig. 27 is a sectional view of a body casting of a suspension very similar to Fig. 26. The insulated bolt *a* is provided

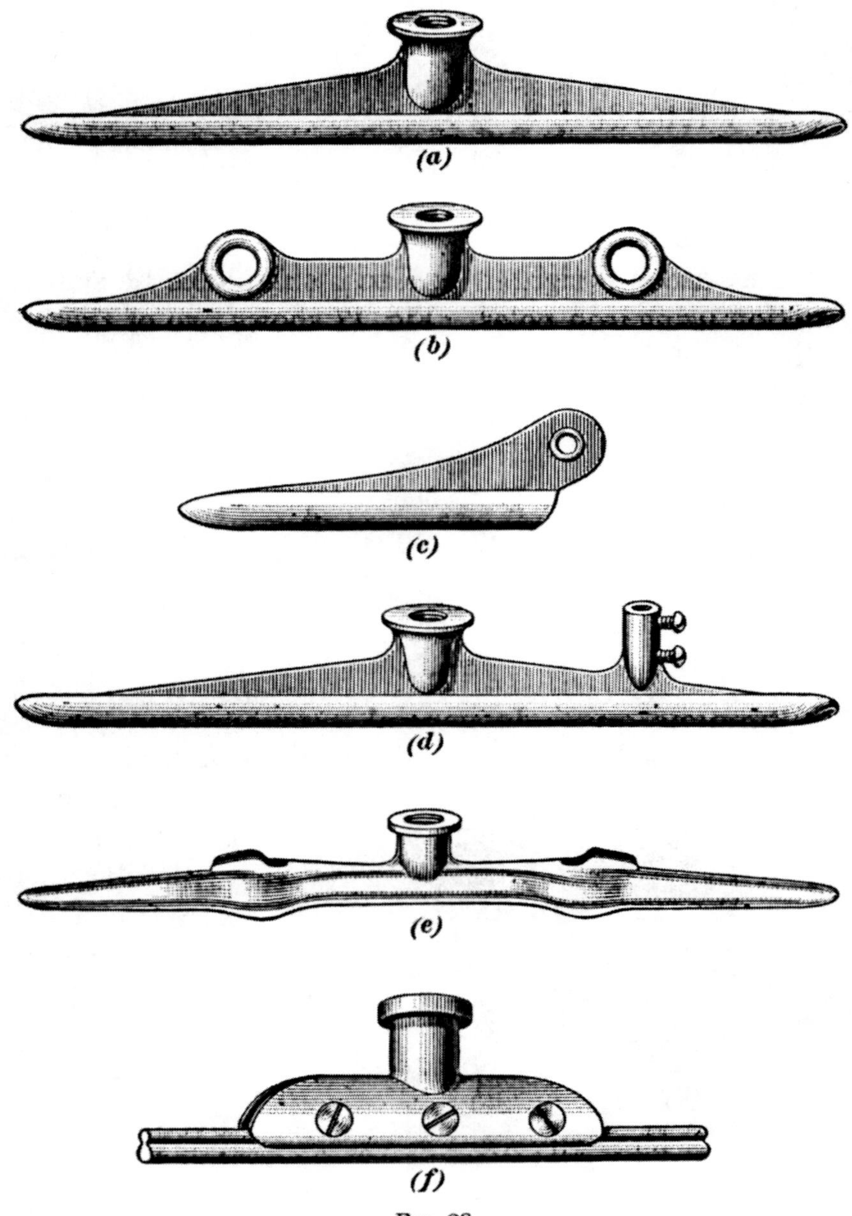

FIG. 28

with molded insulation *b*, and is held in place by the cap *c*; the bolt may be removed by taking off the cap.

34. The ears are made in many forms to suit the special work for which they are intended. Fig. 28 shows a few of these forms. The *plain ear*, Fig. 28 (*a*), is used for straight, or *tangent*, work; the *strain ear*, (*b*), to take anchor wires from both directions as in Fig. 20; the *strain ear*, (*c*), to take anchor wires from one direction as in Fig. 21 (*a*); the *feeder ear*, (*d*), to take the tap wire from the feeder to the trolley hanger; the *splicing ear*, (*e*), to serve simultaneously as a hanger and splice in which the abutting ends of the two trolley wires pass up through the openings and are bent back; the *clamp ear*, (*f*), to clamp the trolley wire and hold it without the use of solder. All of the ears shown, except (*f*), are to be soldered to the trolley wire.

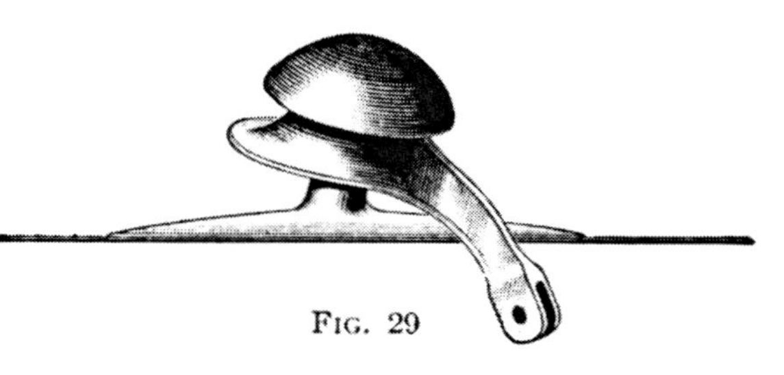

Fig. 29

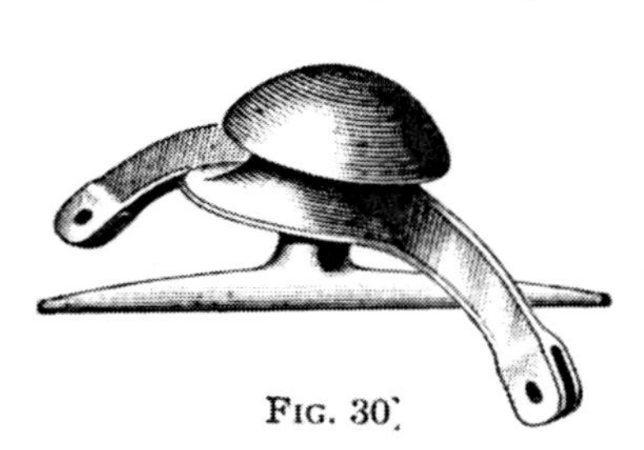

Fig. 30

35. When constructing a curve, suspensions of the type shown in Figs. 29 and 30 are used; the first for construction with a single trolley wire, and the second for construction with two trolley wires in cases where a span wire or a pull-over wire must be attached to each side of the suspension. Methods of using suspensions of this type are indicated in Fig. 21.

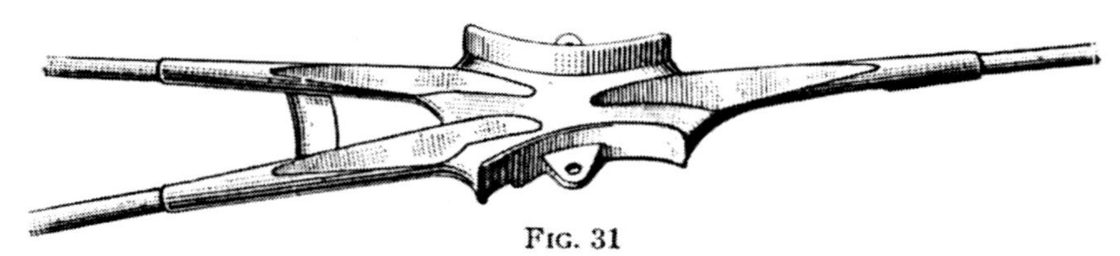

Fig. 31

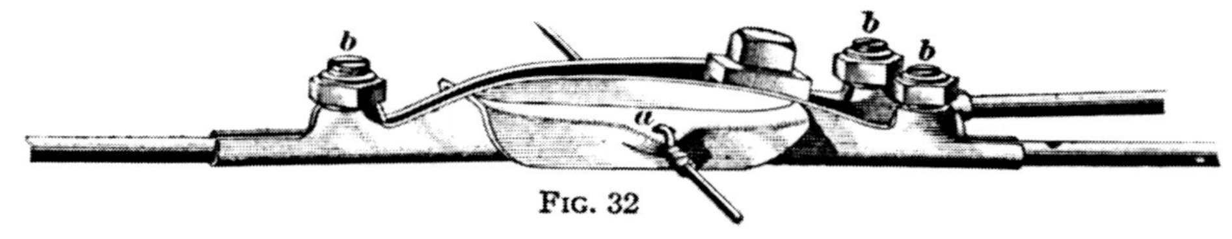

Fig. 32

The metal castings for hangers may be of malleable iron or of brass; soldered ears are of brass; and clamp ears are usually of malleable iron. but may be stampings.

36. Frogs.—Where a line branches or at draw bridges, overhead switches or **frogs** are used to guide the trolley wheel from one trolley wire to another. Fig. 31 shows the under side of a two-way **V** frog and Fig. 32, a frog of the

Fig. 33

V type in its proper position. The span wires are attached at *a* and the trolley wires are held by clamps *b*. Frogs are also made for a right- or a left-hand turnout from a continuous straight track.

Fig. 33 shows a form of frog used at drawbridges. When the bridge is closed, a rib on the underside of the frog is in line with a similar rib on the adjacent frog, and large contact surfaces on the two frogs engage and complete the circuit.

Mechanical fastenings for the trolley wires are desirable, because they allow the frogs to be adjusted by trial for proper position. If a frog is level, the wheel will probably follow the car; if the frog sags to one side, the wheel is liable to be thrown off the wire.

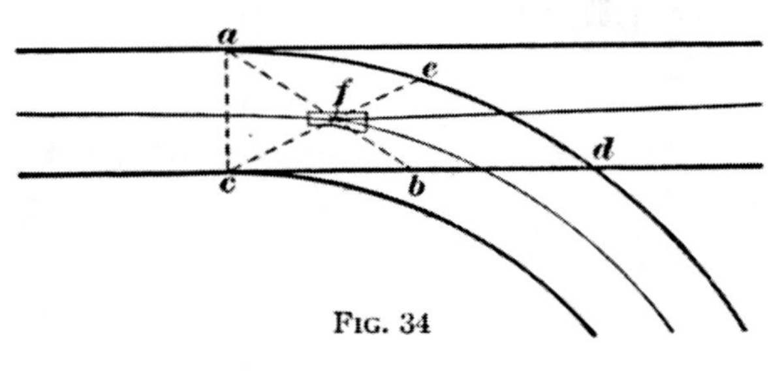

Fig. 34

37. Locating a Frog.—A method for determining the position of a frog in the overhead work, indicated in Fig. 34, is as follows: From switch point *a* draw a line to the center point *b* of frog distance *c d*, and from switch point *c*, draw a line to the center point *e* on arc *a e d*. The intersection of these two lines at *f* indicates the location of the trolley frog. The lines may be laid out on the ground, the intersection found, and a plumb-bob used to locate the frog on the wire above the intersection. Slight adjustment of the position of the frog may be made by the turnbuckles on the span wires.

38. Crossings.—At the intersection of two trolley lines, a device called a **crossing,** or **cross-over,** is installed. Fig. 35

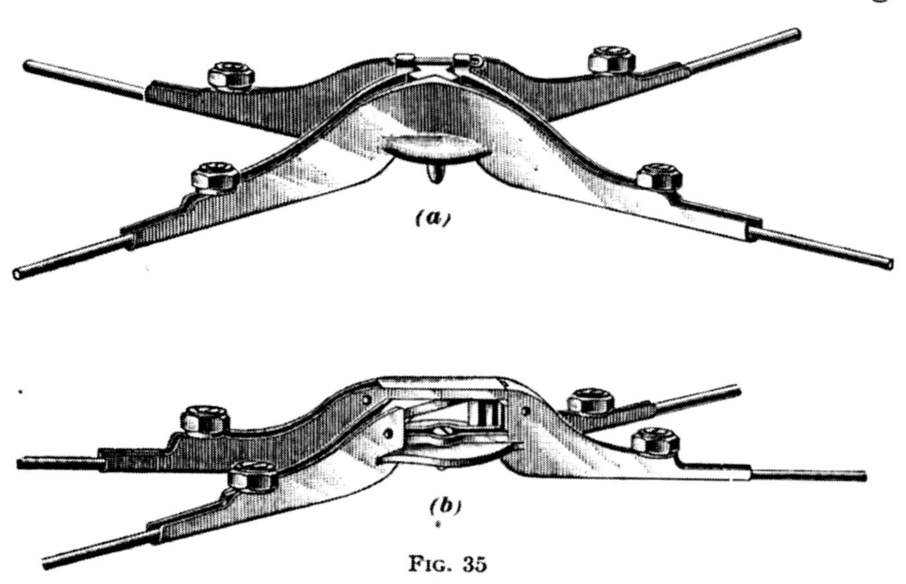

Fig. 35

shows two forms, (*a*) for a right-angle crossing and (*b*) an adjustable crossing for any angle between 30° and 90°. Fig. 36 shows a right-angle crossing in which the conducting circuits are insulated by hardwood; this crossing is used where the two circuits are to be kept electrically separate.

39. Section Insulators.—Section insulators are placed between two trolley wire sections that are fed by separate

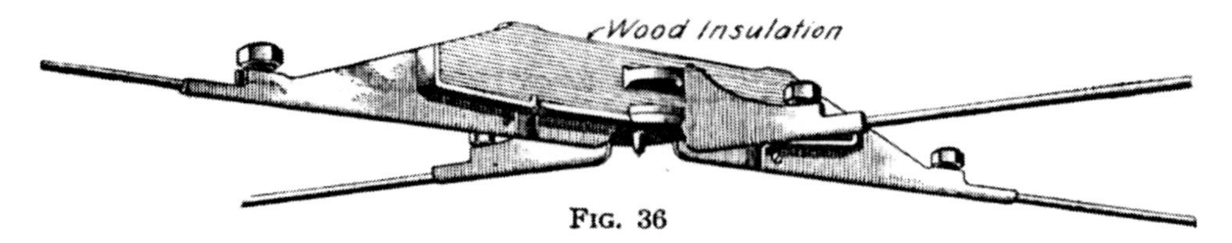

Fig. 36

feeders or from separate taps of the same feeder; they are also called line **circuit-breakers** or **line breakers.**

Fig. 37 shows a type of line breaker suitable for attachment to the body casting of a hanger that is connected to the span wire. The direct line of the trolley wire is maintained by a hardwood runway, which also serves to insulate the two trolley wire sections.

40. Requirements of Line Devices.—The main requirements of all line devices are durability, simplicity, and strength. The line and its devices must be capable of withstanding

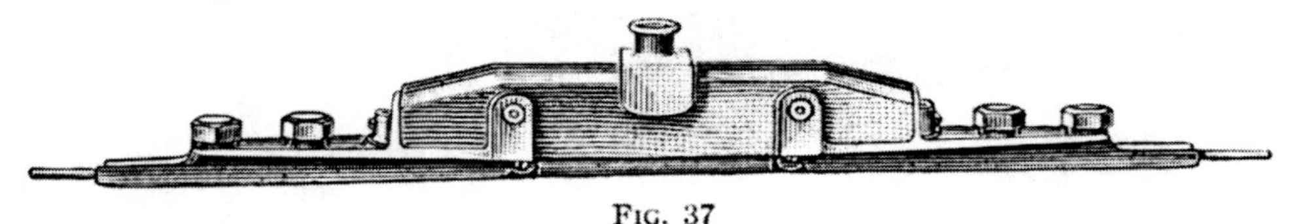

Fig. 37

violent blows from trolley poles that may fly off under a tension of 20 to 30 pounds when the cars are going 25 to 40 miles per hour. All insulation must be good, for while the leakage current over one insulator may be small, that over hundreds may be considerable and the higher the voltage the greater the energy loss corresponding to a given leakage current. The lines should be systematically inspected for minor faults, which can be remedied before they become serious.

TAPPING IN FEEDERS

41. Copper Feeders.—Fig. 38 shows a method commonly used to connect a copper feeder to the trolley wire. One end of a piece of solid or stranded weather-proof wire (No. 00 to 0000) is tapped on to the feeder, and the other end is passed through the eye of strain insulator *a*, given a few

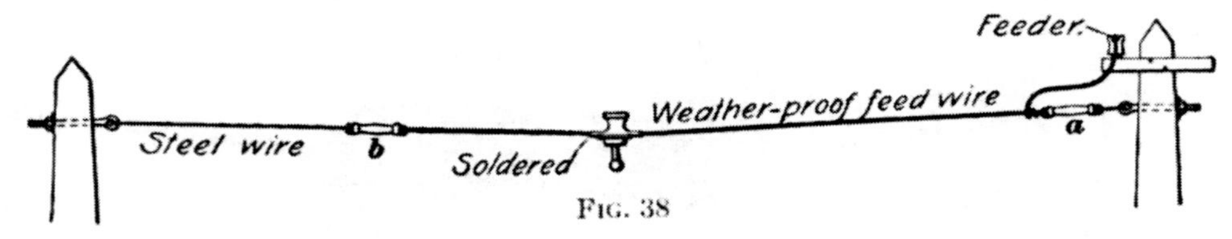

Fig. 38

turns around itself, and attached to one end of strain insulator *b*. Insulator *b* is always placed far enough from the trolley hanger to escape blows in case the trolley wheel flies off the wire, and in some cases is placed near the pole on the opposite side of the street from the feeder. When placed as shown in Fig. 38, ordinary steel wire is used for the balance of the span from *b*. The insulation is removed from the feed-wire at the point where the trolley-wire hanger is attached. The hanger has no insulation between the body casting and the ear and is soldered to the copper span wire. Sometimes an insulated hanger is used and the tap is connected to an ear of the form shown in Fig. 28 (*d*). The feeder and trolley wire are thus electrically connected.

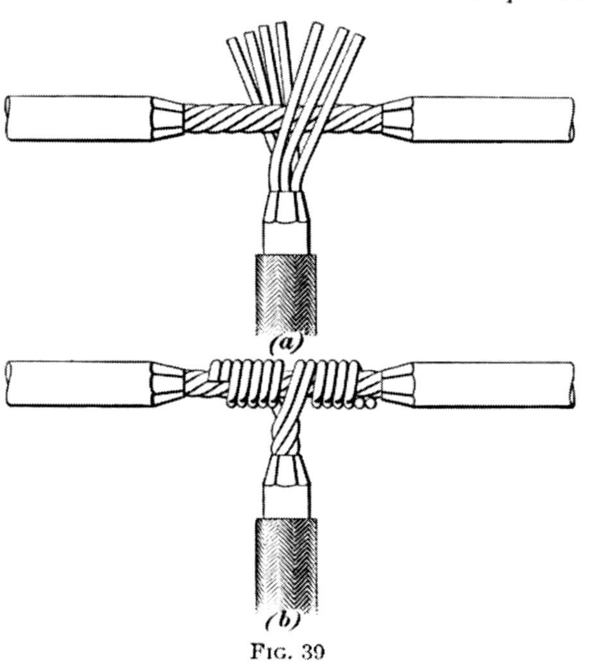

Fig. 39

For connecting the tap wire to the copper feeder, any one of a number of tap joints may be used. In Fig. 39 (*a*) and (*b*), the tap wire is skinned, the strands straightened, cleaned,

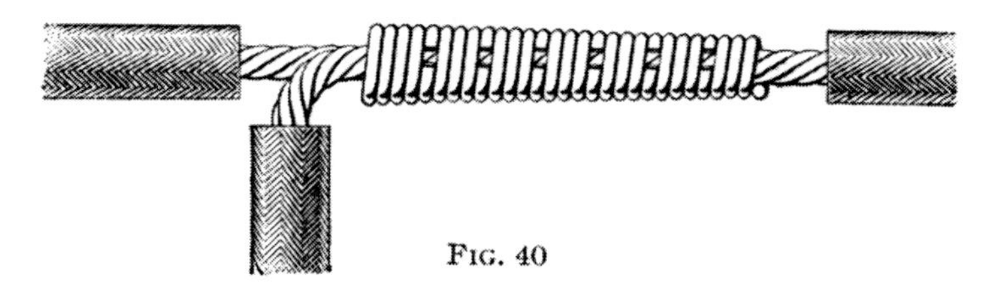
Fig. 40

halved, and the two groups of wires wound in opposite directions around the feeder. In Fig. 40, the strands of the tap

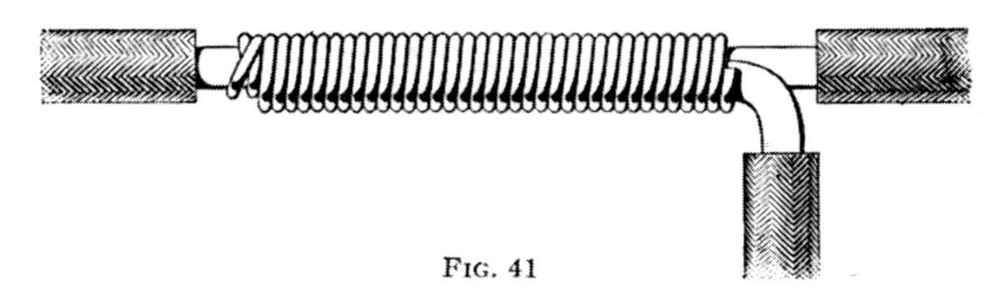
Fig. 41

wire are not separated and are wound on the feeder. In Fig. 41, the tap wire is bound to the feeder by a wrapping wire. The last two methods may be applied to either solid or stranded conductors. In all cases, the joint is soldered, trimmed, tinned, taped, and given a thick coat of weatherproof varnish.

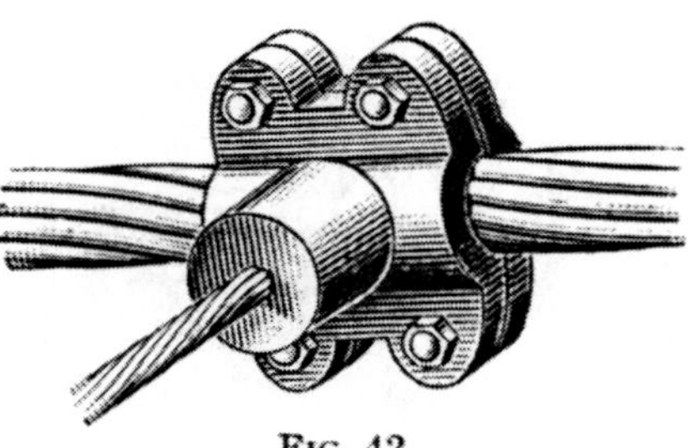

Fig. 42

42. Aluminum Feeders. Copper tap wires are usually employed for aluminum feeders, and are soldered or clamped to them. When soldering, the feeder cable should be kept warm and its surface brightened by scratching.

Fig. 42 shows a form of clamp used to connect an aluminum feeder to a tap wire. All parts are of aluminum and the lug into which the copper tap wire is soldered, or in some cases clamped, forms a part of the clamp.

LINE LIGHTNING ARRESTERS

43. The overhead distributing system of an electric railway should be protected by lightning arresters disposed at least five to the mile and effectively grounded to the rails. Line arresters are similar to those used in station work, except that they must either be housed or adapted to exposure. The connections of a line arrester are indicated in Fig. 19 near *b*.

CATENARY LINE CONSTRUCTION

44. For high-speed, high-potential, railway work, the ordinary overhead construction is not suitable. In such construction, the trolley wire sags between supports and is lifted and bent as the trolley passes under it. Moreover, the insulation of 500- or 600-volt line construction is insufficient for the high voltages sometimes used in alternating-current railway work. To meet the requirement of high-speed railroading and high-voltage insulation, the *catenary line-construction system* was introduced.

SINGLE-CATENARY CONSTRUCTION

45. In the **single-catenary construction,** shown in Fig. 43, a $\frac{7}{16}$-inch, seven-stranded, galvanized-steel cable a, called a *messenger cable,* is supported by bracket construction from poles on one side of the track or from span wires between poles on opposite sides of the track. The trolley wire b is suspended from the messenger cable by means of iron hangers c

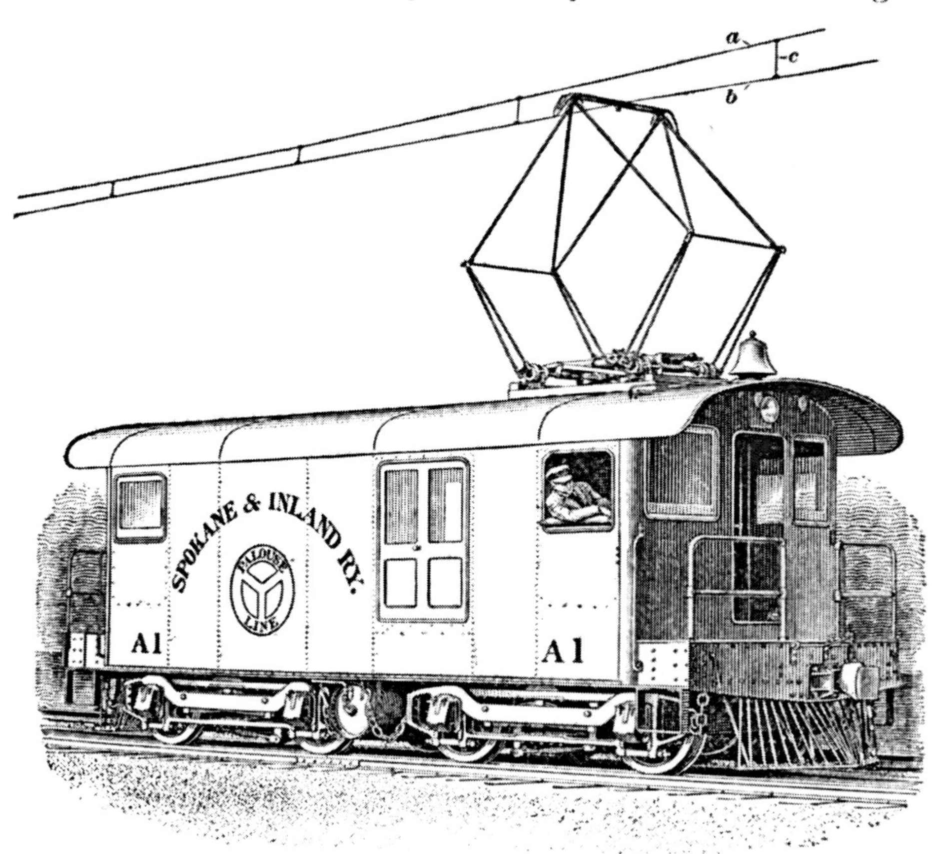

Fig. 43

of proper lengths to maintain the wire at a uniform height above the track. The contact shoe of the pantagraph trolley attached to the roof of the car or the locomotive slides smoothly under the wire and will cause it to move but slightly. Either wheel trolleys or sliding contact shoes are used in this system of overhead construction. The messenger cable, being connected electrically as well as mechanically to the trolley wire, serves as an auxiliary conductor. Both the cable and the trolley wire are insulated from the brackets and poles by means of either porcelain or wooden insulators.

Distances of from 125 to 175 feet between supporting poles are permissible with this type of trolley-wire suspension. In case the trolley wire breaks, it cannot fall to the ground because of the short distance between hangers.

Fig. 44 shows one form of hanger, three or more of which are commonly used between pole supports on single-catenary construction. The bronze ear is clinched on the grooved trolley wire; the hanger rod, which is screwed into the ear, has its upper portion formed into a loop through which the messenger wire passes. When the trolley of a car moves under the hanger, the trolley wire and hanger may move upwards a short distance without lifting the messenger wire. Tendency to bend the trolley wire close to the hangers is thereby lessened.

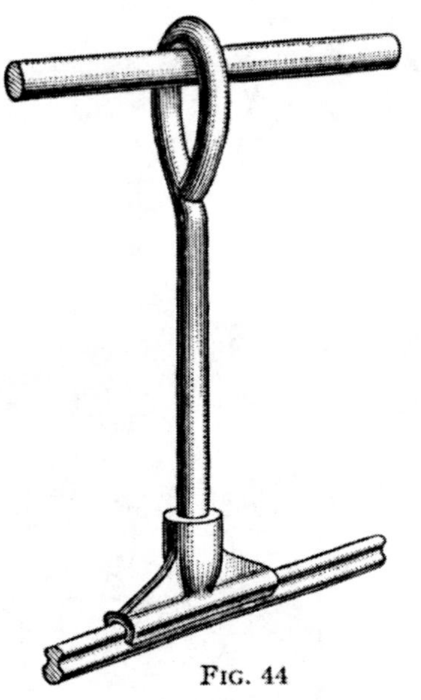

FIG. 44

46. Bracket Construction.—A bracket for straight single track is shown in Fig. 45, in which a bracket arm *a* carries an insulator *b* on which is a messenger wire *c* that supports the trolley wire *d* by means of hangers *e*.

The messenger wire sags between the supporting poles, but the trolley wire, being supported by the hangers, maintains almost the same height above the track throughout its length. The trolley wire is supported about 17 inches below the messenger cable at the poles, and the distance between the messenger cable and the trolley wire gradually decreases toward the center of the span between the poles.

47. Single brackets for curves on single track with the poles on the inside of the curves are shown in Fig. 46. Extensions of the bracket arms support a span wire called a *backbone wire*, to which *pull-off bridles* attached to the top and

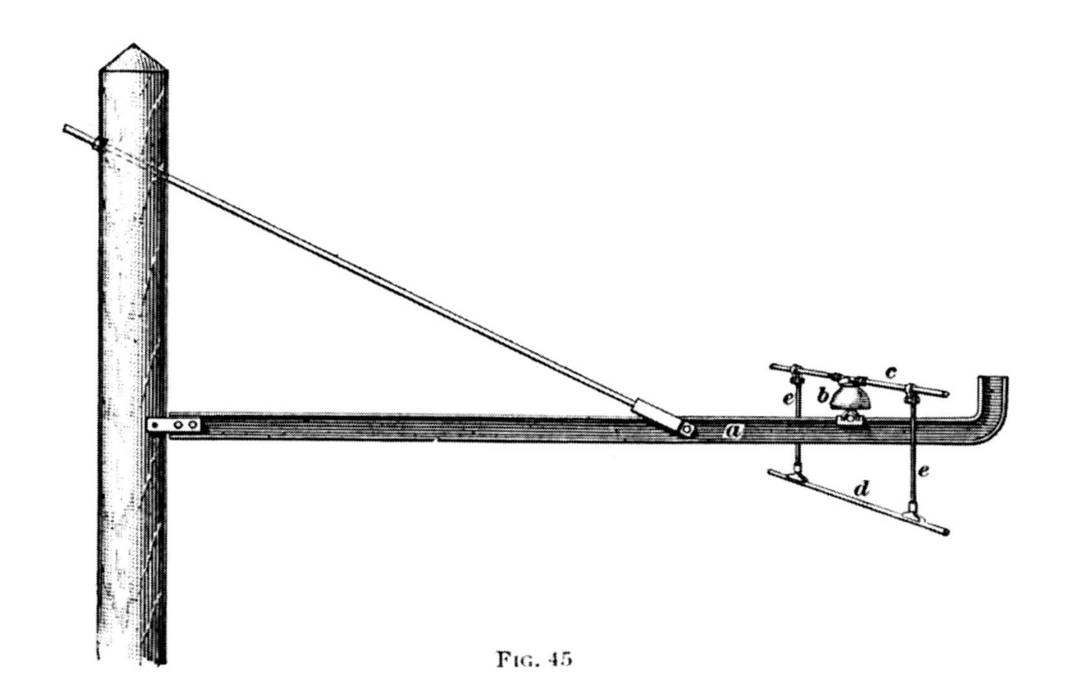

Fig. 45

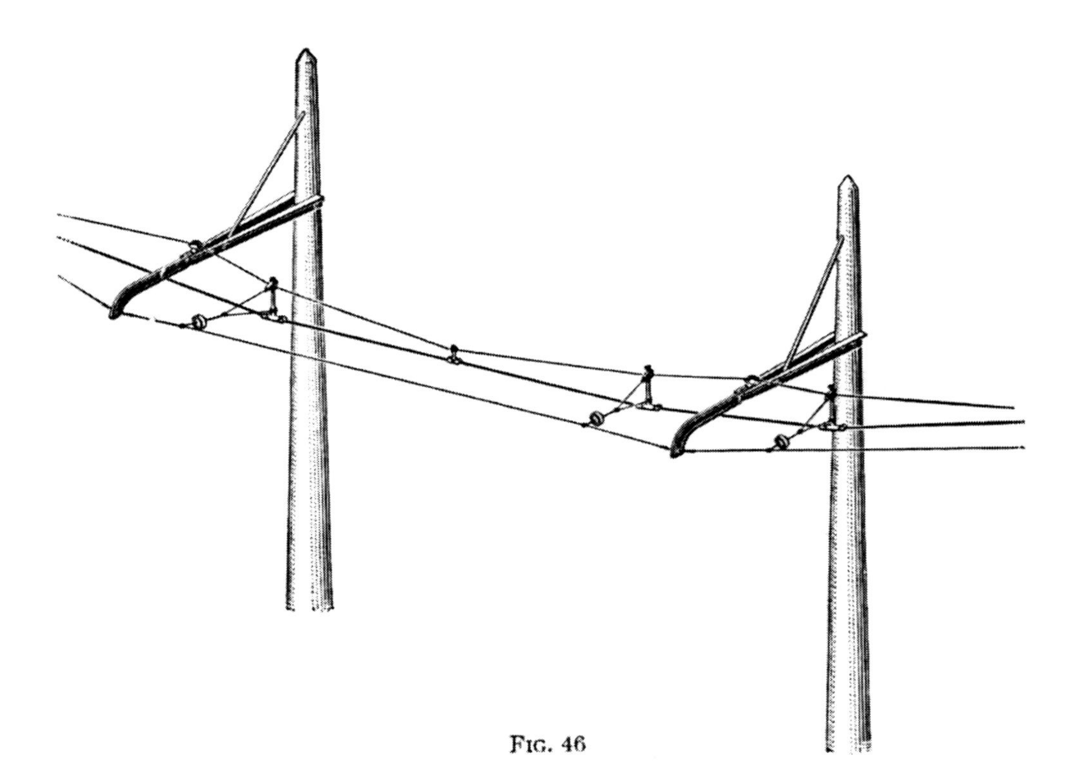
Fig. 46

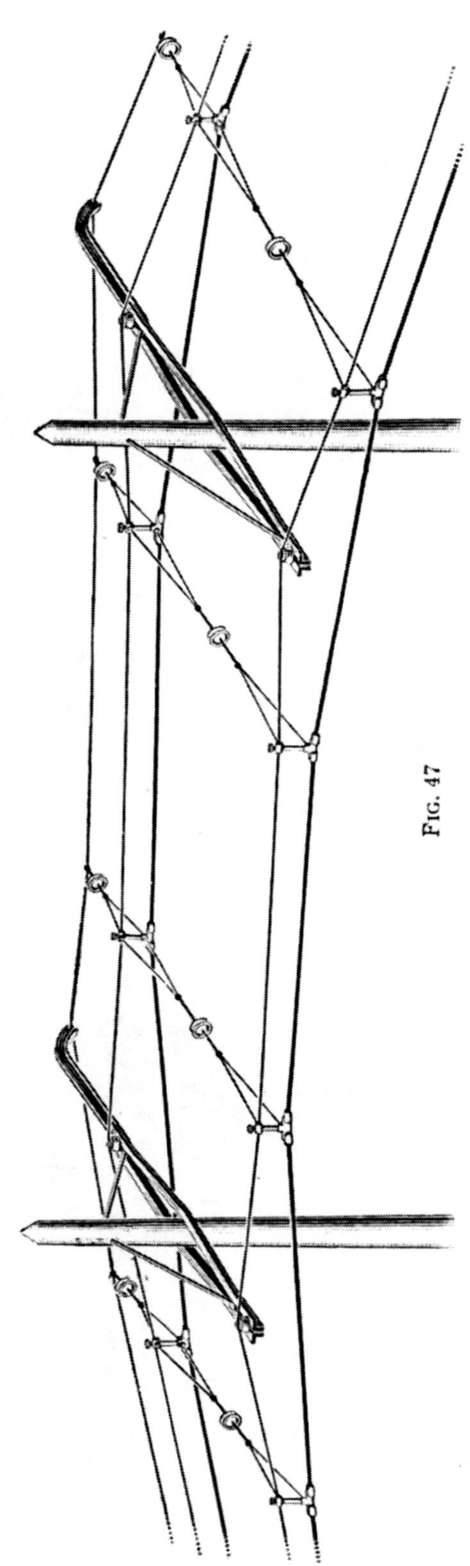

Fig. 47

bottom of the hangers are fastened. The trolley and messenger wires are thus made to conform approximately to the curve of the track.

When the poles are on the outside of the curve, the span wire is supported by the poles, the bracket extensions are omitted, and the pull-off bridles are connected between the hangers and the span wire.

Fig. 47 shows a method of construction for a double-track curve with double brackets provided with extensions at one end that support a backbone wire to which the pull-off bridles from both sets of trolley and messenger wires are attached.

On curves of large radius, and at intervals of about 440 feet on a straight track, a supplemental rod made of hardwood is attached to the bracket arm at one end and to a clamp secured to the trolley wire at the other end, as shown in Fig. 48, thus serving to keep the catenary structure in an upright position. A bracket thus equipped is called a **steady-strain bracket.**

Fig. 49 shows another method of holding the messenger wire and trolley wire in the same vertical plane. The messenger wire *a* is supported on an insulator mounted on the bracket. The trolley wire *b* is supported on a short span wire between the pole and the end of the bracket arm.

48. Feeder Tap for Bracket Construction.—Fig. 50 shows a method of connecting the feeder tap. One end of

Fig. 48

the tap wire is connected to the feeder, the wire supported by insulators on the pole and bracket arm, and its other end connected to a clamp on the trolley wire.

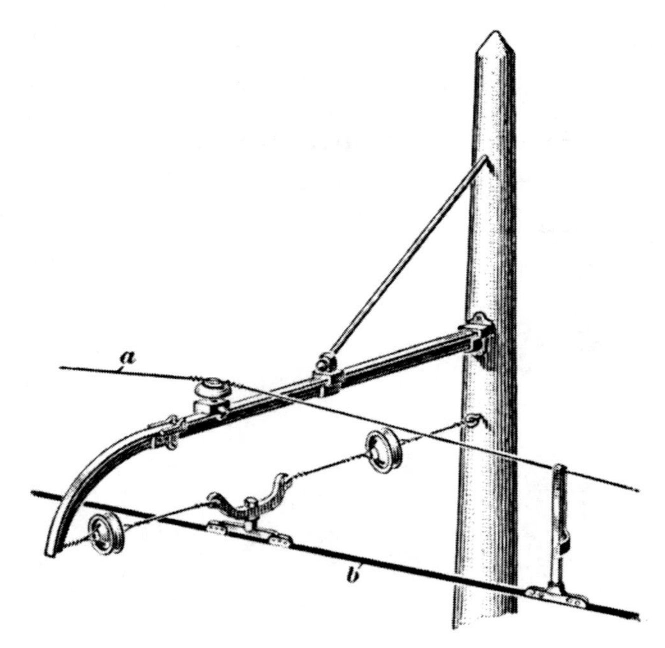

Fig. 49

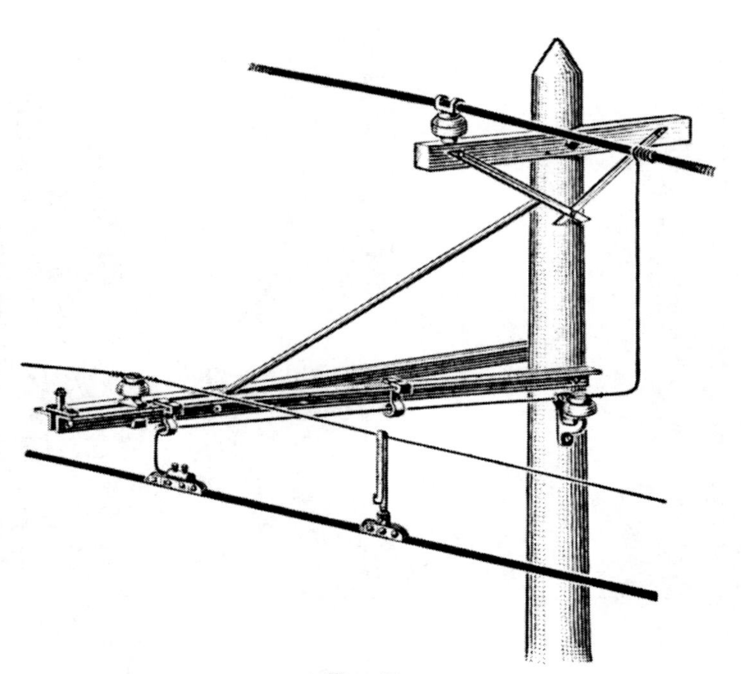

Fig. 50

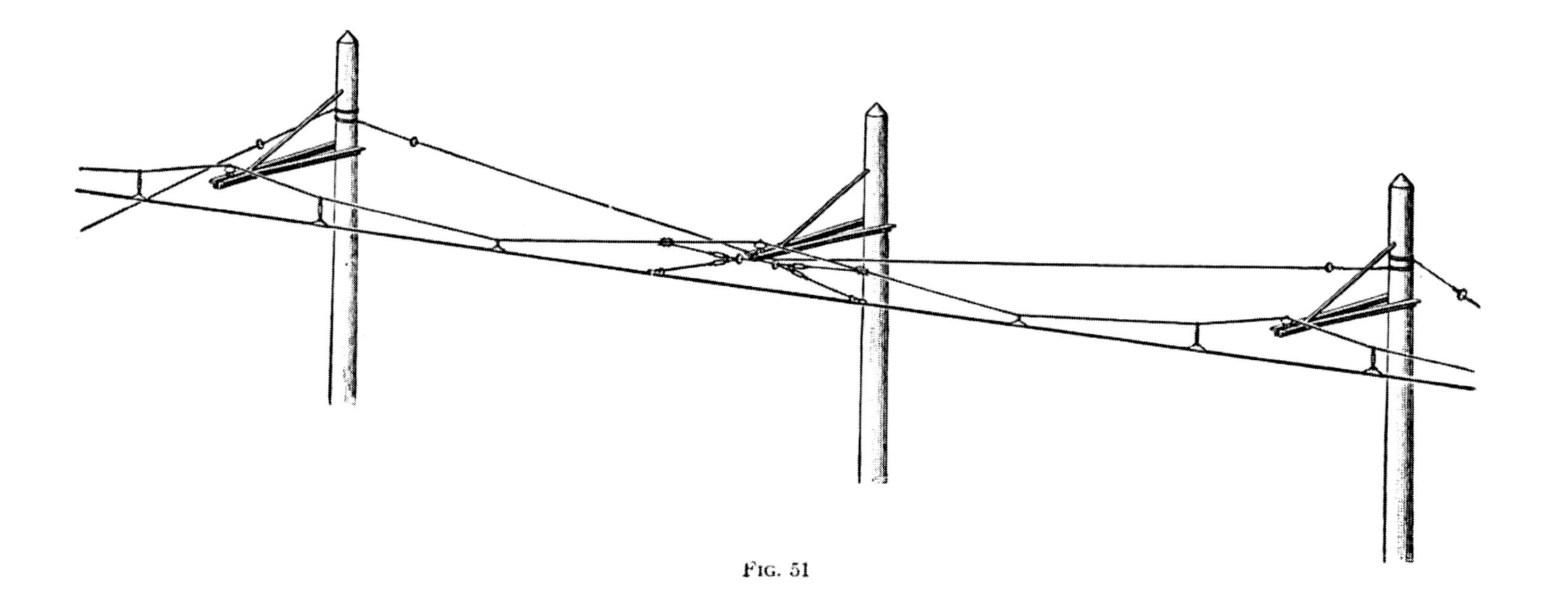

FIG. 51

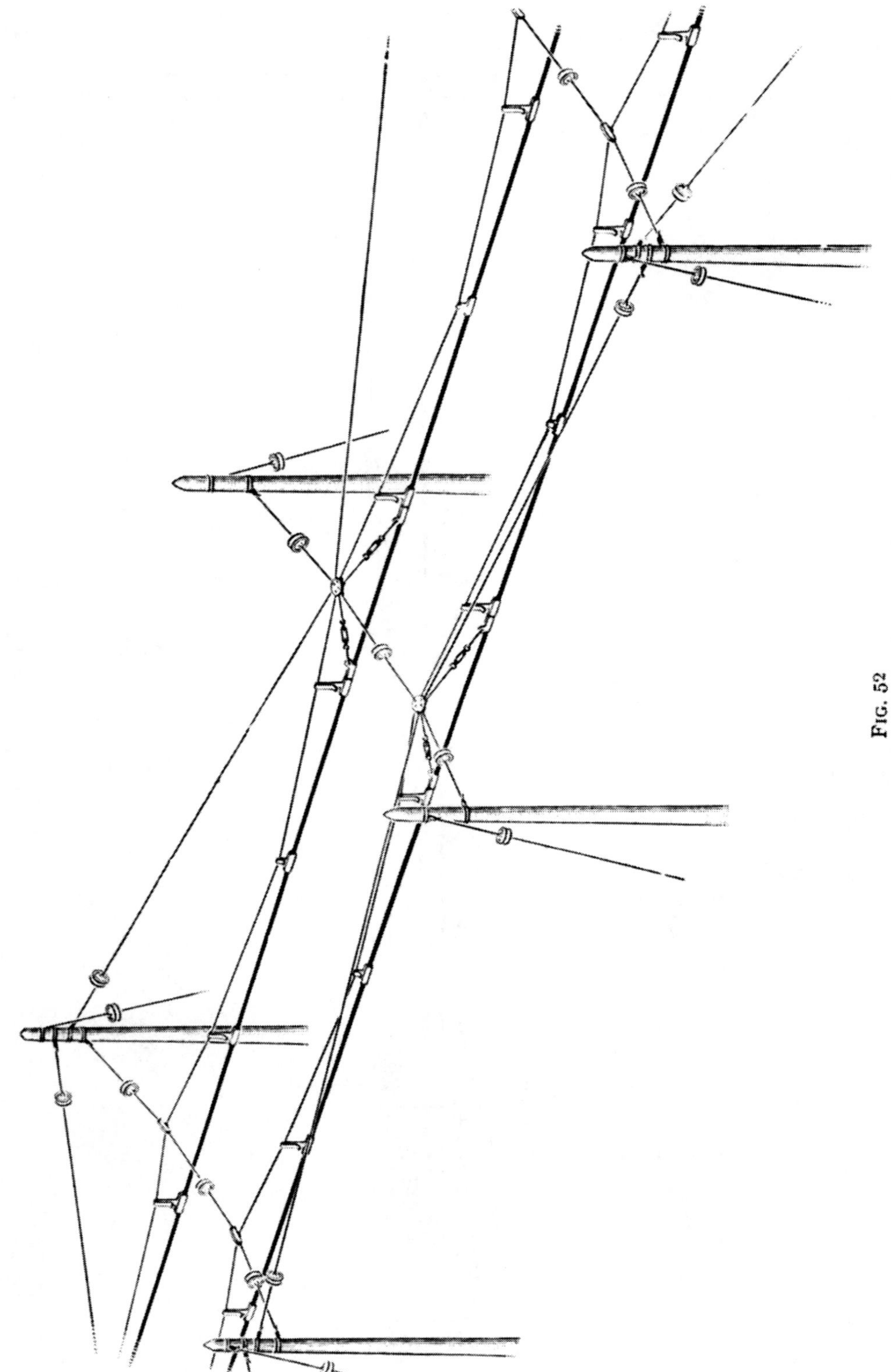

Fig. 52

49. Bracket Anchor Construction.—Both messenger and trolley wires should be anchored every ½ mile for straight track and at all approaches to curves. Fig. 51 shows one method of anchoring for bracket construction. Anchor bridles, attached to the end of a bracket but insulated from it, are clamped to both the trolley wire and the messenger wire, and the bracket is guyed to adjacent poles each way. In some cases, these bridles are fastened to pull-off hangers instead of to separate clamps.

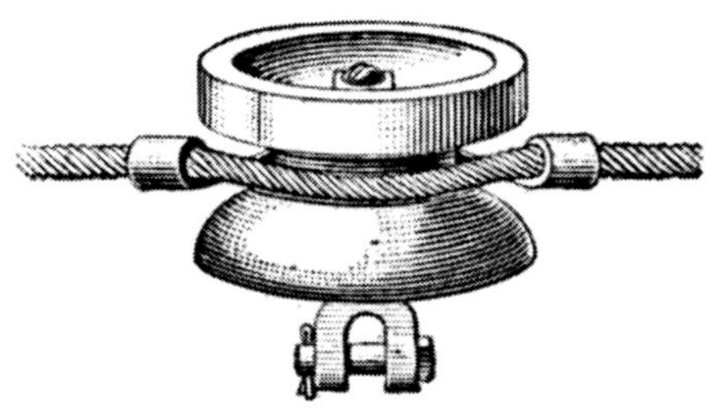

FIG. 53

50. Span-Wire Construction and Anchorage on Straight Track.—Fig. 52 shows a method of span-wire construction and anchorage used for straight tracks. The messenger wire is supported by a loop at the lower end of an insulated hanger, one type of which is shown in Fig. 53. The trolley wire is supported by hangers from the messenger wire as shown in Fig. 52. The method of anchoring the wires is shown near the middle pair of poles, Fig. 52.

In order to preserve the alinement of the messenger and trolley wires, bridles, Fig. 54, are connected at intervals between backbone wires and the top and bottom of pull-off hangers. The backbone wires are connected between adjacent poles.

51. Span-Wire Construction on Curved Track. Fig. 55 shows span-wire construction for curved double tracks.

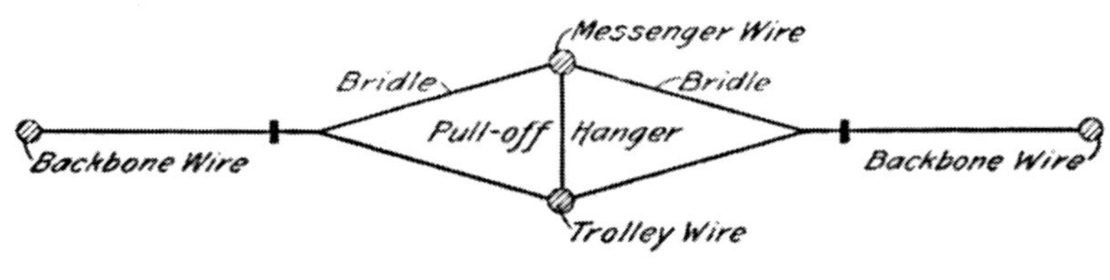

FIG. 54

The poles to which the backbone wires are connected are on the outside of the curve and are guyed, in order to withstand better the stresses of the pull-off bridles.

52. Span-Wire Construction With Bridges on Curved Tracks.—Span-wire construction for curved double tracks where bridges are used instead of poles is shown in Fig. 56. The upper of the two messenger wires for each trolley wire is supported on insulators at the top of the bridges

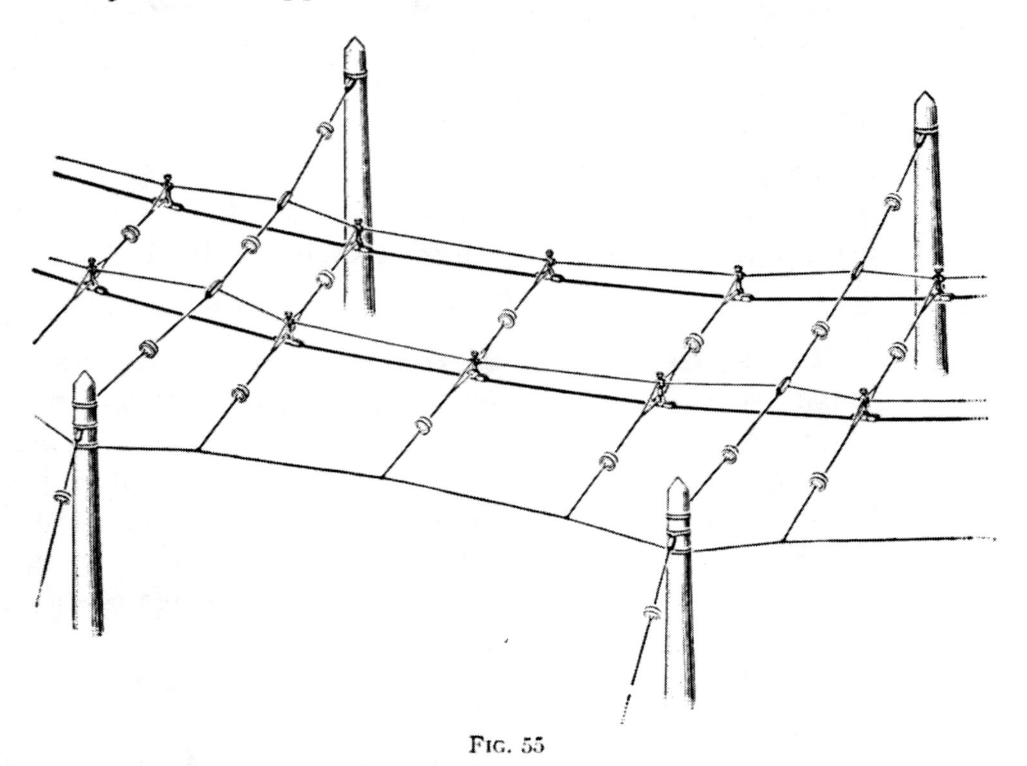

Fig. 55

and is clamped to the lower messenger wire. The trolley wire is supported by span wires at the bridges and by hangers from the lower messenger wire between the bridges. The pull-off bridles are connected to lattice poles placed at points between the bridges and on the outside of the curve.

DOUBLE-CATENARY CONSTRUCTION

53. A form of double-catenary construction is used to a limited extent, notably on a portion of the main line of the New York, New Haven, and Hartford Railroad. It is more expensive to install than the single-catenary system, but holds the trolley wire from swaying better than the single-messenger

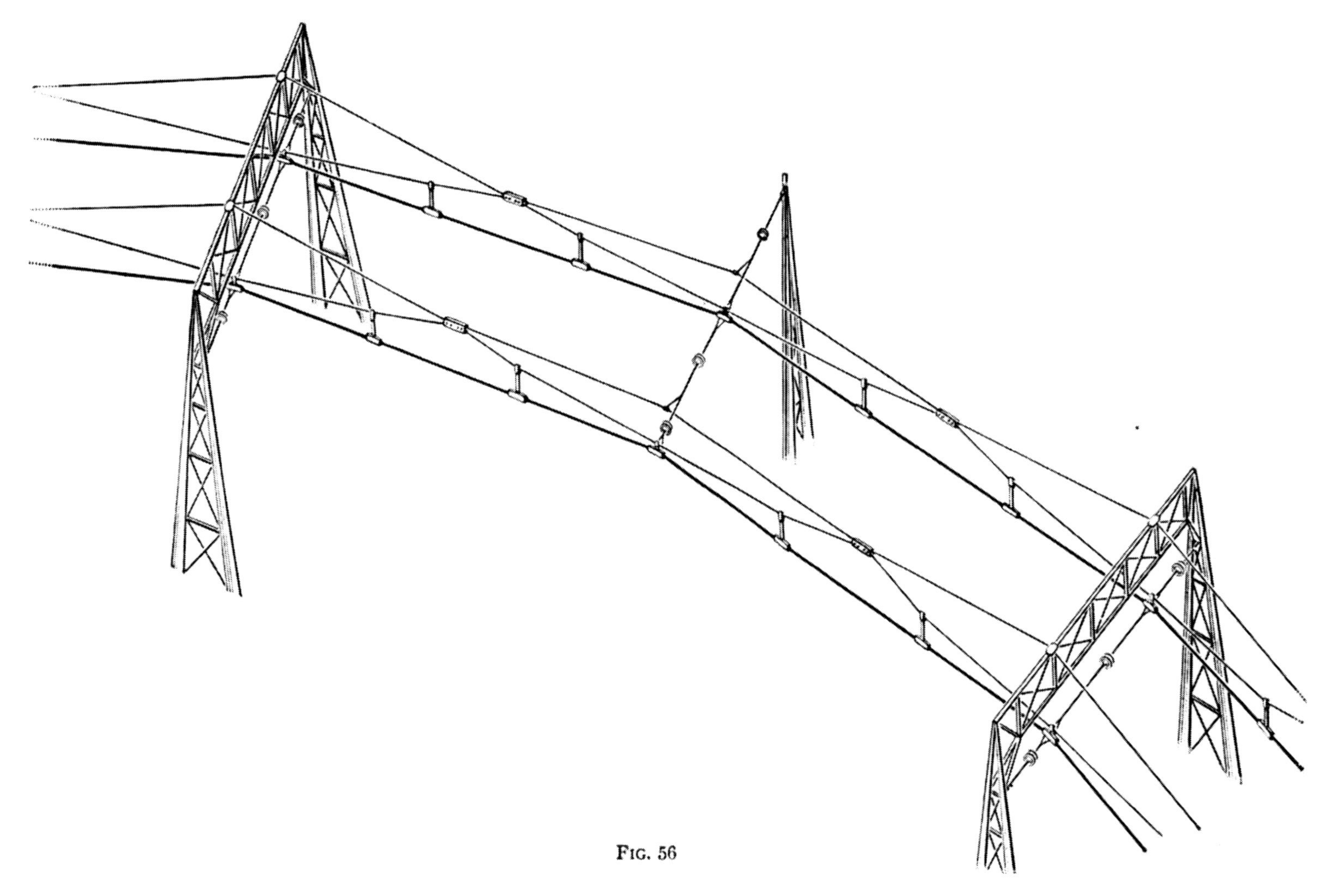

Fig. 56

wire construction. Steel bridges spanning the tracks, or a form of substantial span construction from lattice poles on the sides of the roadbeds, support the messenger wires *a*, Fig. 57. Triangular hangers support conductor *b*, and to this is connected the trolley wire *c* by means of short hangers *d*. The hangers *d* are placed midway between the hangers supporting wire *b*, thus allowing a slight vertical movement of the wires *b* and *c* with but little change in position of the catenary system as a whole.

54. Adjacent to and under low bridges, double-arm clips connected to the lower hangers support the two conductors *b*

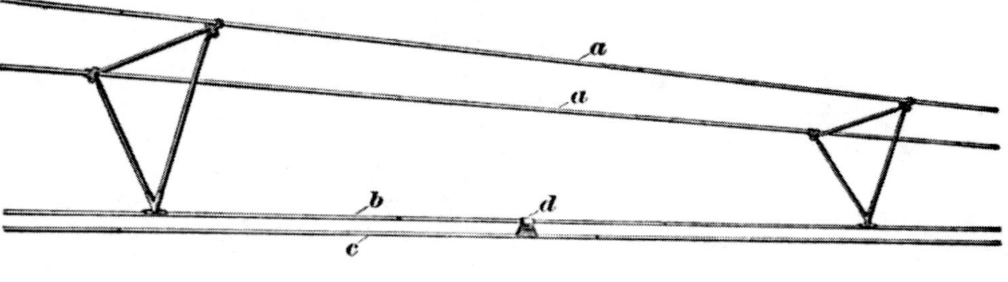

Fig. 57

and *c* of Fig. 57 in the same horizontal plane, as shown in Fig. 58.

At intervals are provided anchor bridges to which the ends of the messenger cables are securely fastened through massive insulators.

55. At the ends of trolley sections, the two trolley wires, about 16 inches apart, are run parallel for a short distance, but are insulated from each other. The broad sliding shoe on the type of trolley used makes contact with the wire of one section before breaking contact with that of the other section.

56. To insure that the sliding trolley shoe will pass smoothly over the junction of the main trolley wire and turnout trolley wire, *deflector wires* are placed in the angle formed by the two wires, as shown in Fig. 59, in which the two trolley wires

Fig. 58

are shown at *a* and the deflector wires at *b*. The illustration shows the junction of the trolley wires of a single- and a double-

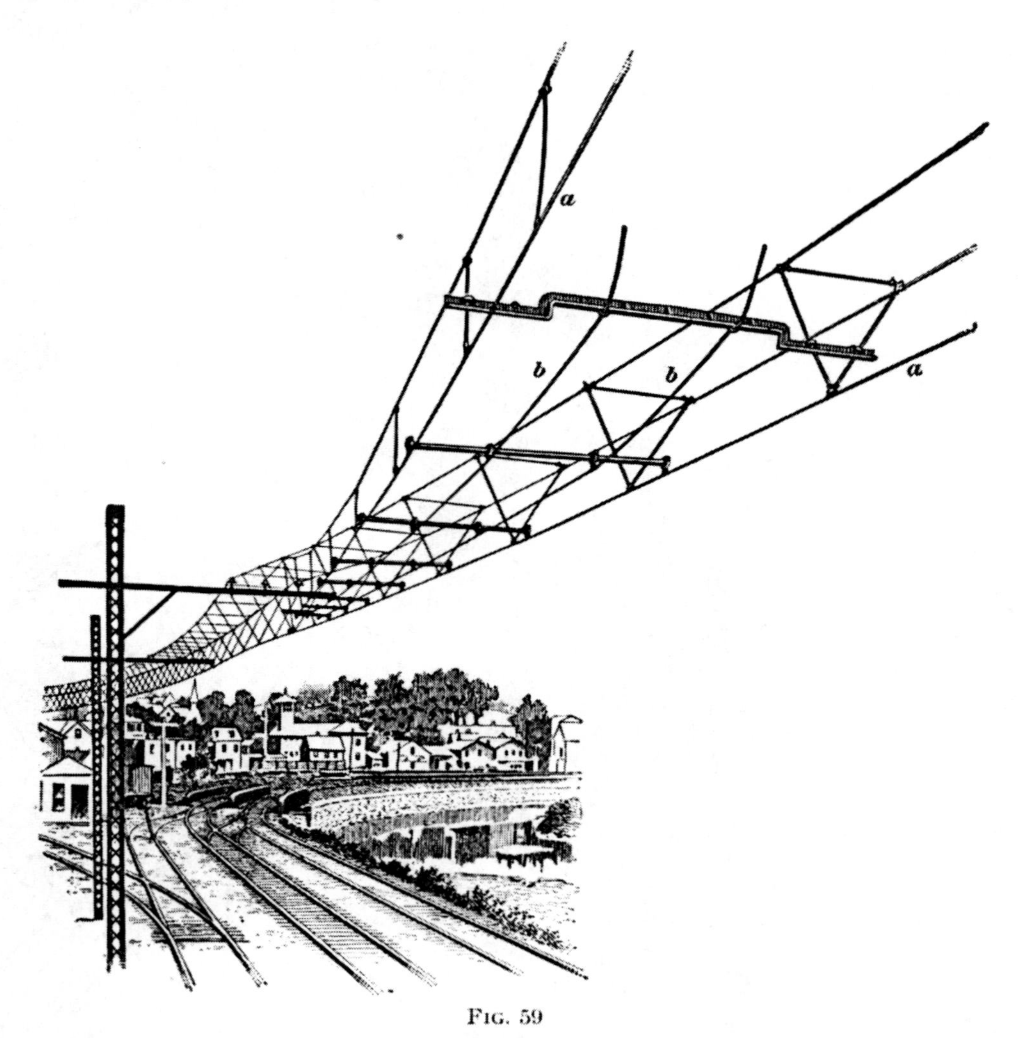

Fig. 59

catenary system. When wheel trolleys are used exclusively, the deflector wires are unnecessary.

HINTS ON INSTALLATION

57. When wheel trolleys are used and the speed of the car is from 40 to 50 miles per hour, three hangers between messenger-wire supports usually result in satisfactory service. The number of hangers may, however, be from two to eleven according to the length of the span, the weight of wire to be supported, the kind of trolley used, and the speed of the car. The greater the number of hangers, the less the sag in the trolley wire, and the straighter the trolley wire for a given tension applied to it.

58. Tension of Wires.—If the messenger and trolley wires in the single-catenary construction are installed with correct tensions, the wires tend to keep in the same vertical plane. With insufficient messenger tension, the upward pressure of the trolley wheel acting through the trolley wire and the hangers

TABLE IV

MESSENGER AND TROLLEY-WIRE TENSIONS

Temperature Degrees F.	Messenger-Wire Tension Pounds	Trolley-Wire Tension Pounds
10	3,400	3,620
20	3,160	3,360
30	2,930	3,100
40	2,680	2,850
50	2,400	2,600
60	2,200	2,340
70	2,000	2,100
80	1,890	1,840
90	1,780	1,600
100	1,690	1,340

will force the messenger wire to one side, thereby allowing an increase in the upward deflection of the trolley wire and making the wheel liable to strike an ear. This may also occur when the trolley wire is too slack, for then the upward pressure of the wheel rotates the trolley wire around the messenger as a center.

Correct tension for messenger and trolley wires, to obtain proper sag in the messenger wire, has been determined for the usual temperature variations and span lengths. Table IV gives data for a 150-foot span and for No. 0000 grooved trolley wire hung from $\frac{7}{16}$-inch steel messenger cable.

59. Lengths of Hangers.—The messenger cable should be installed with uniform tension throughout its length, therefore the short span should have less sag than the long ones. For this reason definite line-pole spacings have been adopted and the lengths of the hangers proportioned accordingly.

Fig. 60 indicates the hanger lengths used in three-point construction for spans varying from 80 to 150 feet. These lengths are based on the assumption that the maximum distance between the messenger cable *c* and the trolley wire *d*, Fig. 45, is 22 inches at the point where the messenger-wire insulator *b* supports the messenger wire.

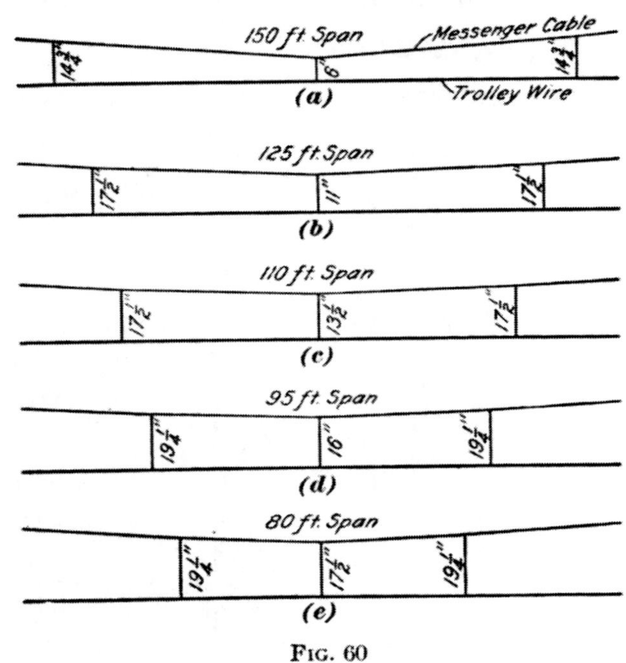

Fig. 60

For shorter spans two-point construction is used and the length of the hangers proportioned to suit the span. For a 70-foot span and two-point suspension, two 20½-inch hangers would be used.

When the standard hangers prescribed for given span lengths are used and the messenger wire adjusted to hold the trolley wire at a uniform distance above the track, the messenger cable will have the correct tension.

60. Size of Wires.—The trolley wires in common use are Nos. 00000 and 0000. For ordinary conditions, the messenger wire is $\frac{7}{16}$-inch, extra-galvanized, steel cable; the anchor wires, $\frac{3}{8}$-inch steel cable; and the pull-off wires, $\frac{1}{4}$-inch steel cable.

61. Insulation of Wires.—Voltages from 600 to 11,000 are employed on roads equipped with the catenary trolley construction. The messenger cable must, therefore, be very thoroughly insulated from the supporting poles or bridges, and massive insulators are used for this purpose. Smaller insulators are used with the span wires, anchor wires, guy wire, etc. Both porcelain insulators and wooden rods are employed for insulating purposes.

62. Installing the Wires.—The bracket arm should be located 18 inches above the desired position of the trolley wire in single-track construction, and 16 inches above the trolley in double-track construction; the additional 2 inches for the single bracket is allowed for sag of the end of the bracket due to the yielding of the pole when the bracket is loaded.

Generally, this construction does not require back guys on straight work; but on curves and at all anchor points, all

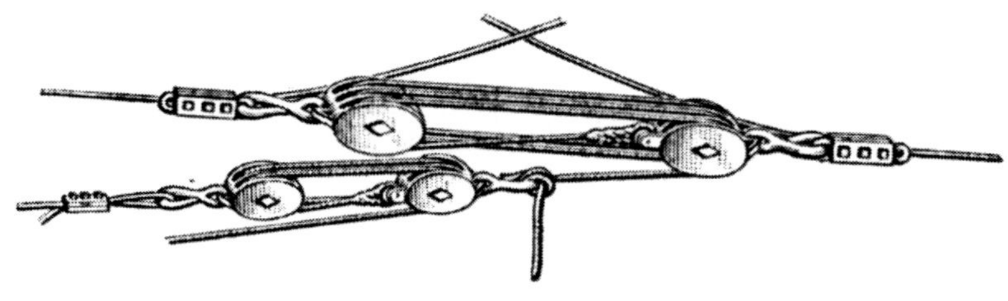

Fig. 61

poles should be guyed. It is recommended that strain insulators be used in all guy wires.

With brackets and messenger insulators in place, the trolley and messenger wires are both run out and hung over the brackets, except at curves where the messenger wire is run over the bracket arms and the trolley wire supported below them; the trolley is then pulled tight and temporarily anchored.

It is generally inconvenient to measure the tension of the trolley wire in course of installation. The desired tension of about 1,000 pounds for 0000 wire can be secured if the pull is made with a pair of three-sheave blocks, Fig. 61, and a "luff," or purchase, with a pair of two-sheave blocks. Two strong men can pull the wire to about the right tension with this arrangement of blocks. The messenger wire should next be adjusted for tension to give the sag *a*, Fig. 62, of 9 inches at 30° F.; 10 inches at 60° F.; and 11 inches at 85° F.; after which it may be lifted on to the messenger insulators and tied. The trolley wire should then be dropped off the ends of the bracket arms and temporarily supported by hooks from the brackets and from the messenger wire at the center of the span; the line will then be ready for the hangers to be installed. Both messenger and trolley wires should be anchored every ½ mile on straight track and at all approaches to curves. Sufficient slack should be left in the curves to allow the messenger and trolley wires to be pulled to proper position over the track. Where bridles for pull offs and anchor wires are used, care should be taken to see that no wires are allowed within a space 6 inches above the horizontal plane of the

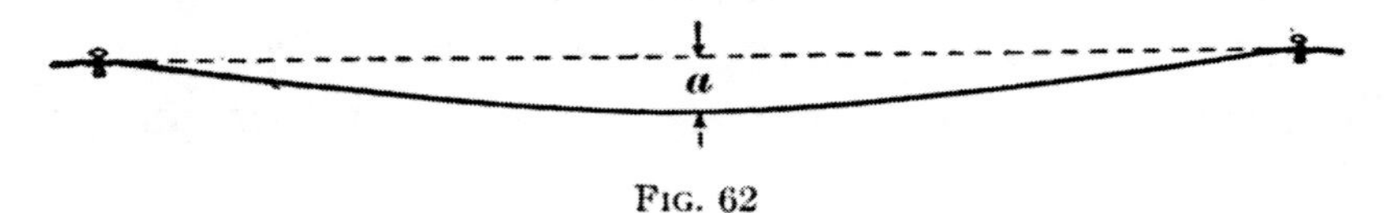

Fig. 62

trolley wire at a distance of 3 feet either side; this clearance is necessary to avoid interference should sliding contacts be used.

63. In span construction, the span wire should be so installed that when the weight of the messenger and trolley is put on it, there will be a sag of about 1 foot for each 20 feet of span. The back guy wires that run from the tops of the poles to the ground, thus preventing the tops from being pulled toward each other, should be insulated for the full line potential. After the poles are guyed and the spans are in place, the messenger and trolley wires are run out and are hung temporarily from the span wires by hooks. Tension is then applied to the trolley and messenger wires and the installation of the hangers may then proceed as in bracket construction.

THIRD-RAIL SYSTEMS

64. Rail Location.—The conductor, called the *third rail*, or the *contact rail*, from which current is taken for the cars in the third-rail system, is an iron rail mounted a short distance from the ground and to one side of the track rails. A **T** rail is commonly used, but rectangular and **U**-shaped rails are employed to a limited extent.

At grade crossings, the third rail is omitted, as the momentum of the car is sufficient to carry it over the gap in the conductor rail. In the case of a train, the shoes on the first car will make contact with the forward section of the third rail before the shoes on the rear cars have left contact with the other section of the rail.

On double-track roads, the two contact, or third, rails are usually placed between the tracks; on single-track roads the third rail may be placed first on one side and then on the other of the track, according to convenience; since all cars have collector shoes on both sides this is permissible. The relative position of track and third rail is largely determined by the clearance required by rolling stock. Where steam locomotives also operate on the road, their cylinders govern the limiting clearances. One recommendation for position is that the vertical center line of the third rail and the gauge line of the nearest track rail, which is a vertical line drawn on the inside edge of the head of the rail, should be 2 feet 3 inches apart and that the top of the third rail should be $3\frac{1}{2}$ inches above the plane of the tops of the two track rails. These relations are, however, not always observed.

Fig. 63 shows a typical third-rail construction. A plan view is shown in (*a*), an end view in (*b*), and a side view in (*c*). The third rail *r* is, in this case, an ordinary **T** rail weighing 80 pounds per yard. It is supported on reconstructed granite insulators *a* located on every fifth tie, which is about 2 feet longer than the other ties.

65. Rail Weights.—The weight per yard and the contact surface of the third rail depend somewhat on the current to be collected. Weights of less than 60 pounds per yard are seldom used except for light traffic; more often the weight of the conductor rail is about the same as that of the track

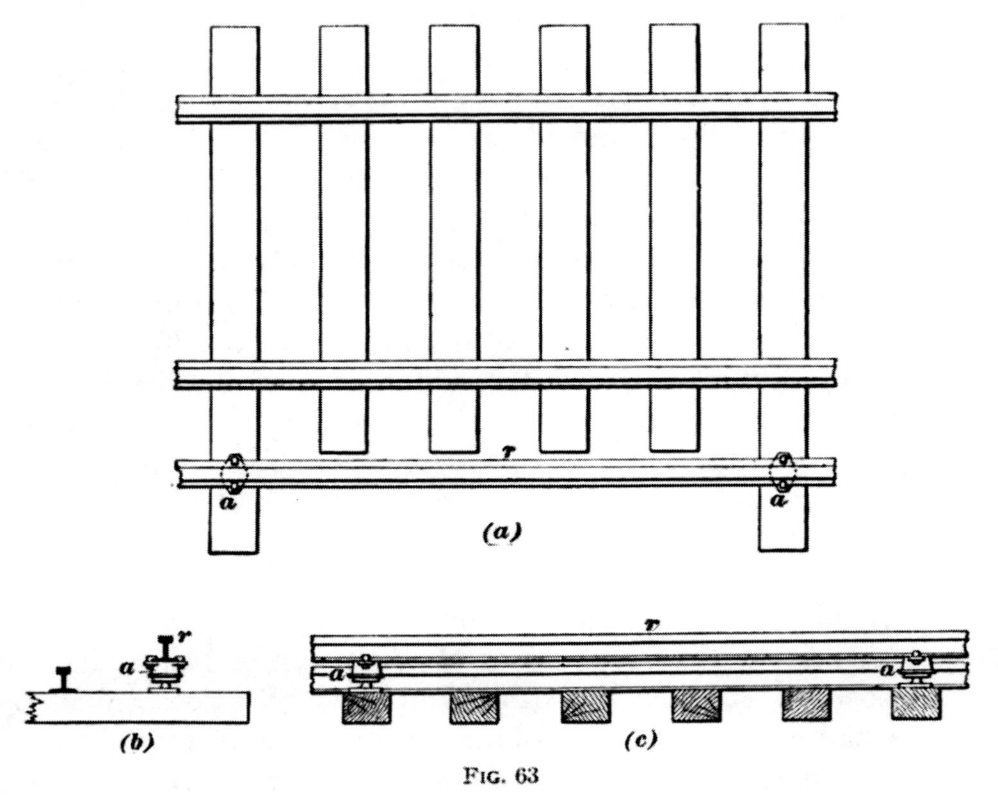

Fig. 63

rails. Third rails are often made in 60-foot lengths, in order to lessen the number of bonded joints that are required.

66. Rail Insulators.—The third rail is supported on insulators that allow for expansion or contraction of the rail. Specially treated wooden blocks provided with iron caps to hold the rail have been used, but in the best construction reconstructed granite or porcelain has been adopted.

Fig. 64 shows three granite insulators; in (*a*) and (*b*), the rail rests on an iron cap and is held by the lugs; in (*c*), it rests on the granite block and is kept alined by special castings held by a bolt passing through the granite body. The castings should allow some vertical play to avoid undue stresses on the insulators. Third rails for 1,200-volt, direct-current, railway systems require better insulation and protection, but in general the method of mounting is similar to the ordinary construction.

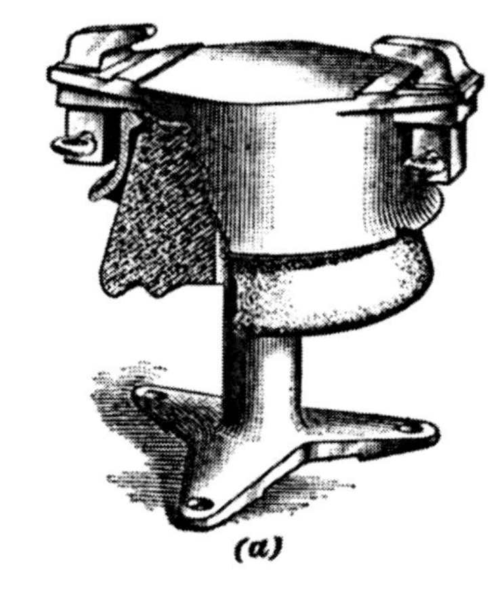

(a)

(b)

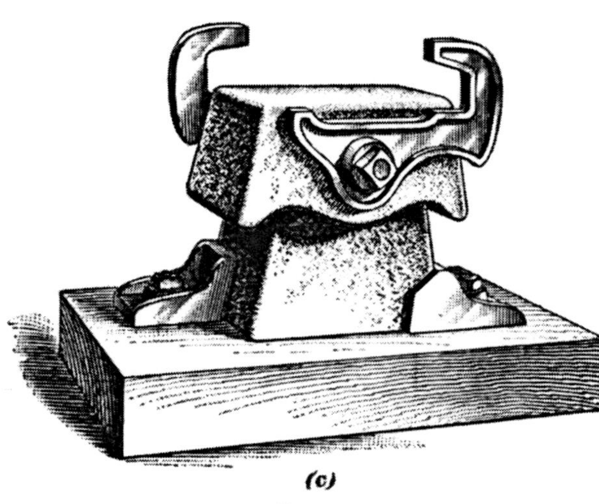

(c)

Fig. 64

67. Rail Protection.—On many roads the third rail is exposed, but the present trend is toward protection that will prevent accidental contact between the third rail and ground or track rails and at the same time prevent the accumulation of snow or sleet on the rail.

Fig. 65 shows a method that has been used on elevated roads; planks parallel to the rail and projecting about 2 inches above it, prevent contact by anything accidentally dropped across it; but this arrangement does not prevent accumulation of sleet and snow.

In Fig. 66 is shown a type of substantial third-rail protection that is sometimes used in subways.

The type of protected third rail used in the electric zones of some large steam railroads is shown in Fig. 67. The rail is of the bullhead, under-running type. The car shoes are pressed upwards against the lower surface of the rail by shoe springs.

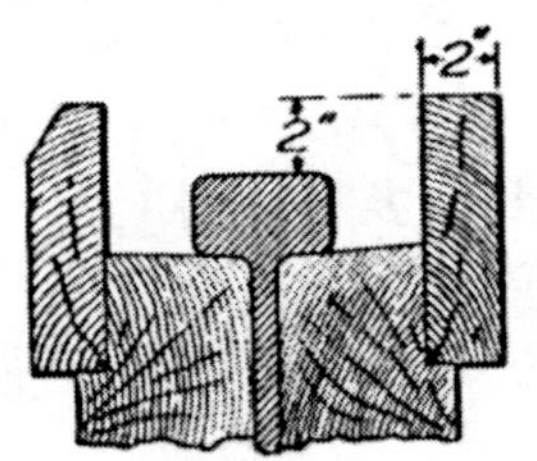

Fig. 65

The method of supporting the third rail is shown in greater detail in Fig. 68. The iron bracket *a*, located at intervals of 10 feet, supports the semiporcelain insulator blocks *b* that fit around the third rail *c*. A steel hook bolt *d* passes around the insulator and through a lug on the top of the bracket. Lugs on both the bolt and the bracket fit into vertical and horizontal grooves in the insulator blocks and thus hold these blocks in position.

The rail between the brackets is first covered with flexible-fiber sheathing and further protected by means of a top strip and two side strips of yellow-pine boards. With this thorough protection, there is but slight chance of an accidental connection between the third rail and the ground or the track rail. The rail is also protected from snow and sleet.

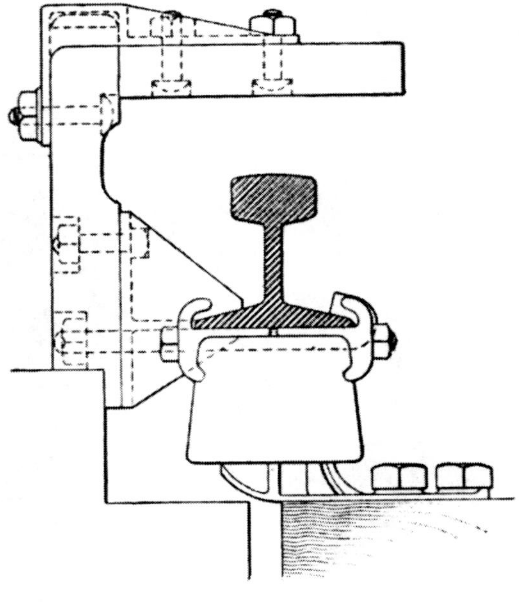

Fig. 66

The third rail is anchored at frequent intervals by means of bolts, Fig. 69, that pass through the rail and wooden insulator blocks mounted on each side of a regular semiporcelain insulator, similar to *b*, Fig. 68.

68. Connections of Third Rails at Highway Crossings.—At public highway grade crossings and at turnouts in the track system, the line of contact rail must be broken and connections must be arranged between the two sections of rail thus formed by means of either an overhead or an underground cable, usually called a **jumper.** The connection is nearly always made by means of an under-

Fig. 67

ground cable, which can be made short and direct. Fig. 70 shows a typical crossing. The contact rail *a* is provided with

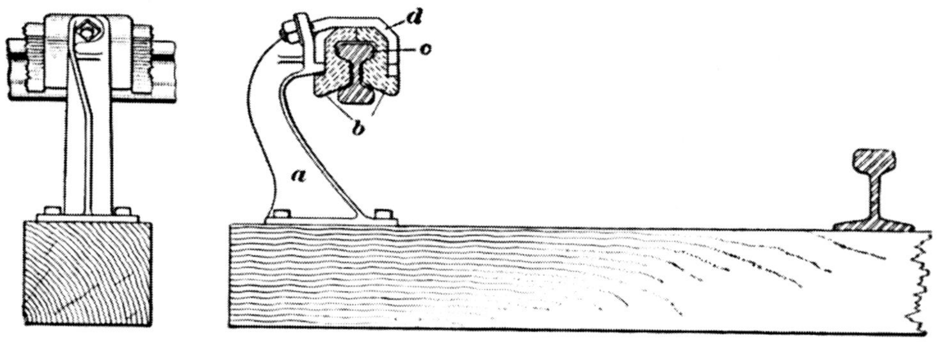

Fig. 68

cast steel, inclined approach blocks *b* that allow the shoes to glide on to the rails without shock. A cable is attached to

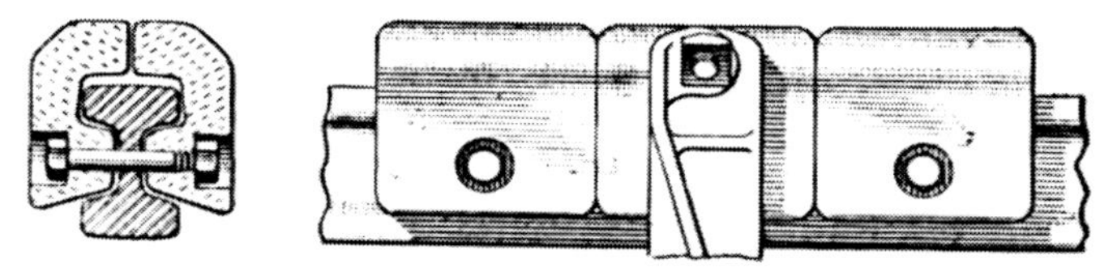

Fig. 69

the rail at *c*, and it is carried underground to *d*, where it connects to the rail again, thus preserving the continuity of the conducting system. The cable should have a cross-section equivalent, in carrying capacity, to that of the rail. For this purpose, 1,000,000-circular-mil, lead-covered, paper-insulated cable is often used.

69. The jumper cable must be thoroughly insulated from the ground, in order to prevent leakage current. There are various methods of installing jumpers, one of which is shown in Figs. 71 and 72. In this case the cable is drawn into a

Fig. 70

black, bituminized, fiber tube, which is laid in solid concrete. A concrete or a terra-cotta cap protects the top of the cable, and flexible leads connect the cable to the third rail. In case it is desired to have independent jumper cables for the two sets of third rails on a double-track road, another cable can be installed in the second fiber tube, Fig. 71.

70. Sleet Troubles.—When sleet gathers on the third rail or when rain freezes as it strikes the cold rail, there is formed on the surface a non-conducting film of ice that often results in delays in the operating schedule. The sleet is difficult to remove, but is easy to prevent if rail cleaning devices are

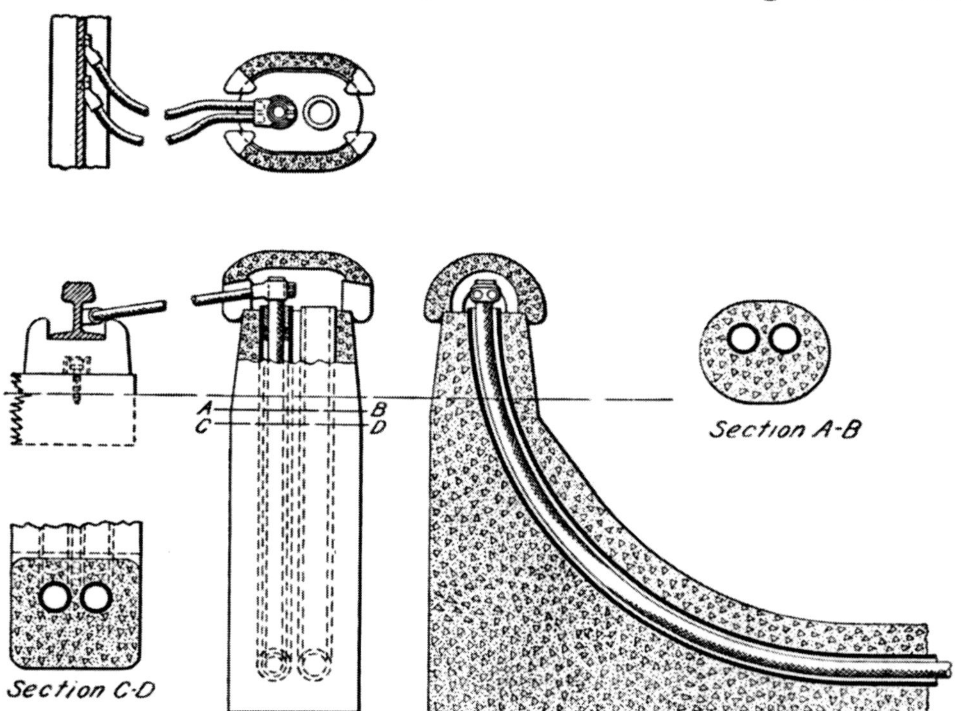

Fig. 71

promptly applied. At complicated special work, it is customary to use salt freely, and in some cases large special blow torches burning crude oil have been used to heat the rail; on straight rail some form of sleet brush hung from the car trucks is depended on to keep the rail surface clean. At the least prospect of trouble, the rail should be swept frequently, even if extra cars must be run. Fig. 73 shows one type of sleet-cutting device. It is fastened to an extension of the collecting-shoe beam; the cutter consists of steel plates *a*

Fig. 72

cast into a block *b*, and set at an angle so that they will slide over projections at joints. A flat spring *c* presses the cutter against the rail and a cam *d* operated by lever *e* holds the shoe up when it is not in use; by moving lever *e* to one side the shoe can be lowered by the motorman.

71. Third-Rail Leakage.—The current leakage per mile of rail is negligible where precautions are taken to avoid it; the leakage from a poorly insulated underground cable at a crossing may easily exceed that of several miles of third rail.

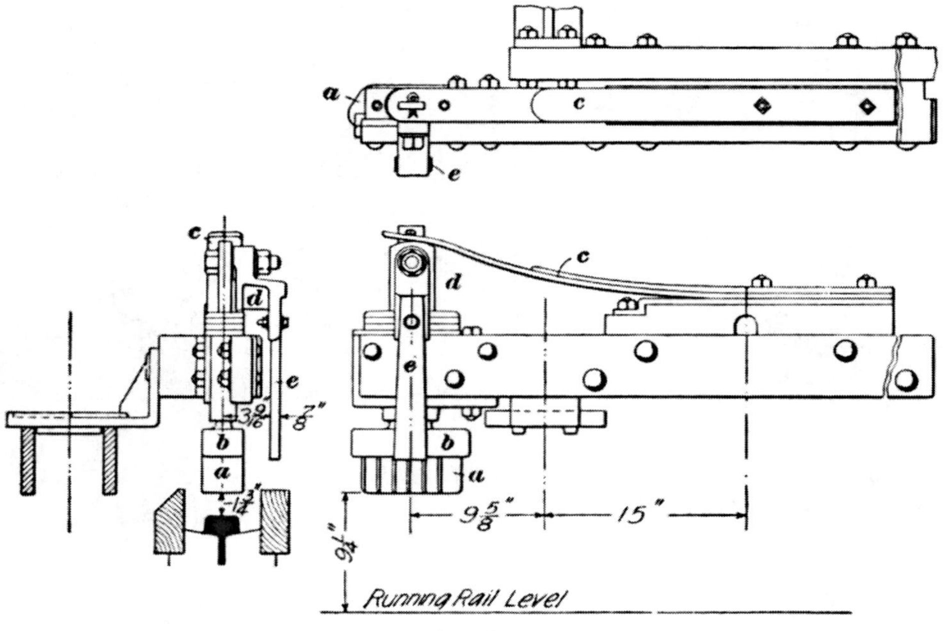

FIG. 73

Tests on a section of third rail disconnected from all crossing cables showed leakage varying from .057 ampere per mile, after a 20-hour rain, to .023 ampere per mile in hard freezing weather with a light snow on the track. Tests of the whole road, including crossing cables, showed an average leakage of about .5 ampere per mile, and an investigation located the greater part of this in defective insulation of crossing cables. The third-rail insulators in the case under test were of specially prepared wood with iron caps.

CONDUIT SYSTEMS

72. Construction.—Owing to their great first cost, conduit systems are installed only where city ordinances prohibit any other form of construction. Theoretically, both sides of the line are insulated from the ground; as a matter of fact, one side of the line or the other is generally grounded enough to nullify this advantage.

Fig. 74 shows the conduit construction used in New York, which may be taken as typical of this class of construction. The rails are supported on heavy cast-iron yokes *a* spaced 5 feet apart; every third yoke has handholes and carries the insulators *b* that support the conductor rails *c* every 15 feet. Fig. 75 shows a section through one of the handhole yokes and illustrates the method of supporting the conductor rails. The conduit between yokes is made of concrete filled in around a sheet-iron form that is afterwards removed. The conduit may be lined with steel plates or may be constructed of concrete alone. Each manhole connects to the sewer through a 6-inch drain pipe. The outgoing and return feeders are run in terra-cotta ducts *d*, Fig. 74. To facilitate the installation of new feeders or the repair of old ones, duct manholes are provided every 400 feet. The yokes are designed to stand the pressure of the earth (packed down by the heavy traffic) and also the pressure due to freezing of the soil in cold climates. Cast-iron, steel, and wrought iron have been used, but cast iron is the most common. Light-weight yokes gave much trouble from breakage, so castings weighing from 200 to 400 pounds have been adopted. In some cases, the metal yokes have been replaced with concrete, but the best construction calls for the heavy metal yokes.

73. Operating Features.—Mud accumulates in the main conduit, which must be cleaned about once a month in the summer and oftener in the winter. With special scrapers the

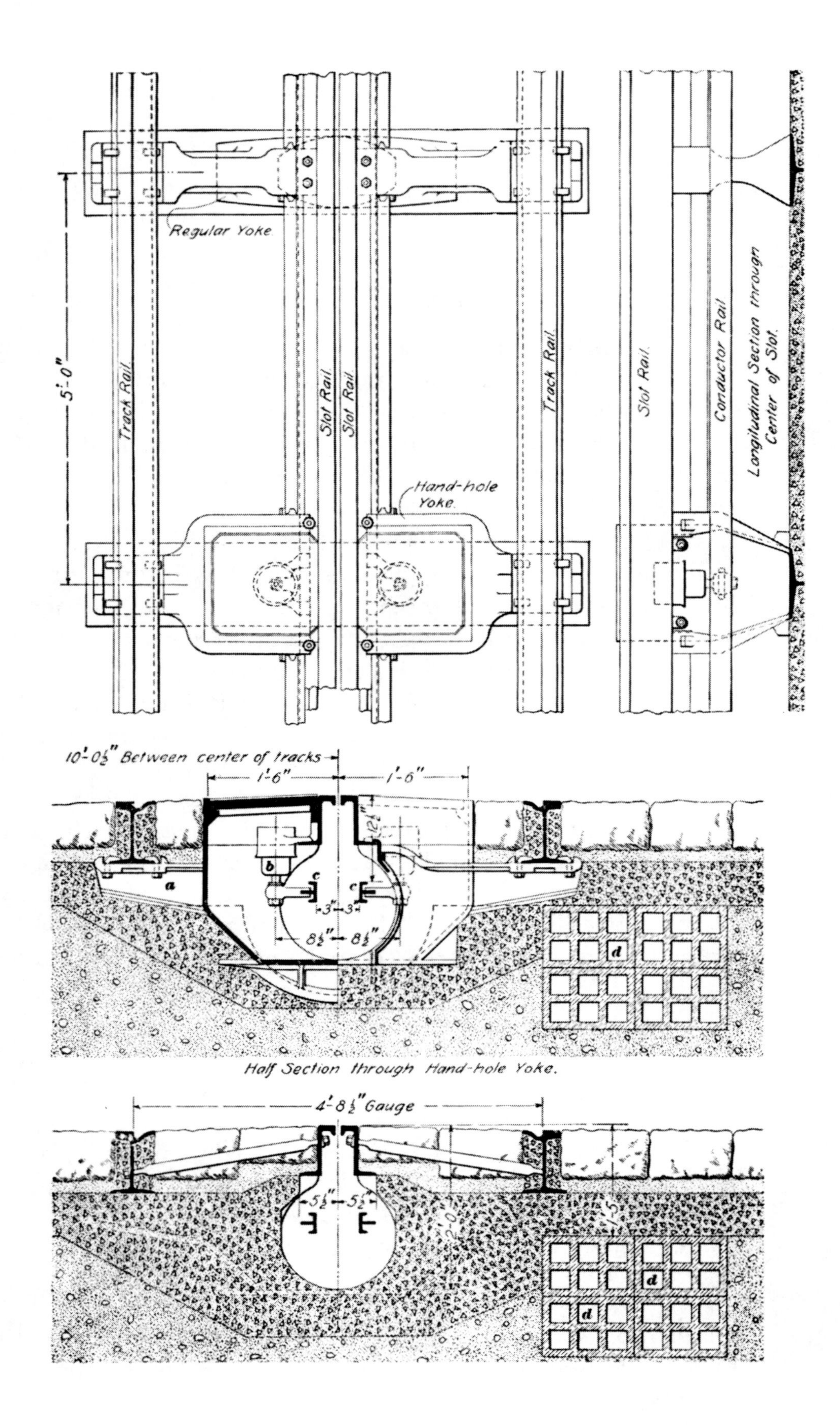

Half Section through Hand-hole Yoke.

mud is drawn into the yoke manholes from which it is removed. The conductor rails are divided into insulated sections about 1 mile long and each section has its own feeder, so that trouble on one section does not interfere with traffic on others. As each feeder includes its own switch and automatic circuit-breaker, if two grounds occur in a section, its circuit-breaker opens and the power-house attendant can locate the defective section. Separate sections supplied by individual feeders have the advantage that in case of a block on the road, the simultaneous efforts of all motormen to start their cars cannot overload the station, because the switchboard attendant has all sections under his control and can compel the starting of the cars one section at a time.

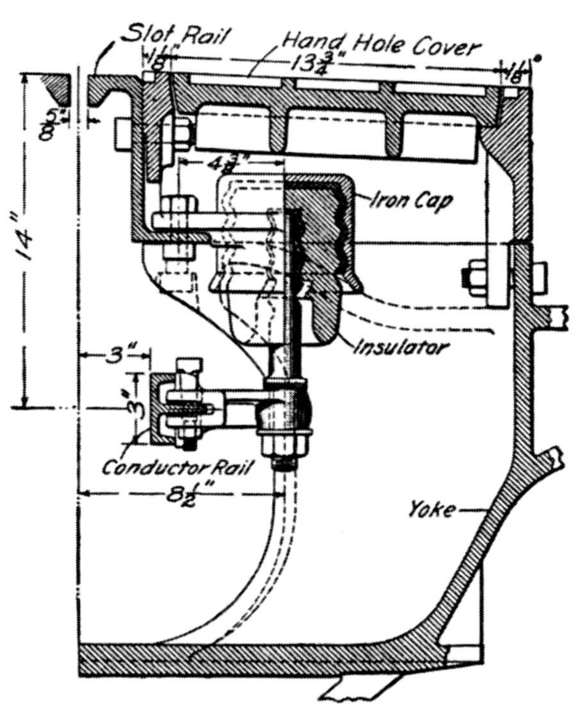

Fig. 75

TRACK CONSTRUCTION

ROADBEDS

HINTS ON CONSTRUCTION

1. Definitions.—A **railroad** consists of the *foundation*, the *ballast*, and the *track*. The foundation consists of the earth support, the top of which is called the *subgrade*. The ballast rests on the subgrade and consists of broken stone, gravel, sand, cinders, or slag. The ballast serves to hold the ties, rails, and connecting-rods that constitute the track in position and helps to drain the moisture from the surface of the roadbed. The name *roadbed* is given to the arrangement of ballast, concrete, paving blocks, etc. that is built on the subgrade and holds the track.

2. City Roads.—Through improved city districts, the location of the roadbed is fixed by street limits and city ordinances. The tracks should be so located that the maximum car overhang on straight track or on curves will not cause interference between cars or with any fixed object near the tracks.

On paved streets the rails are generally supported by wooden ties though metal ties are sometimes used, and in rare cases ties are omitted and the rails supported along their length by stringers of wood or of concrete.

3. Interurban Roads.—On interurban roads in undeveloped districts, steam-railroad construction of the roadbed can be followed closely. Some roads require very expensive construction and others, on account of natural conditions, may be spared most of this expense. When crossing swamps and marshes, it may be necessary to effect thorough drainage in order to obtain a solid foundation that will not yield under heavy cars; on a yielding track a heavy car pushes a wave of rail ahead of it, thereby increasing the resistance to train motion. Before building across marshes or lowlands, soundings should be taken to determine whether the foundation must be supported on piles. In general, where the subsoil is yielding, the substructure must be more substantial and have more area exposed to subsoil. In steam-road construction, instances are recorded in which sections of improperly constructed road disappeared under water.

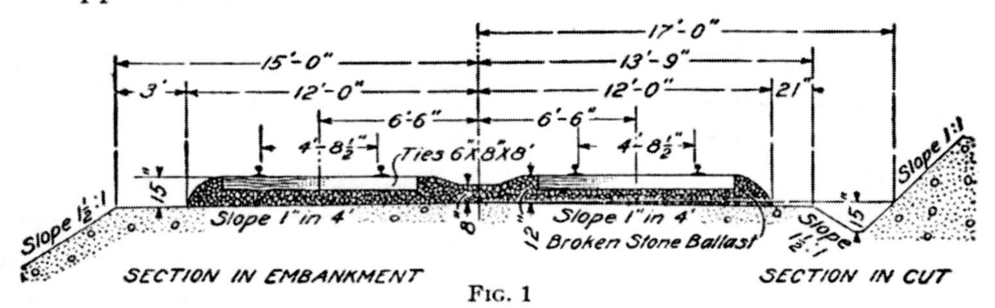

Fig. 1

The subgrades should be crowned, or graded downwards from the center to the sides, to help drainage, and should be sufficiently wide at the top to support ballast around the ends of the ties. The lines of the grades in undeveloped districts in the suburbs of cities should be given by the city engineers, to avoid the expense of later raising or lowering the tracks.

In allowing clearances between the cars and the walls and roof of tunnels and bridges, the height of the car, including trolley stands, ventilators, and stove pipes must be considered; and where margins are close, measurements must allow for possible increase in the sizes of car wheels to be used. Failure to consider such points may cause great expense for later changes.

TYPICAL ROADBEDS

INTERURBAN ROAD

4. In Fig. 1 is shown a cross-section of a roadbed construction for an interurban double-track road. The right-hand half of the figure represents the construction used for a cut or on the side of a hill. The construction for an embankment is shown at the left. The subgrade is crowned to promote drainage, and ample provision for ballast is provided at the ends of the wooden ties. Most of the measurements refer to the center line of the double roadbed.

CITY ROADS

5. For city construction, a trench is excavated in the street wide enough for the single or double track. In the construction shown in Fig. 2, only one-half of a double-track roadbed is

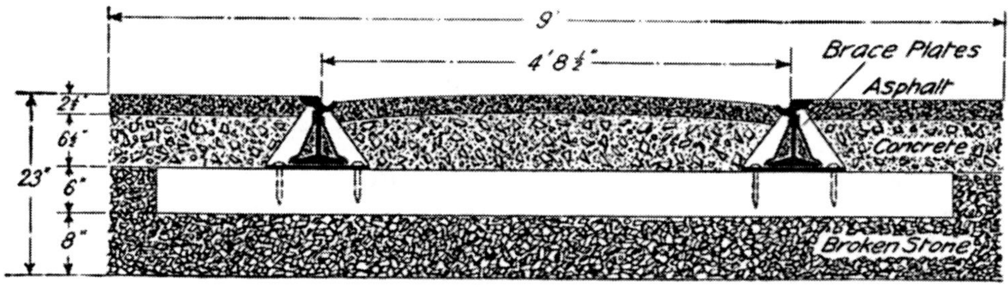

FIG. 2

indicated. The trench for both tracks is 18 feet wide and 23 inches deep. The bottom is rolled and the trench partly filled with 2-inch broken stone; soft spots in the rolled surface are dug out and also filled with stone or other solid material. The stone is rolled until it is firm at a depth of 8 inches. On this ballast are laid the wooden ties, 6 in. × 7 in. × 7 ft. 6 in., a little less than 2 feet between centers, except at the rail joints, which are supported by three ties about 15 inches between centers; 60-foot rails are then laid on the ties, ends butted and joints staggered; before joining, the ends of the rails and the joint plates are well cleaned. The rails are then coupled, the plates bolted tight, brace plates installed every other tie, the ties lined up and spiked to the rail. The track is then lined and surfaced and the space between the ties filled with broken stone well tamped to the top of the tie. The rail is then finally lined, the joints secured, and the broken stone or concrete brought up to the proper surface for the asphalt pave.

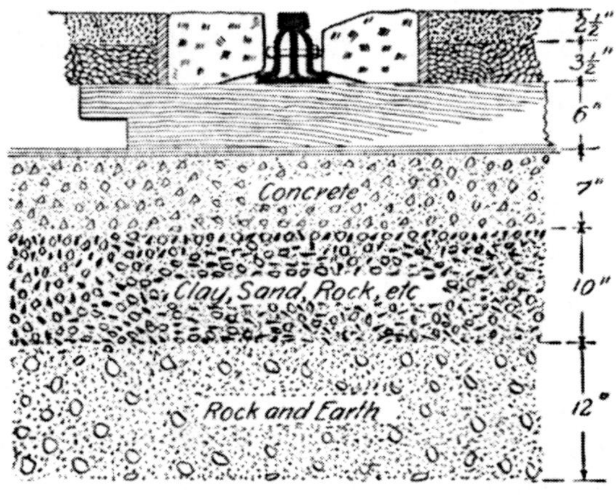

Fig. 3

6. Fig. 3 shows a roadbed construction for a weak subsoil. A trench 36 inches deep and the width of the tracks is dug and filled to a depth of 29 inches with successive layers of 12 inches of hard earth and rock well beaten down; 10 inches of earth, pebbles, clay, sand, and rocks, well tamped, and 7 inches of concrete. Hard pine ties 6 in. × 8 in. × 8 ft., treated with asphalt, are laid on the concrete, and these support 80-pound **T** rails. More concrete is then added to form the surface for the asphalt pave.

7. Fig. 4 shows a roadbed construction in which granite blocks are used for the street surface. The wooden ties, spaced 3 feet between centers, are embedded in concrete. A layer of sand on the concrete serves as a bed for the granite paving

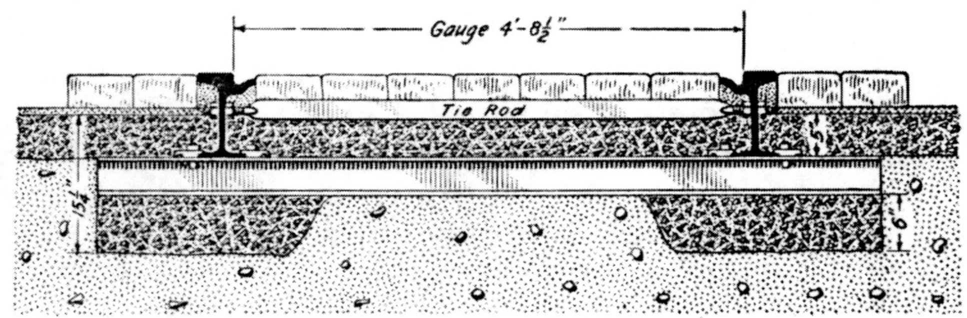

Fig. 5

blocks. The spaces near the web of the rails are filled with cement mortar.

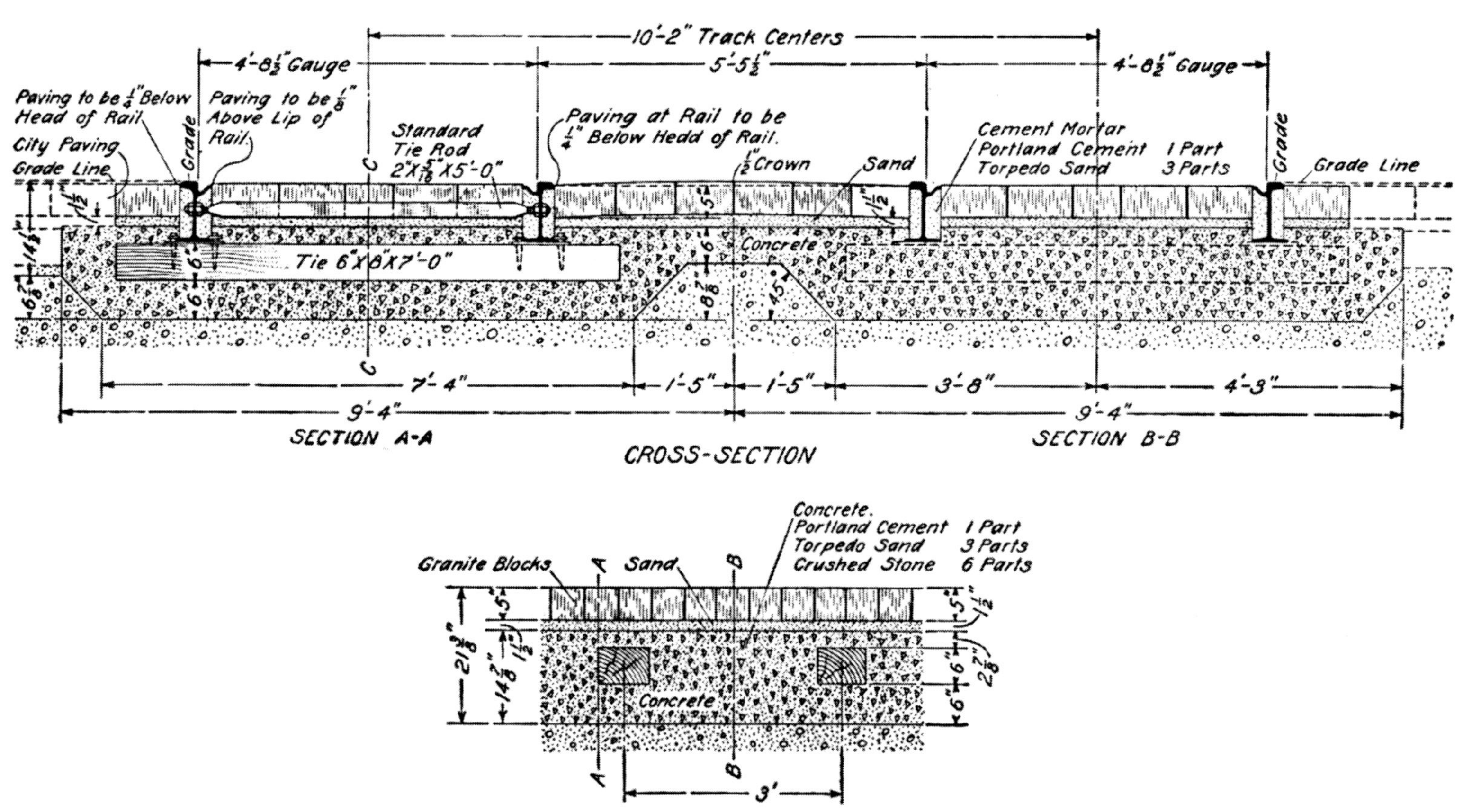
10'-2" Track Centers
4'-8½" Gauge
5'-5½"
4'-8½" Gauge
Paving to be ¼" Below Head of Rail
Paving to be ⅛" Above Lip of Rail
Standard Tie Rod 2"X5/16"X5'-0"
Paving at Rail to be ¼" Below Head of Rail.
½" Crown
Sand
Cement Mortar
Portland Cement 1 Part
Torpedo Sand 3 Parts
Grade
Grade Line
City Paving Grade Line
Tie 6"X8"X7'-0"
Concrete
7'-4"
1'-5"
1'-5"
3'-8"
4'-3"
9'-4"
9'-4"
SECTION A-A
CROSS-SECTION
SECTION B-B
Concrete.
Portland Cement 1 Part
Torpedo Sand 3 Parts
Crushed Stone 6 Parts
Granite Blocks
Sand
Concrete
3'
LONGITUDINAL SECTION C-C

FIG. 4

In other construction, bricks or blocks are shaped so that they fit against the web and lie partly under the head of the rail. Cement mortar is poured over the brick or block paving to hold all parts together.

Fig. 5 shows a construction in which steel ties are used. Concrete is used under the ends of the ties but is not necessary under their middle portions.

THE TRACK

TIES

8. Wooden Ties.—Ties are the wooden or metal supports to which both rails of a track are fastened. The ties are laid at right angles to the rails. Wooden ties are shown and their position indicated in Fig. 4. The woods most used for ties are black locust, red cedar, cypress, oak, chestnut, pine, hemlock, and spruce, here given approximately in the order of their life untreated and under average conditions. The life of untreated ties varies from 4 to 10 years according to the wood and climatic conditions.

Ties for standard gauge road (4 feet $8\frac{1}{2}$ inches) are usually 6 in. × 8 in. × 8 ft.; in third-rail construction, the insulator ties are about 2 feet longer. Wooden ties are generally spaced from 2 to 3 feet between centers. Heavy T-rail construction requires the closer spacing and roads designed for light traffic the wider spacing. Economy in tie spacing, however, may under some conditions be false economy because of excessive wear and tear on rails and rolling stock. Completely embedded ties deteriorate more rapidly than ties partly exposed. If stone ballasted, their life is longer than when buried in soil, because of better drainage.

9. Preservation of Wooden Ties.—The life of a tie can be prolonged by treatment with preservatives applied by dipping, boiling, or vacuum impregnating; in any case the tie should be heated before being treated, to dry it and to expel part of the air. Zinc chloride or creosote are the preservatives generally used.

One method of impregnating ties with creosote is as follows: The ties are placed in an iron tank, from which the air is then partly exhausted by means of a vacuum pump, thus withdrawing air from the pores of the ties. When the gauge shows a constant degree of vacuum, creosote is admitted to the tank. The creosote fills the empty pores and penetrates the wood, the distance depending on the hardness and condition of the ties, but always far enough to do much good. The process is helped if a pressure of 50 to 100 pounds per square inch is applied after the creosote is admitted to the tank. The pores being filled, moisture is excluded from the wood, and decay is thus retarded.

10. Steel Ties.—Increasing scarcity and the cost of suitable timber for cross-ties has led to many experiments with concrete, concrete and metal, and all-metal ties. Steel ties have proved most satisfactory and many miles of track with such ties have been installed. Among the advantages of steel ties are:

1. Ability to maintain correct track gauge. Steel ties can be punched accurately, and bolt clips hold the rails permanently in place, provided the nuts on the bolts are tightened occasionally. With wooden ties, frequent spiking is necessary to prevent rail spreading.

2. The useful life of steel ties is nearly three times that of wooden ties; renewals, including new ties and disturbance of the roadbed to place them are correspondingly less frequent.

3. Steel ties and rails are practically integral and the lower tie flanges are sometimes crimped so as to minimize creeping of tracks.

4. After serving approximately three times the useful life of wooden ties, steel ties have a scrap value of from 40 to 50 per cent. of their original value, resulting in some financial gain by the use of steel ties.

5. Uniformity of spans of rails between steel ties equalizes rail deflections, resulting in smoother riding than with the unequal spans over wooden ties.

11. Steel ties are made of open-hearth steel of about structural-steel grade. They can be obtained of manufacturers

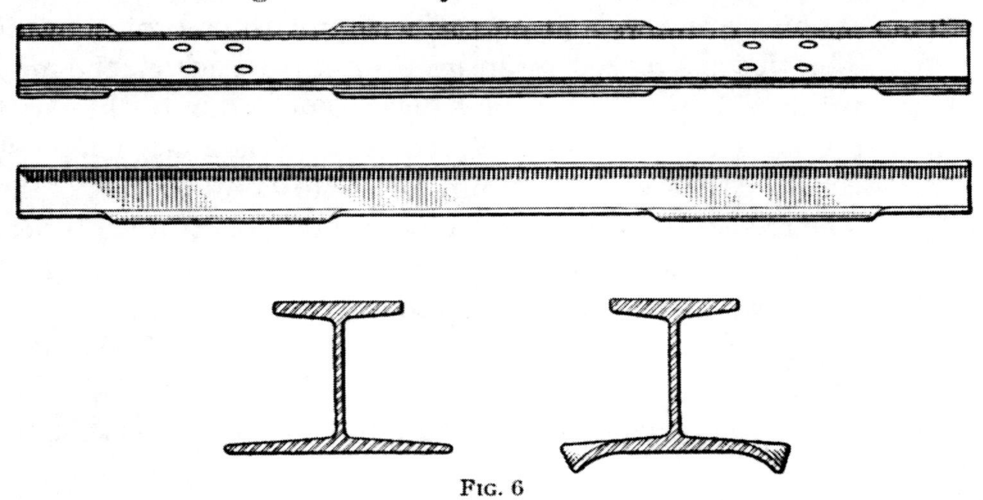

Fig. 6

who furnish the ties in any lengths, accurately punched to gauge if desired, and also dipped in coal tar to prevent rusting if such treatment is deemed advisable.

Fig. 6 shows a steel tie punched to support an 80- or a 100-pound rail; the inner pair of holes is forthe lighter weight rail, and the outer pair for the heavier. The left-hand section is taken through the part of the tie where the lower flange is straight, and the right-hand section through the part where the lower flange is crimped. The crimped portion in conjunction with the ballast tends to prevent the rails and ties from creeping. Under heavy traffic conditions and on grades, the tracks used by the descending trains should be equipped with the crimped ties, since when the brakes are applied, the moving train tends to drag the track along with it.

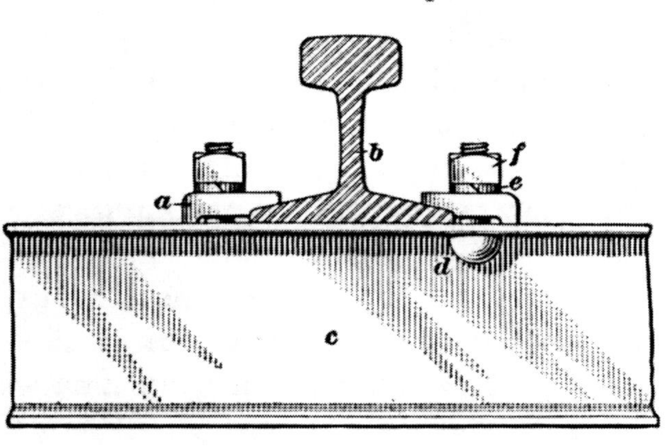

Fig. 7

12. Methods of Fastening Rails and Steel Ties. Fig. 7 shows the fastening devices for intermediate ties, and

Fig. 8 those for ties near a joint. Similar letters refer to corresponding parts in the figures. The shoulders on steel clips *a* bear against the rail *b* and the clips are secured to the tie *c* by bolts *d*, lock washers *e*, and nuts *f*.

In Fig. 8, the two splice bars that fasten the ends of the rails together are shown at *g*. Some types of splice bars are stengthened by a projection under the rail beneath the joint, as shown by the dotted lines in view (*a*). In order to avoid drilling special bolt holes in the ties near rail joints, slots are provided in the splice bars permitting the clips to be placed the same as on intermediate ties between joints; these slots aid in preventing creepage of the rails. In the absence of such slots, special clips more widely spaced than normal may be used to hold the joints in position on the adjacent tie.

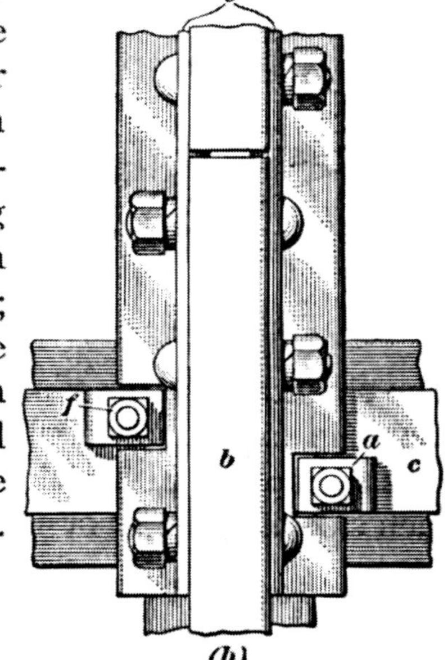

(b)

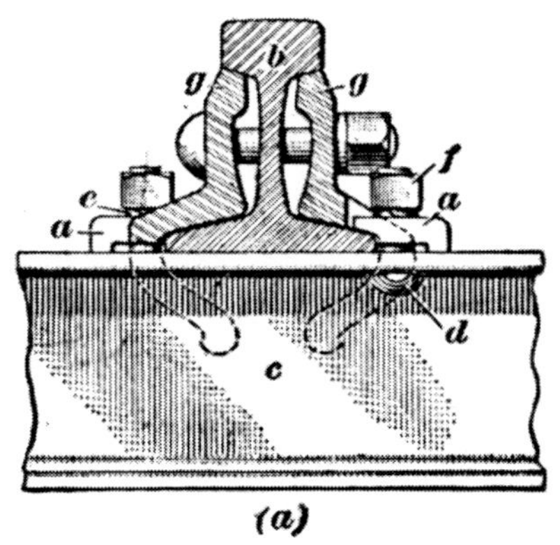

(a)

Fig. 8

RAIL AND TIE ACCESSORIES

13. Tie-Plates.—Pounding, expansion, contraction, and creeping of rails tend to wear wooden ties immediately under the rail. As tie-plates increase the wearing surfaces and decrease the wear, thereby adding to the life of the tie and to the permanency of the track, they are desirable, especially for comparatively soft-wood ties. Fig. 9 shows a common form of plain tie-plate and also shows its position on the rail.

14. Tie-Rods.—Tie-rods are used to hold the two rails of a track to gauge. The method of installing them is indicated in Fig. 5. In some cases, a tie-rod is placed every 4 feet of single track and in other cases the tie-rods are 10 feet or more apart, depending on the type of roadbed. In general, the farther apart the ties are, the closer should be the tie-rods.

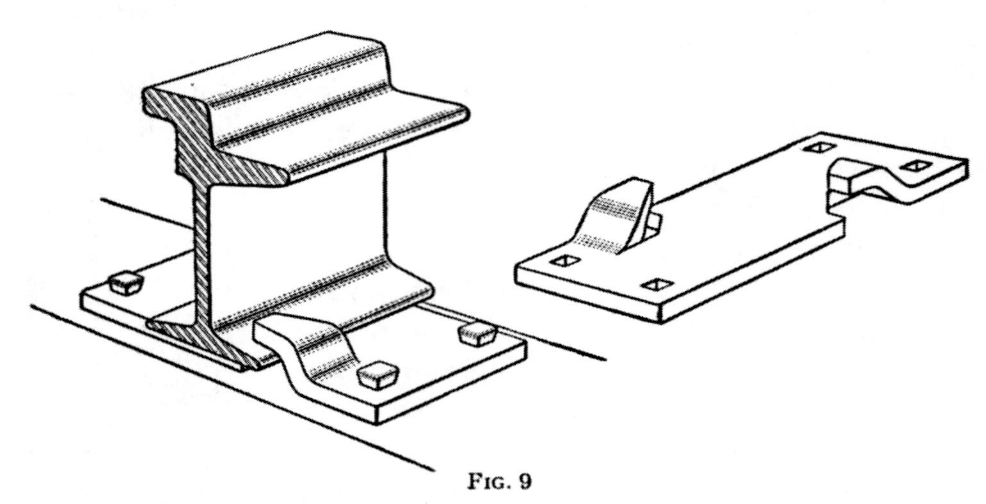

FIG. 9

The shoulders of the tie-rods are such a distance from the ends that the rods may be placed in the holes in the rails after the rails are spiked in position, and nuts are then tightened against each side of the web of each rail.

15. Rail Chairs.—In some cases, rails that are too low to accommodate paving are raised to the desired height by rail chairs, one type of which is shown in Fig. 10; the chairs are spiked to the ties.

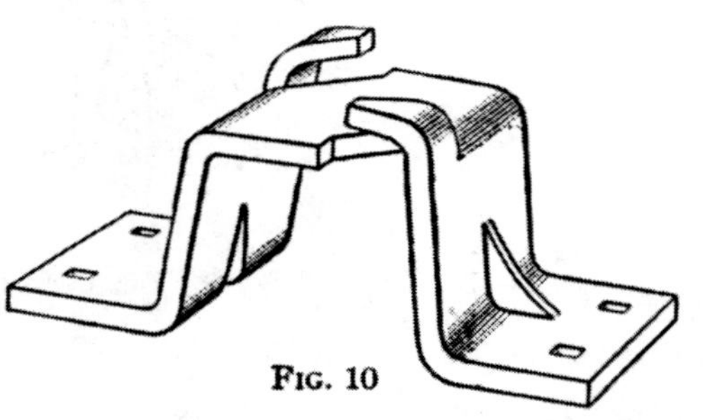

FIG. 10

16. Brace Tie-Plate. Another device to keep the rails to gauge, especially on curves, is the brace tie-plate, types of which are shown in Fig. 11. View (*a*) shows a brace for one side of the rail, (*b*) shows a double brace, and (*c*) shows types designed for **T** rails. Since the top of the braces bears against the head of the rail, the effect of any force tending to turn the rail is lessened.

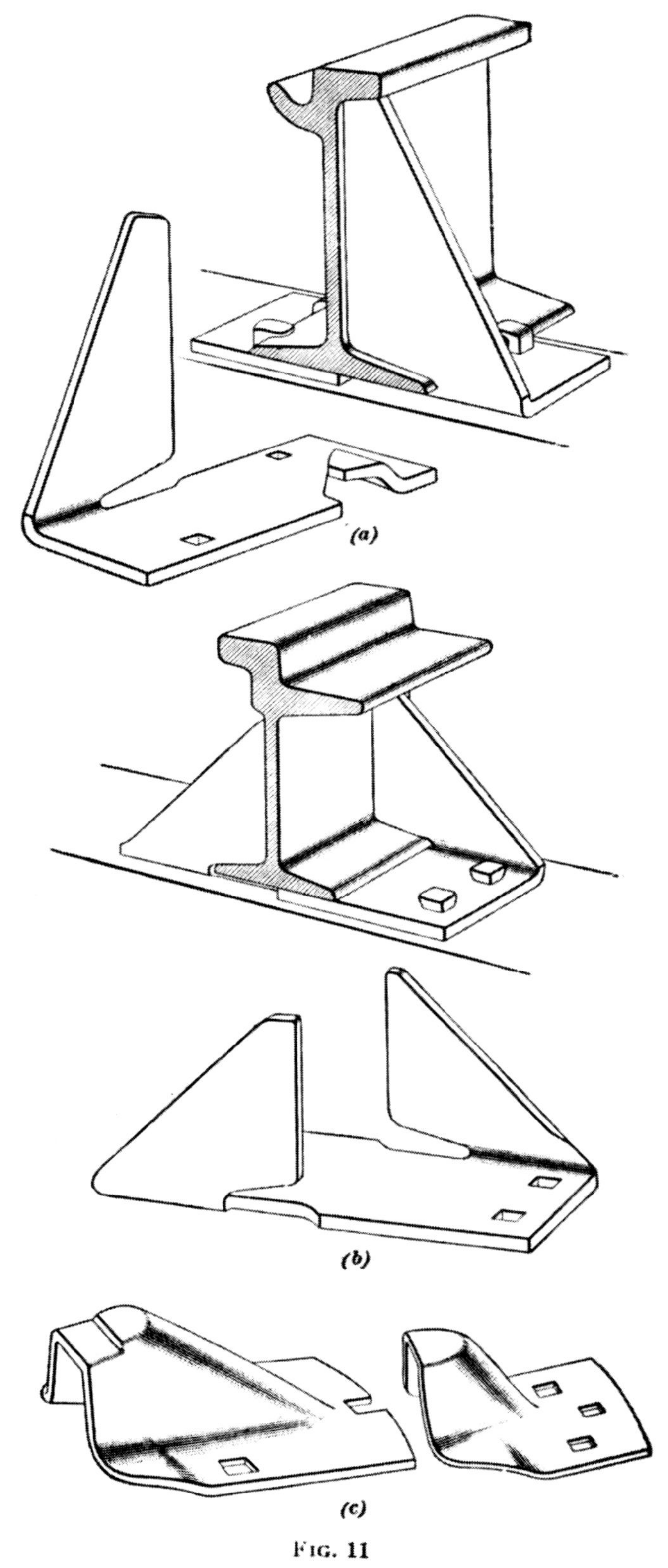

FIG. 11

RAILS

RAIL COMPOSITION

17. Track rails are of mild steel that contains low percentages of carbon, manganese, and silicon and lower percentages of sulphur and phosphorus. If the percentages of carbon and manganese are too low, the rails are soft and of poor wearing quality; if too high, the rails are brittle and their electric conductivity low. The percentages as used by a large steel company for rails intended for electric service are indicated in Table I.

TABLE I

MODIFYING CONSTITUENTS OF STEEL RAILS

	Rails Weighing From 50 to 60 Pounds a Yard	Rails Weighing From 61 to 70 Pounds a Yard	Rails Weighing From 71 to 80 Pounds a Yard	Rails Weighing From 81 to 90 Pounds a Yard	Rails Weighing From 91 to 100 Pounds a Yard
Carbon	.35 to .45	.35 to .45	.40 to .50	.43 to .53	.45 to .55
Phosphorus, not over	.10	.10	.10	.10	.10
Silicon, not over	.20	.20	.20	.20	.20
Manganese	.70 to 1.00	.70 to 1.00	.75 to 1.05	.80 to 1.10	.84 to 1.14

DROP TEST

18. Sections of rail made from a given lot of steel and not less than 4 feet or more than 6 feet in length are in some cases subjected to the drop test. The rails are placed head upwards on supports 3 feet apart and a weight of 22,000 pounds with a 5-inch face allowed to fall on them. The temperature of the test pieces should not be lower than

60° F. nor higher than 120° F. The distance, in feet, that the weight drops is 16 feet for 61- to 80-pound rails; 17 feet for 81- to 90-pound rails; and 18 feet for 91- to 100-pound rails. If the falling weight breaks the rail, tests on two additional pieces are made from the same lot of steel. If either of these rails breaks, all of the rails made of that lot of steel are usually rejected. If the two rails pass the test satisfactorily, the lot of rails is usually accepted as far as this test feature is concerned.

RAIL WEIGHT

19. Rails are made in many sizes to suit widely differing operating conditions, but all are rated in pounds per yard of rail. For electric service, rail weights generally range from 60 pounds to 141 pounds per yard; 129-pound rails are used in Chicago, Illinois, and 141-pound rails in Columbus, Ohio.

RAIL SECTIONS

20. Rails for electric service are rolled in many shapes, but may be divided into three general classes: The **T** rail, Fig. 12 (*a*); the *girder* rail, view (*b*); and the *groove* rail, view (*c*).

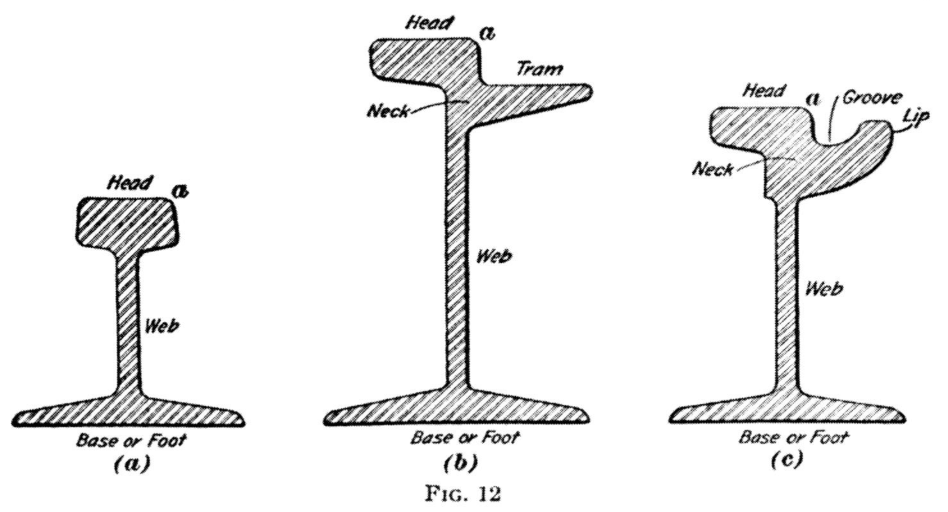

Fig. 12

The names *head*, *web*, *base*, *tram*, *groove*, *lip*, and *neck* have been given to the different parts of the rails as indicated in the three views. The gauge of the track is measured from the inside edge *a* of the head of each rail, called the *gauge line*. The tendency in electric-railway practice is to use long rails in order to reduce the number of joints. Ordinarily, rails are 30 feet in length, but some roads use 60-foot rails.

21. T Sections.—**T** rails of much the same sections as used on steam roads are employed on the unpaved portions of suburban and interurban railroads. Somewhat higher sections are quite frequently used on paved streets for city roads. Fig. 13 shows a section that has a wider than normal tread in order that interurban cars with 3-inch wheel tread may operate on city streets without interference with the paving. The splice bars are shown unshaded.

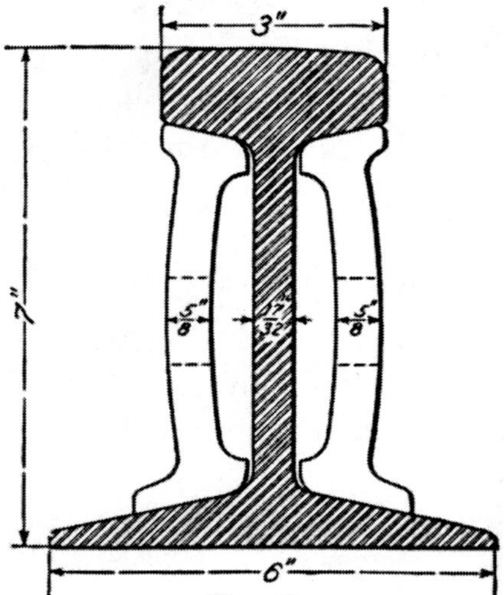

FIG. 13

The **T** section gives maximum strength and stiffness with minimum weight and has no groove nor tram to collect dirt; the head remains clean, thus lessening the power required to propel cars. **T** rails are cheaper than groove or tram rails, are easier to lay and have as good, and in some cases better running qualities. An old objection to their use was the difficulty in paving to their head in a manner to keep the street surface unbroken; with the special paving blocks and bricks now used, this objection does not exist, and **T** sections, as listed in Table II, are advocated by good authorities.

22. Girder and Groove Sections.—Rails with either a tram or a groove offer minimum interference with wagon traffic. The groove rail leaves the street surface near the track practically level. The tram of the girder rail offers a good path for vehicle wheels which results in a very considerable wear of the rails; also, wagons have difficulty in turning out from such a rail; therefore, in places of dense traffic, groove rails or **T** rails with special paving blocks are usually to be preferred.

To obtain satisfactory service with a groove rail, dirt must not be allowed to accumulate in the groove because of the danger of derailment. With some sections of groove rail the lip is ½ inch lower than the head and the groove so shaped that the flange of the wheels tends to push the dirt out of the groove.

There is always a given shape of car-wheel flange best suited to a groove of given form. In buying car wheels the shape of groove in the rail on which they are to operate should be considered in order to lessen the wear on both the flange and the groove.

TABLE II

WEIGHTS AND DIMENSIONS OF STANDARD T RAILS

(A. S. C. E. Sections)

Weight per Yard Pounds	Area of Cross-Section Square Inches	Width of Base and Height Inches	Thickness of Web Inch	Width of Head Inches
100	9.8	$5\frac{3}{4}$	$\frac{9}{16}$	$2\frac{3}{4}$
95	9.3	$5\frac{9}{16}$	$\frac{9}{16}$	$2\frac{11}{16}$
90	8.8	$5\frac{3}{8}$	$\frac{9}{16}$	$2\frac{5}{8}$
85	8.3	$5\frac{3}{16}$	$\frac{9}{16}$	$2\frac{9}{16}$
80	7.8	5	$\frac{35}{64}$	$2\frac{1}{2}$
75	7.4	$4\frac{13}{16}$	$\frac{17}{32}$	$2\frac{15}{32}$
70	6.9	$4\frac{5}{8}$	$\frac{33}{64}$	$2\frac{7}{16}$
65	6.4	$4\frac{7}{16}$	$\frac{1}{2}$	$2\frac{13}{32}$
60	5.9	$4\frac{1}{4}$	$\frac{31}{64}$	$2\frac{3}{8}$
55	5.4	$4\frac{1}{16}$	$\frac{15}{32}$	$2\frac{1}{4}$
50	4.9	$3\frac{7}{8}$	$\frac{7}{16}$	$2\frac{1}{8}$
45	4.4	$3\frac{11}{16}$	$\frac{27}{64}$	2

Groove rails must be kept accurately to gauge in order to prevent the wheel flange from binding on the rail head or lip when the car is operating on a straight section of track or when the car is rounding a curve. The groove must also be of sufficient depth so that the wheel flange will not ride on the rail.

TRACK AND WHEEL GAUGES

23. The standard track gauge in the United States is 4 feet $8\frac{1}{2}$ inches. There are, however, many roads with a track gauge other than standard. The device for testing the distance apart of the rails is also called a *track gauge;* one of these is indicated in Fig. 14 (*a*). When in service, the shoulders should engage the rail heads at the gauge line on each rail when the testing device is laid at right angles to the track.

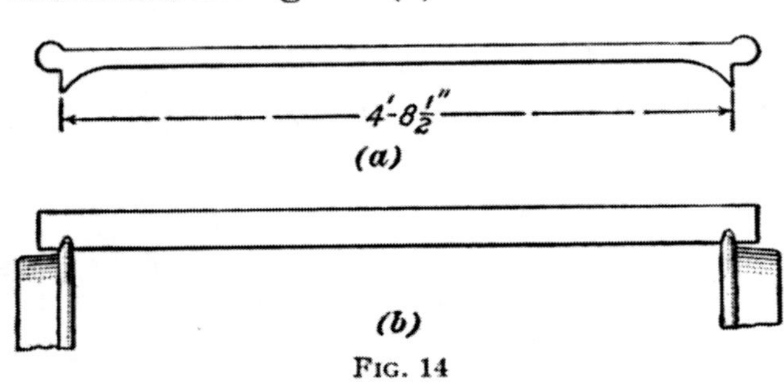

Fig. 14

The car wheels are pressed on the axle so that the outside of the wheel flanges are 4 feet $8\frac{1}{4}$ inches apart, this distance being tested by a device called a *wheel gauge* shown in Fig. 14 (*b*). This gauge should fit over the flanges so that if one end is held pivoted over the flange, the other end may be moved laterally about $2\frac{1}{2}$ inches.

RAILS WITH CONICAL TREAD

24. Car-wheel tread diameters are greater next to the flange than they are at the outside edges, thus permitting the car to center itself on the track when the two wheels on the same axle differ slightly in diameter. This so-called *coning* of the treads can satisfactorily perform its service when the difference in the diameters of the two wheels on the same axle is not enough to cause differences of more than $\frac{3}{8}$ inch in the circumferences of the two wheels. It was formerly customary to make the rail top level; under this condition, until there is some wear in either the rail tread or in the wheel tread, the

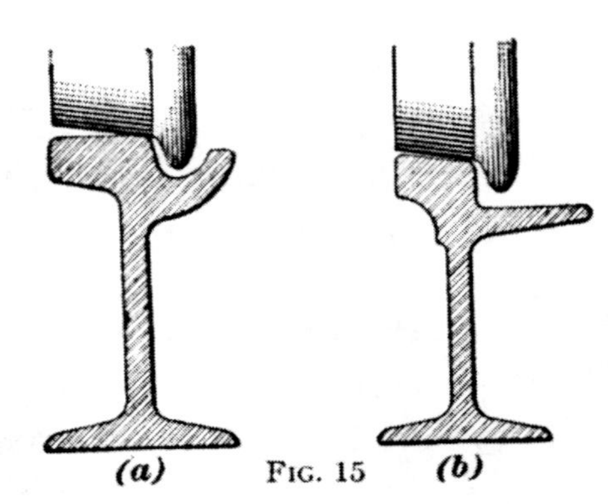

Fig. 15

bearing surface between the two is practically a line, as shown by Fig. 15 (*a*).

Girder and groove rails are now rolled with a conical tread, as in Fig. 15 (*b*), thus providing good traction surface between wheel and rail and increasing the life of both. T rails are coned in both directions from the center and to such an extent that when a fair life has been realized from one side of the rail head, the rails can be turned end for end and the other side of the head used.

GUARD-RAILS

25. When a car rounds a curve, the flange of the forward outer wheel presses against the gauge line of the outer rail. The contact of these two surfaces guides the car around the curve and causes a tendency for a wheel to climb the rail; the sharper the curve and the higher the speed, the greater is the climbing tendency, which may be dangerously aggravated by chipped wheel flanges or open rail joints. To prevent climbing, curves are laid with a *guard* feature which may be either a part of the running rail itself or a separate rail laid alongside the running rail. In some cases both running rails are provided with guards, but this practice increases the cost as well as the amount of power required to propel the car against the increased friction. The best authorities are agreed that a guard to the inner rail of the curve affords ample protection. The groove rail guard is generally a part of the running rail itself, as indicated in Fig. 16; this rail is similar to a groove rail except that the guard lip is heavier and extends above the rail tread. The dotted line indicates the contour to which the guard wears in time.

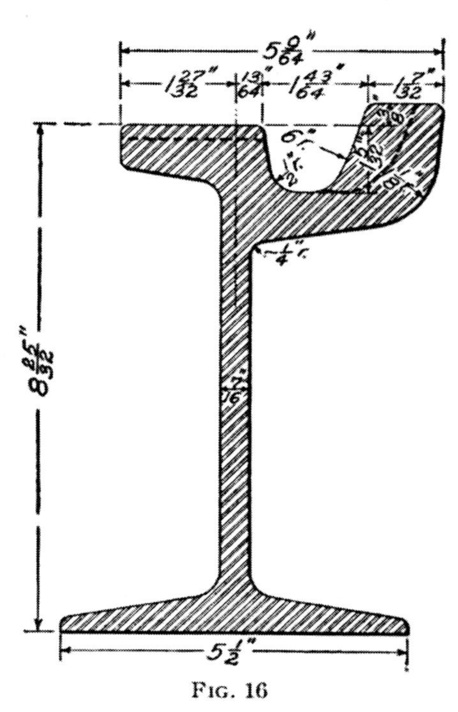

Fig. 16

T-guard rails for city work are usually fastened to the **T** rail as indicated in Fig. 17 (*a*) for high rails and in (*b*) for low rails.

In open-country work, a second line of **T** rails is sometimes

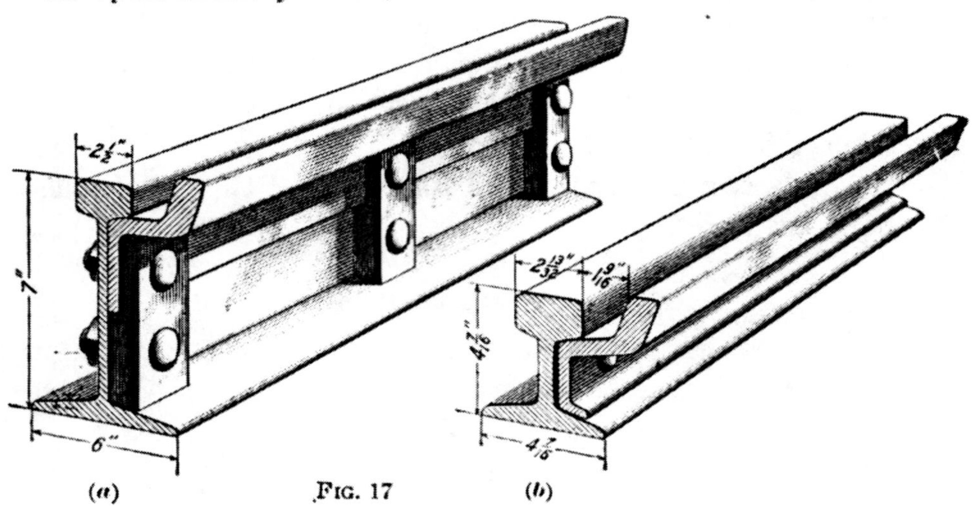

(*a*) Fig. 17 (*b*)

laid near the inner rail of the curve and between the two track rails. On bridges, a guard-rail is laid near each track rail and between them.

RAIL JOINTS

26. Necessity of Good Joints.—Substantially constructed rail joints are essential on any railroad, and especially on electric roads, in order to obtain durability of track and rolling stock and to maintain good conductance of track circuit. A portion of the weight of the car motors is supported directly by the car axle without the intervention of springs, so if the rail ends are slightly uneven or loose, the car wheels strike them a heavy blow thus tending to flatten the heads of the rails. The blows become heavier as the rail heads flatten, and unless the joints are promptly made even and secure, they soon get in such bad condition as to decrease greatly the conductance of the track and to rack the rolling stock. The remedies are to cut off the damaged ends of the rails, or if the rails are not too much damaged, to grind the rail treads so as to slope the flat back some distance from the joint. The better plan is to make the joints perfect and rigidly secure when the track is laid. A few of the bolted joints will be described.

27. Standard Channel-Plate Joint.—All forms of bolted joints depend for their fastening and stiffening on the *channel plates*, sometimes called *fish-plates*, *splice bars*, or *joint*

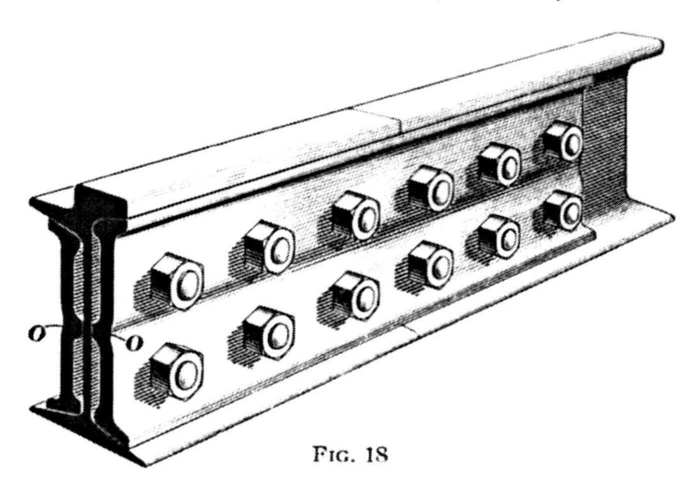

Fig. 18

plates. The most common form of bolted joint is made by bolting a channel plate on each side of the rail. The holes in the channel plates are oblong as are also a portion of the shanks of the button-headed bolts so that the bolts cannot turn when they are being tightened.

Fig. 18 shows a standard twelve-bolt joint made with two channel plates; the plates have projecting ribs *O*, to prevent buckling when the bolts are tightened. The channel-plate flanges bear against the under side of the rail head and tram and the upper side of the rail foot, thereby stiffening the joint.

In Fig. 18 the bolts have hexagon heads; ordinarily, square-headed nuts are to be preferred, but with girder rails and high **T** rails where two rows of bolts are used in the plates and it is desirable to have the bolts close to the edges of the plate, the hexagon headed nuts are used because they are easier to get at with wrenches. Square-headed nuts have more bearing surface on the rail.

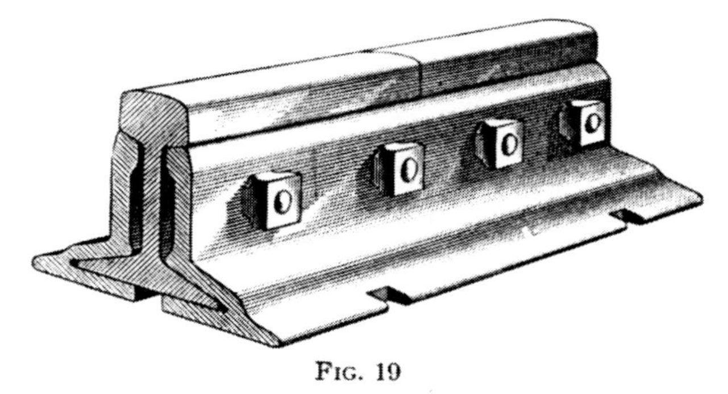

Fig. 19

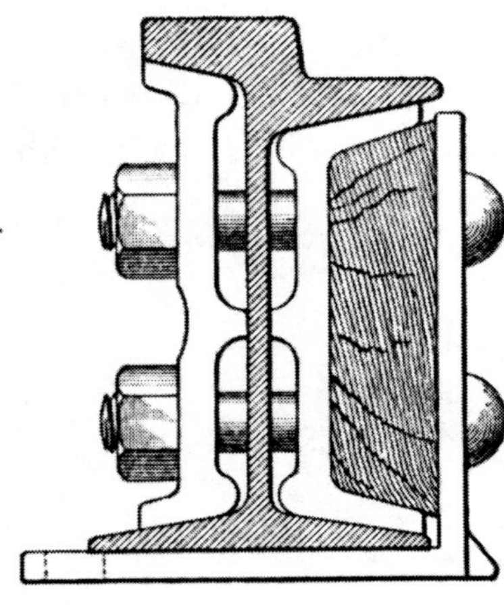

Fig. 20

28. Continuous Joints.—Fig. 19 shows a form of continuous rail joint made of sections so shaped that a flange projects under the foot of the rail from each side. The abutting rail ends are thus held firmly in line. The slots, Fig. 19, in the splice bars made for the rail spikes or bolts also serve to prevent rail creeping, as explained in Art. **12.** In another form of base-supporting joint, the under portions are bolted together below the rail.

Fig. 20 shows a base-supporting joint made up of two channel plates, a block of wood, and a rolled angle of steel, all of which are bolted to the two abutting rail ends.

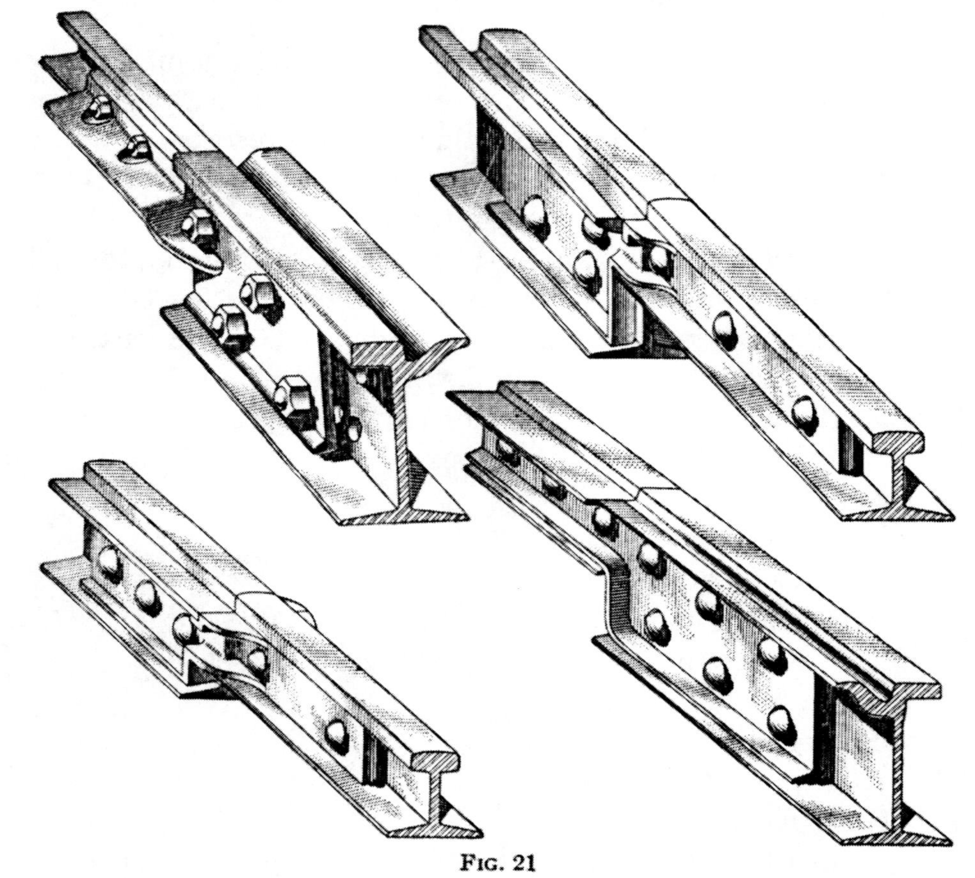

Fig. 21

29. Joining Rails of Different Sections.—For joining rails of unlike sections, combination joints can be made with special splice bars that directly connect the rail ends together, or special sections of cast-steel rails may be used and these joined to the two rail ends by ordinary splices. Fig. 21 shows four combination joints made with special splice bars and Fig. 22 shows a form of rail casting that may be inserted in the line to join a groove rail to a **T** rail.

30. Use of Welded Joints.—Rails laid to their full depth in paving expand and contract but little with ordinary changes of atmospheric temperature, because the paving tends to equalize the temperature of the rail and the earth. The joints of such rails can therefore be welded so as to form a long length of rail and no allowance need be made for expansion and contraction.

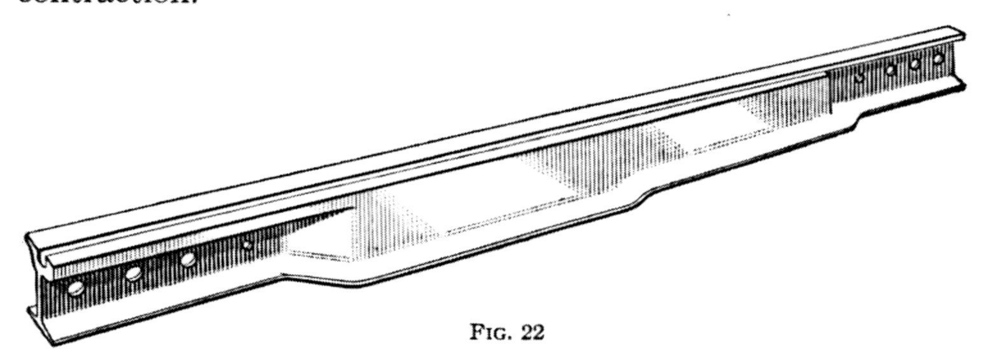

Fig. 22

If the rails are exposed, as in open-country work and on elevated structures, they may be welded together if joints that allow a slight movement of one rail toward or away from the other rail are installed at intervals of about 1,000 feet on straight track and at the ends of curves.

31. Lorain Welded Joint.—In the Lorain method, mild-steel plates are electrically welded to each side of the webs of the two abutting rails by passing a very large current through the plates and the rails.

The welding outfit consists of a sand blast, a synchronous converter, a transformer, a welder, and grinding apparatus, all mounted on four cars. The surfaces to be welded are cleaned by directing a blast of sand against them. The converter serves to change direct current from the trolley circuit to alternating current and the transformer provides the proper value of voltage for the welder. The tops of the rails at the welded joint are smoothed by the grinder.

Fig. 23 shows a splice plate, one of which is welded to each

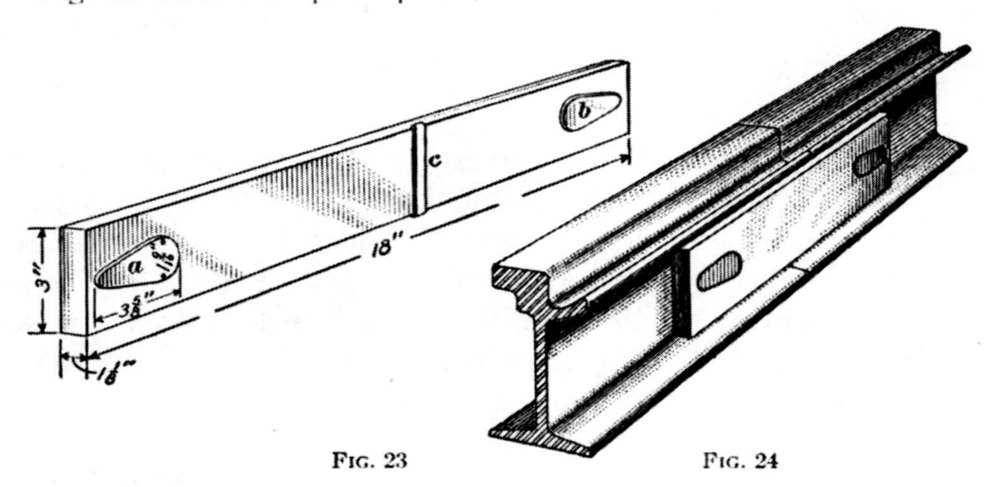

Fig. 23 Fig. 24

side of the rail web as shown in Fig. 24. Each plate is welded at points *a*, *b*, and *c*, Fig. 23, the total area affected being about $10\frac{1}{2}$ square inches on each plate.

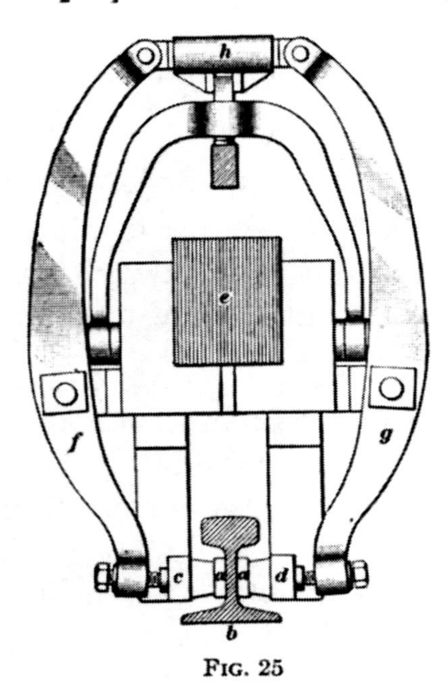

Fig. 25

32. The welder is shown in Fig. 25 and is suspended by a crane from the front of one of the cars so that it can easily be placed in position on either rail of the track. After cleaning the rail surfaces, the splice bars *a* are clamped in position on the rails *b*. The terminals *c* and *d* of the welding transformer *e* are pressed against the plates over the portions to be welded by means of levers *f* and *g* operated by pistons in a hydraulic cylinder *h*. The primary coil of the welding transformer *e* is supplied with alternating current. The secondary circuit of this transformer consists of a single loop of very large conductor and includes in the circuit, plates *a* and rails *b*. A secondary current of about 25,000 amperes at 7 volts melts the surfaces to be welded; the current is then cut off and the surfaces firmly pressed together and the joint allowed to cool. The central weld of the joint is made first, followed by the welds at the ends of the plates.

33. Fig. 26 (*a*) shows a horizontal section through a completed joint; (*b*) a section of the center weld; and (*c*) of an end weld. The resistance of a joint of this kind is less than the corresponding length of rail.

For welds on old rails, a special section of plate on the tread side of the joint is sometimes used; this section is provided with a lug that fits under the rail head, and the small space between

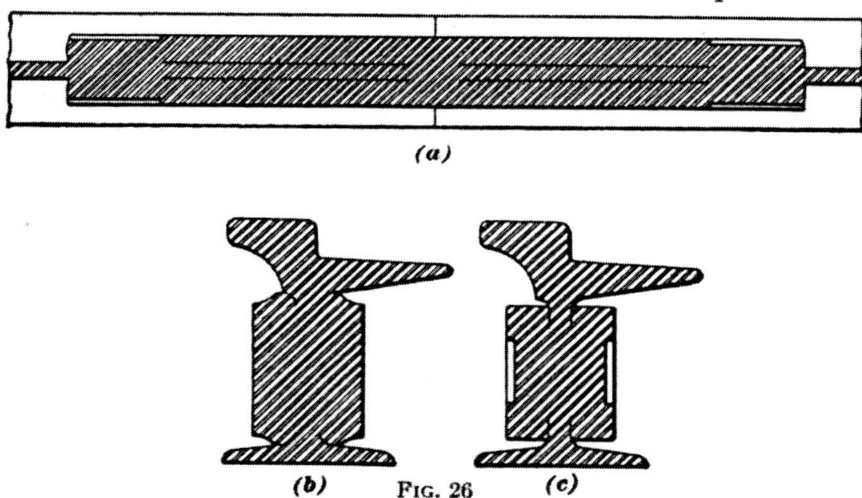

Fig. 26

the lug and the rail head is poured full of melted zinc, thus improving the support of the rail head.

34. Arc Welded Joint.—The welding of rail joints by the arc method is accomplished by clamping plates on each side of the webs of the rails and then causing metal to flow from a soft-steel electrode along the junction of the plates and the rails. The steel electrode is a rod about $\frac{1}{4}$ inch in diameter and 30 inches long and is connected through an adjustable rheostat with the trolley wire. The rail forms the negative side of the circuit. When the electrode is touched and then withdrawn from the plates, or rail, an arc is formed and the melted steel from the rod lodges against the rail and welds the edges of the plates to the rails.

35. Cast Welded Joint.—Cast welded joints are made by pouring molten iron into a cast-iron mold placed around the abutting rail ends, which are first sand-blasted for a distance of 6 or 8 inches from the center of the joint.

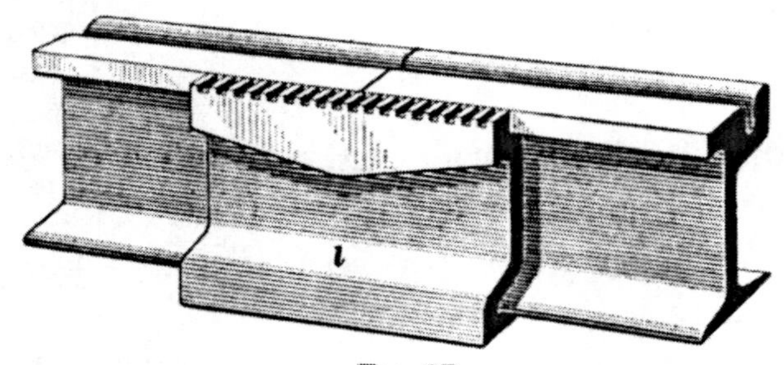

Fig. 27

Fig. 27 shows a cast-iron weld. The added iron which is indicated at *l* may weigh from 75 to 225 pounds, depending on the shape of the joint. The apparatus used with this method consists of a sand-blast machine for cleaning the rails, rail molds, and a portable cupola for melting the iron.

Fig. 28

36. Thermit Joint. Thermit is a powdered mixture consisting mostly of finely divided iron oxide and aluminum. The mixture is placed in a crucible over the joint. The metal aluminum has a great affinity for oxygen, and if the mixture is suddenly heated by burning some ignition powder on its surface, a very rapid chemical action takes place that melts

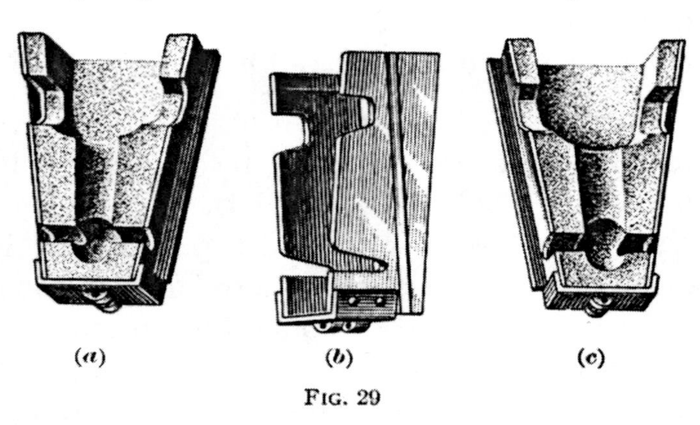

Fig. 29

the whole mass in the crucible. The iron oxide and other ingredients are reduced into steel and this is poured into molds attached to the rail ends. The melted metal is so hot that it welds itself into the structure of the rails. A scratch brush is used to clean the rails before making the joint.

Fig. 28 shows a thermit joint and Fig. 29 the molds. The sheet-iron case (*b*) is placed over a model of the joint and tamped full of China clay and loam. The molds for the two sides of the rail thus formed are shown at (*a*) and (*c*) and these are clamped to the rails.

37. Zinc Joint.—Fig. 30 shows a type of zinc joint. The rails are sand-blasted and the rolled joint plates are riveted to the rails. There are open spaces left at the top of the plates and around the foot of the rail. The whole joint is heated to a temperature of 300° or 400° F. by means of portable oil burners and molten zinc is then poured in the spaces near the foot of the rail and between the plates and head of the rail. The zinc fills all irregularities of the surfaces of plates and rails and offers a firm support for the joint.

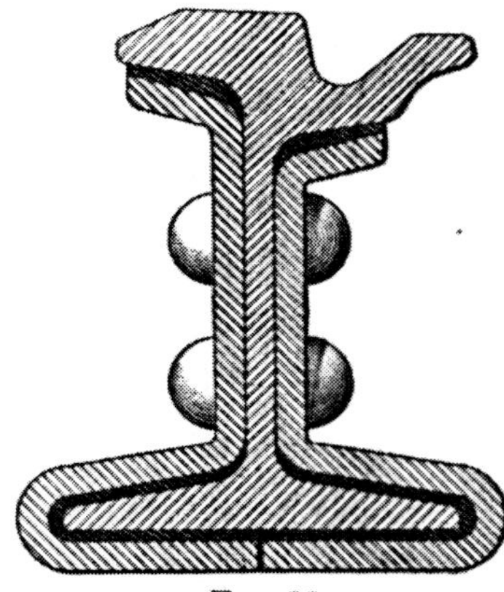

Fig. 30

38. Expansion Joint.—In exposed rails with welded joints, expansion joints are usually installed at intervals of 800 to 1,000 feet on straight track and at the approaches of curves. Each expansion joint allows slight endwise movements of one of the rails, and thus lessens the danger of the rails breaking or the curves being distorted due to changes in length of the rail caused by temperature changes.

Fig. 31 shows one type of expansion joint with the rails in the positions assumed in warm weather. Rail *a* is fixed to the base *b* which is spiked to the ties. Rail *c* is free to move endwise to the right through the guides *d* as the rail contracts as the result of cold weather. Rail *a* is the movable rail at the next expansion joint on the left.

The extent of rail movement depends on conditions such as the amount of temperature change and of rail surface exposed.

39. Disposition of Joints.—Some engineers advocate placing the joints in the two rails of a track opposite each

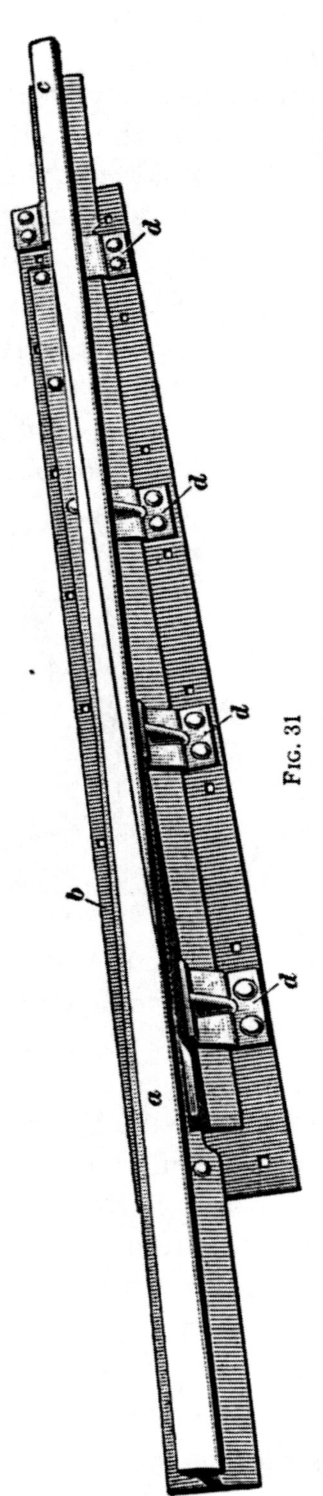

Fig. 31

other, while others prefer to locate the joint in one rail opposite the center of the rail length on the other side of the track. In the first method, the track is said to be *even jointed* and in the second method it is said to be *broken jointed*.

When the track is so laid that the ends of the rails meet on a tie, the joints are called *supported joints;* when the rails meet between ties, the joints are called *suspended joints*. Both methods are in general use.

RAIL BONDS

40. Use of Rail Bonds.—Rail bonds usually consist of large bare copper conductors, the terminals of which are firmly attached to points near the ends of the abutting rails of a joint. As the sole purpose of rail bonds is to improve the track conductance they are not usually employed with rails having the joints welded; the chief use of bonds is with bolted joints, where loosening bolts and rusting surfaces would otherwise reduce conductance.

No matter what type of bond is used, the conductance of the joint depends on the mechanical excellence of the bolted connections and of the contacts between the bond and the rails, because continual movement of the rails is likely to loosen the bond contacts and increase the resistance of the joint. It is highly important, therefore, that the track should be inspected systematically and that the joints be kept tightly bolted.

41. The maintenance of high rail conductance keeps the energy loss low and helps in maintaining good car operation.

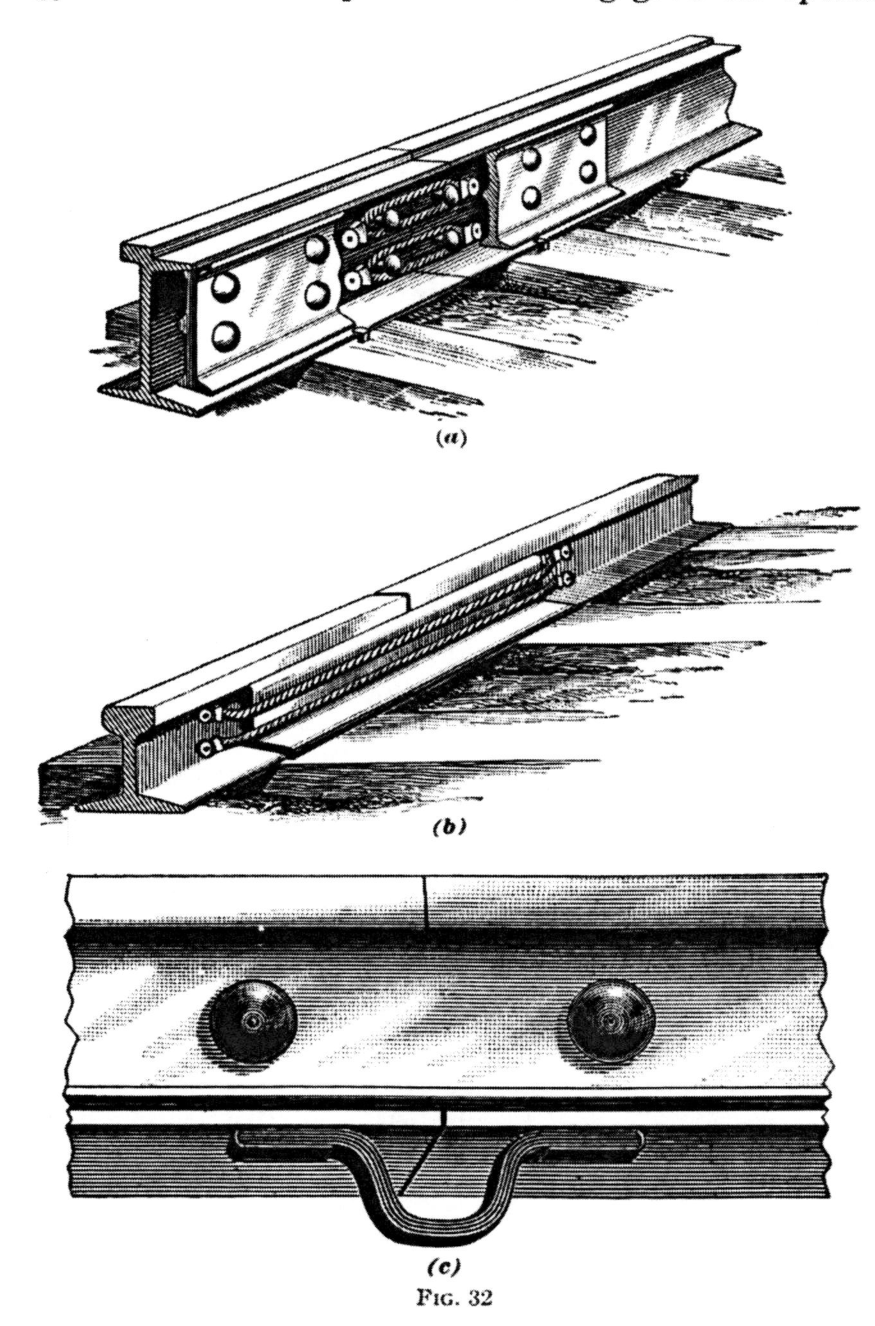

(a)

(b)

(c)

Fig. 32

High track conductance also minimizes leakage of the return current to adjacent pipes with possible resulting damage to them.

Bonds placed between the channel plates and the rail to protect them from mechanical injury and from thieves are known as **protected bonds**; those not so covered are known as **unprotected bonds.** Fig. 32 (*a*) shows a protected bond and (*b*) and (*c*) two types of unprotected bonds.

42. Compressed Stud Terminal Bond.—It is important that the contact between the copper conductors of the bond and the iron rail be so made as to secure a permanent low-resistance connection between the two devices.

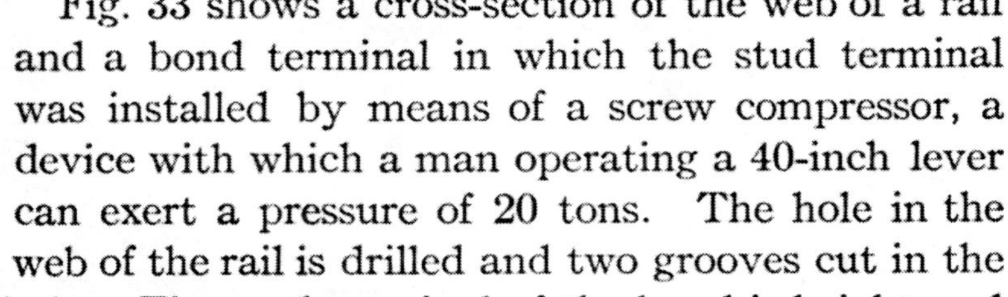

Fig. 33

Fig. 33 shows a cross-section of the web of a rail and a bond terminal in which the stud terminal was installed by means of a screw compressor, a device with which a man operating a 40-inch lever can exert a pressure of 20 tons. The hole in the web of the rail is drilled and two grooves cut in the walls of the hole. The stud terminal of the bond is brightened by emery paper, then inserted in the hole and pressure applied. The copper flows into the grooves and further pressure forms the rivet head as shown. Soda and water or plain water should be used instead of oil to lubricate the drill when boring the holes in the web. The surface of the walls of the hole should be bright and free from oil or moisture and the bonds should not be installed in damp weather. A solution of red lead and linseed oil may be applied to the bond terminals and adjacent surface of the rail after installation, to seal the joint from moisture.

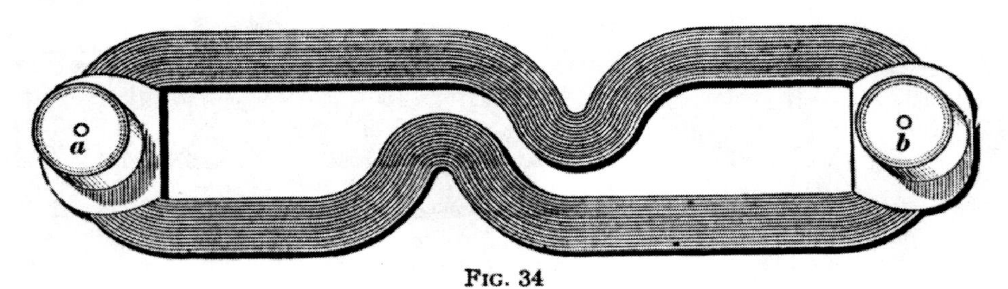

Fig. 34

Fig. 34 shows a protected bond of this type made of copper ribbon and cast copper terminals *a* and *b*. Fig. 32 (*a*) shows the position on the rails of two bonds of this general type but made of flexible cable. When the channel plate is in position the bond is covered.

The main portion of the bond may be formed of solid copper, flexible cable, or ribbon. The stud terminals may be formed by welding together a portion of the wires of the cable or ribbon conductors, or made separately and welded or soldered to the conductors. The conductors are usually bent so as not to interfere with the bolts of the joint and to allow for slight contraction or elongation of the rails, due to temperature changes, without causing serious stresses on the contacts of the bond terminals.

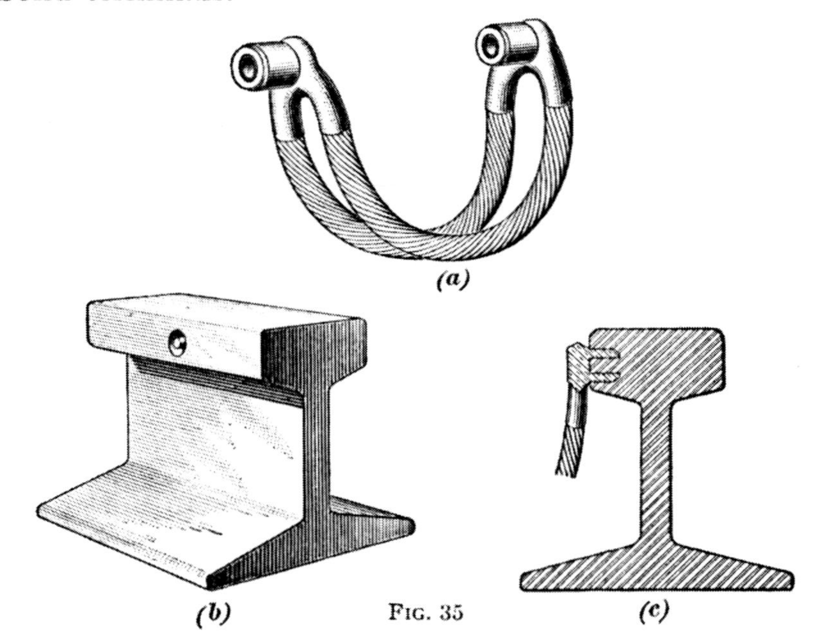

Fig. 35

Bonds may be installed on one side of the rails or on both sides; in the latter case, the method is called *double bonding*. One or more bonds on each side may be used. Sufficient copper is used to cause the conductance of the joint as a whole to be nearly the same as the conductance of an equal length of rail.

43. Hammered Stud Terminal Bond.—Fig. 35 (*a*) shows a bond provided with terminal studs that may be hammered into place after the holes in the rail heads, one of which is shown in (*b*), have been drilled by a special milling cutter. The holes in the studs indicated in (*a*) fit over the pins formed in the rail heads as indicated in (*c*), thus obtaining large contact surface.

44. Soldered Bonds.—A bond that is soldered to the base of the rails is shown in Fig. 32 (*c*) and one soldered to the head of the rails in Fig. 36. A special application of these bonds is for temporary work or for bonding old rails where the cost of removing channel plates and renewing bolts would be prohibitive.

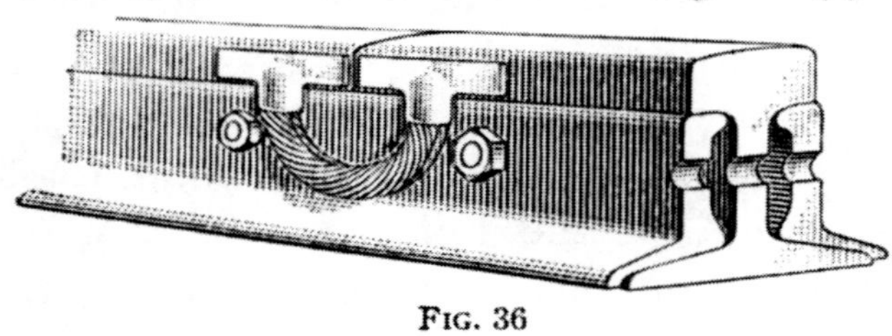

Fig. 36

The installation of soldered bonds requires the utmost care to insure good electrical and mechanical union between the copper and steel. All rust and scale must be removed from the steel surfaces and the rails heated until the cleaned surfaces show a light blue color. Soldering flux, preferably zinc chloride, is then applied and a bar of solder rubbed on the cleaned surfaces until they are thoroughly tinned. The bond is then clamped on the rails and the rails again heated sufficiently to melt wire solder applied to them; the clamp is then tightened and wire solder melted on the edges of the terminals as the joint cools. After the joint has cooled it should be painted with waterproof paint.

45. Electrically Welded Bonds.—Fig. 37 shows the installation of two welded bonds at a joint of a conductor rail. These bonds may be applied either to the head or foot of the track or conductor rails.

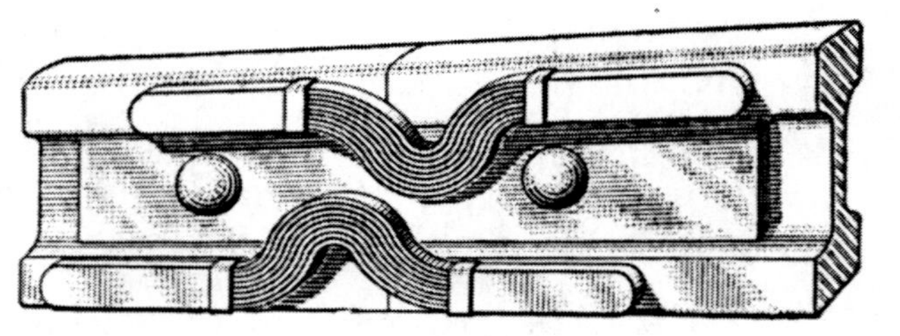

Fig. 37

Fig. 38 shows a welded bond applied to the conductor rails of a conduit system. The two parts of the bond are separately welded to the rails and when the rails are in place, the free ends of the bond conductors are clamped or soldered together.

A special car on which is mounted a rotary converter, transformer, and welding clamps is used to weld the bonds onto the rail.

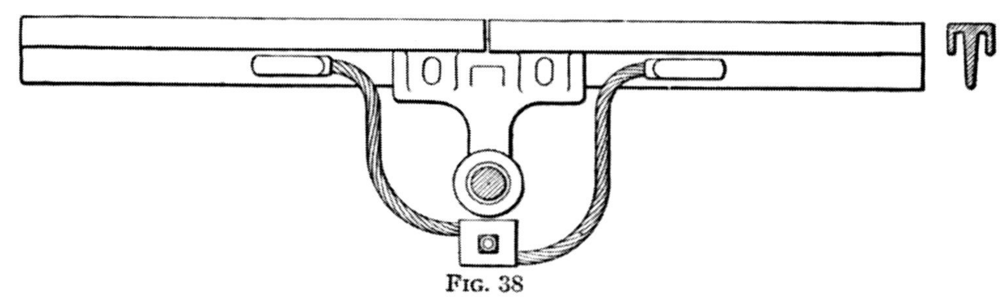

FIG. 38

46. Copper-Welded Bonds.—In some cases the copper conductor of a bond is welded to the rail by pouring melted copper into a mold clamped against the rail and into which the end of the bond conductor is placed. Fig. 39 shows the mold in position. The interior includes two chambers connected by a narrow neck. Melted copper, from a crucible carried on a special car, poured into one of these chambers surrounds the copper conductor and comes in contact with the cleaned rail surface. Excess copper flows into the second chamber and assists in raising the temperature of the rail to a point where the copper in the first chamber is welded to the rail. The joint is then allowed to cool, the mold removed, and the excess copper cut loose at the narrow neck, leaving the bond terminal as indicated in Fig. 40. Copper cables may be welded to the third rail or to track rails by this method.

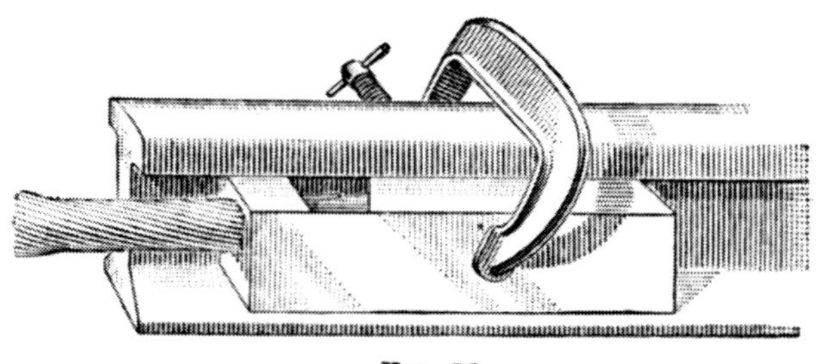

FIG. 39

47. Bonds Using Amalgam. — The conductance of the contacts between the rails and the channel plates is greatly improved if the contact surfaces are cleaned and coated with an amalgam of copper or other metal. In some cases the

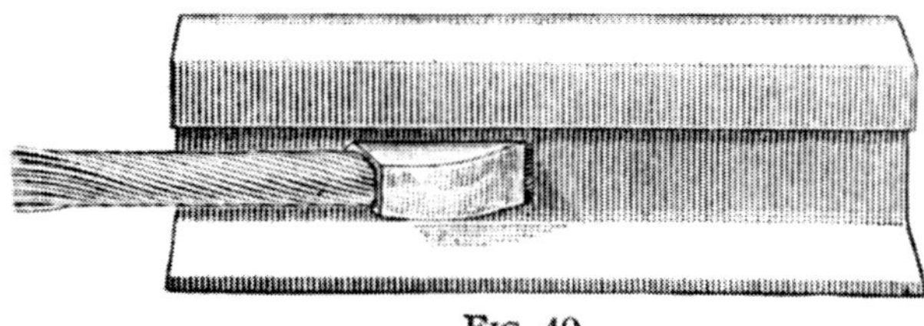

FIG. 40

amalgam is held in cork cases and these are compressed between the rails and the channel plate. In joints of this kind, the channel plate serves as the bond conductor and the amalgam affords a low-resistance path between the plate and the rail.

48. Cross-Bonds.—The two track rails should be connected together at intervals of 500 feet approximately by cross-bonds, thus affording a low-resistance path around a poor joint in either rail.

On double track the four rails should be cross-bonded at about the same intervals as for single track. Parts of crossings and switches that are bolted together are in many cases electically connected together by long copper cables.

In order to increase track conductance, bare copper cable is sometimes laid along the track and connected at intervals to the joint bonds. The rails at a point near the station are connected to the negative bus-bar by large cables. The cables are usually welded or brazed to the rail.

SPECIAL WORK

49. Designation.—Track construction relating to curves, branch-offs, crossings, etc., is usually known as **special work.** Some of the more common forms of special work are shown in Fig. 41. A *plain curve* is shown at (*a*); it may be right-hand or left-hand, simple, or compound. A *left-hand branch-off* is shown at (*b*) and a *right-hand branch-off* at (*c*); these are used where a branch road leaves the main line. When determining whether left or right, face the point of departure of the branch.

A *connecting curve and crossing* are shown at (*d*); a *plain* **Y** at (*e*); a *three-part* **Y** at (*f*); and a *through* **Y** at (*g*). The three-part **Y** can be used in place of a loop to turn cars at the end of a line. A *reverse curve* is shown at (*h*); it is used where a cross-street is broken at the main street. A *right-hand cross-over* is shown at (*k*) and a *left-hand cross-over* at (*l*); they are installed at intervals in the track so that a car may cross from one track to the other, thus shortening the regular run in case of the disablement of the car or to make up time. A *diamond turnout* is shown at (*m*); an *ordinary siding* at (*n*); and a *thrown-over turnout* often used for temporary work to avoid interference with track repair at (*o*). Other than the names just given to these parts are used in some localities.

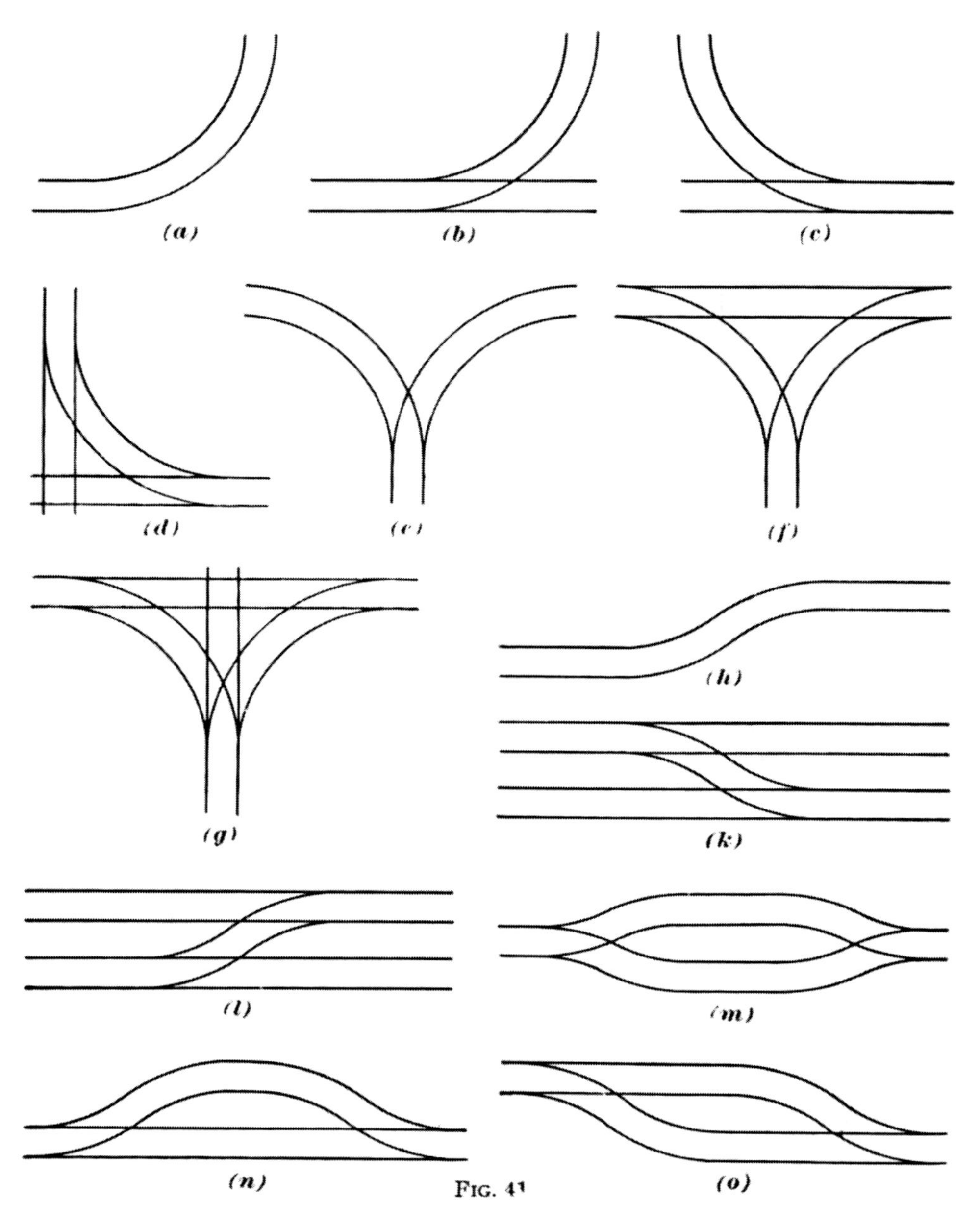

FIG. 41

Fig. 42 shows a layout of special work and indicates the names given to some of the more detailed parts as recommended by a large steel company.

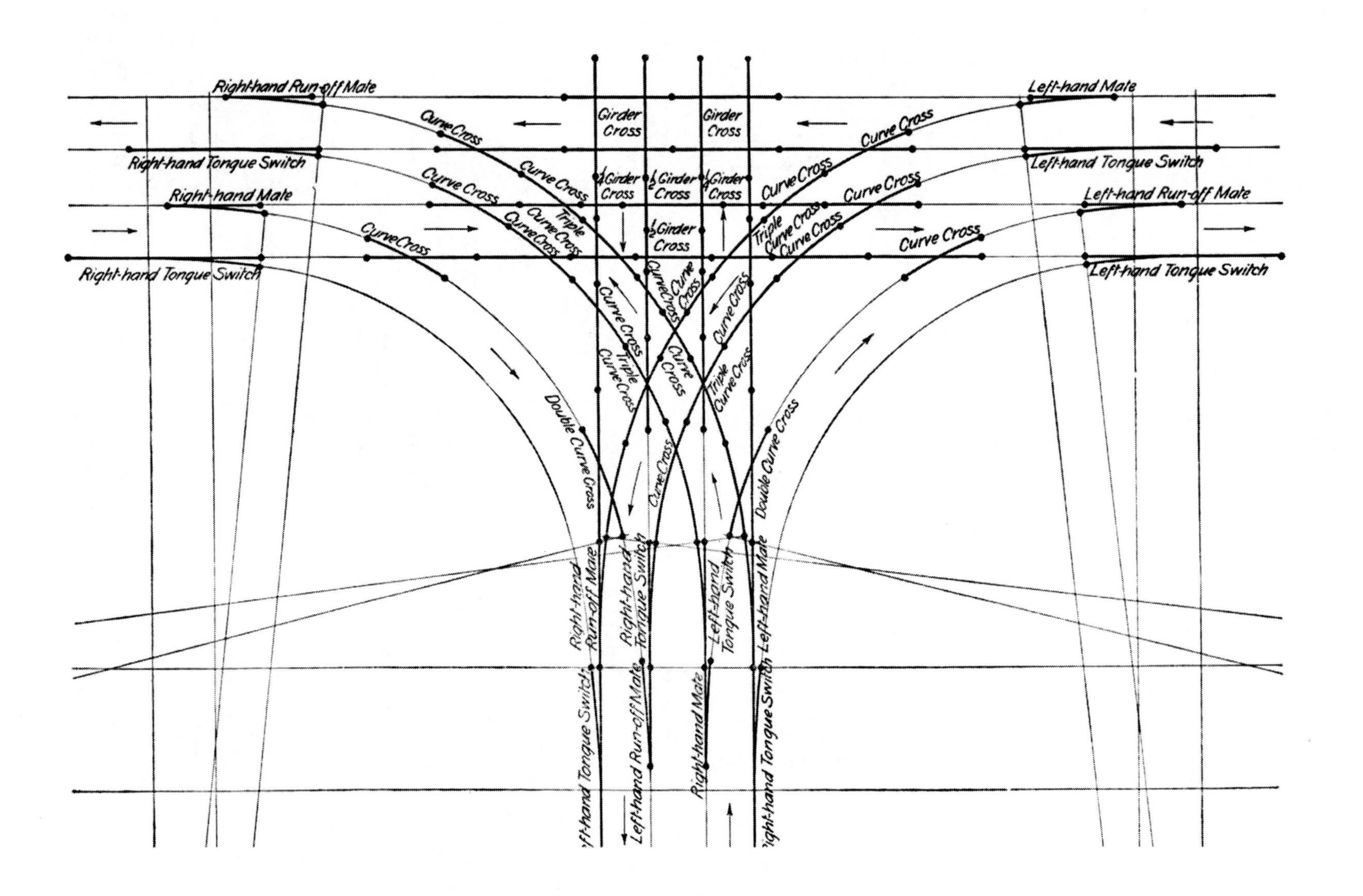
Right-hand Run-off Mate
Right-hand Tongue Switch
Right-hand Mate
Right-hand Tongue Switch
Left-hand Mate
Left-hand Tongue Switch
Left-hand Run-off Mate
Left-hand Tongue Switch
Curve Cross
Girder Cross
½ Girder Cross
Triple Curve Cross
Double Curve Cross
Right-hand Run-off Mate
Right-hand Tongue Switch
Left-hand Tongue Switch
Left-hand Mate
Left-hand Run-off Mate
Right-hand Mate
Right-hand Tongue Switch

50. Construction.—Important special work is made up at the steel works and shipped ready to install. The work of construction must be carried out with great precision. The site of the proposed work is surveyed and a drawing made; this drawing is checked and from it the work is laid out in actual size with chalk on a hard smooth maple laying-out floor; if the layout checks correctly, the lines on the floor are used as guides for making wooden templets for the patternmakers and rail benders. When the separate parts of the job are complete, it is assembled in the laying-out yard where any mistakes that may have escaped preceding inspections or any inaccuracies due to uneven shrinkage of the cast parts or to carelessness in the bending may be detected and rectified. The system of checking and rechecking the job in the various stages of its progression from the surveyor to the shipper is so thorough that the chances of error are a minimum. Special work on electric railways generally lasts longer than on steam railways, because subjected to lower tonnage and speed; also on electric railways the rules generally require stopping or slowing at crossings and intersections.

51. In switches, frogs, and crossings, the greatest wear takes place at the points of the switches and the breaks in the tread of the rail, places subjected directly to the pounding of the wheels. Various methods have been adopted for inserting hard steel at these places. One make of special work, called *manganese*, takes its name from special plates of hard manganese steel that are placed at the intersections; these are held by special bolts or fastenings so that worn plates can be readily renewed. Another class of special work, known as *guarantee work*, is guaranteed to last as long as the abutting rails; in it, tempered-steel wearing plates are held by keys and zinc poured in around the inserted piece. In a third class of special work, known as *adamantine work*, the crossings are made of steel castings.

Fig. 43 shows a crossing of the guarantee type. Renewable hardened-steel plates *a* are set in as indicated; the joints are stiffened by cast welding the rail ends at the crossings. The piece as a whole is in two parts fastened together by means of standard joints. The crossing illustrated is for groove rails, but similar construction is adaptable for use with high **T** rails.

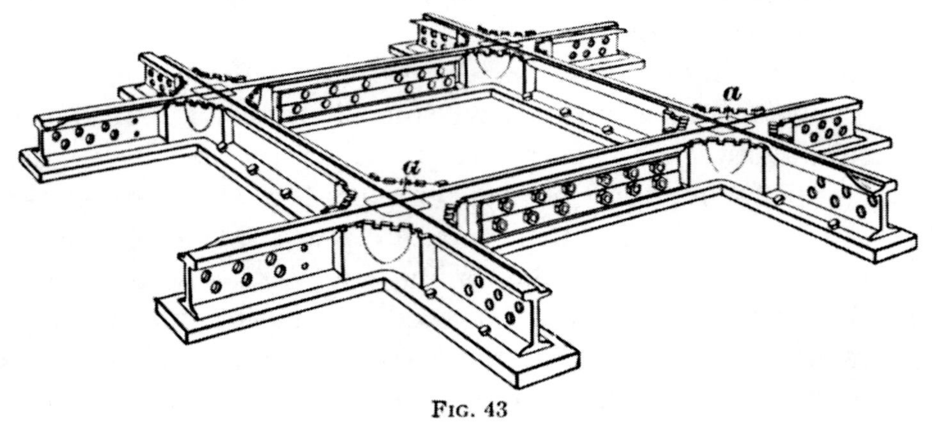

Fig. 43

Fig. 44 shows a cast manganese-steel crossing; it is made in two pieces and connected together by standard joints.

Fig. 45 indicates the arrangement of the renewable parts in a guarantee-curve crossing. The hardened-steel plate *a* is held by the wedges *b* and *c*, which are embedded in zinc to prevent them from working loose. To remove the plate, the wedges are driven down.

52. Gauntlet Track.—For double track through a narrow street or tunnel or over a narrow bridge, the *gauntlet*

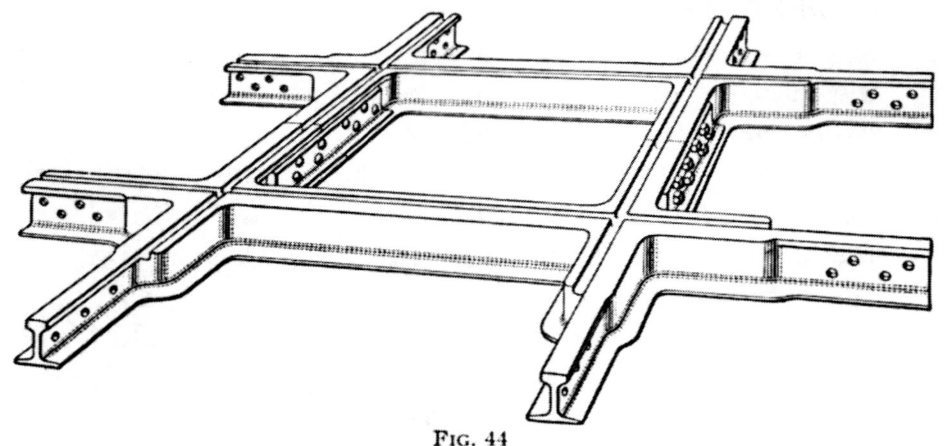

Fig. 44

construction indicated in Fig. 46 may be used. The gauntlet construction is cheap and simple, the special work being limited to two plain ordinary frogs.

53. Curves.—Curves are of two kinds, *simple* and *compound*, or *easement*, curves. A **simple curve** is described with but one radius throughout its length, while a compound curve

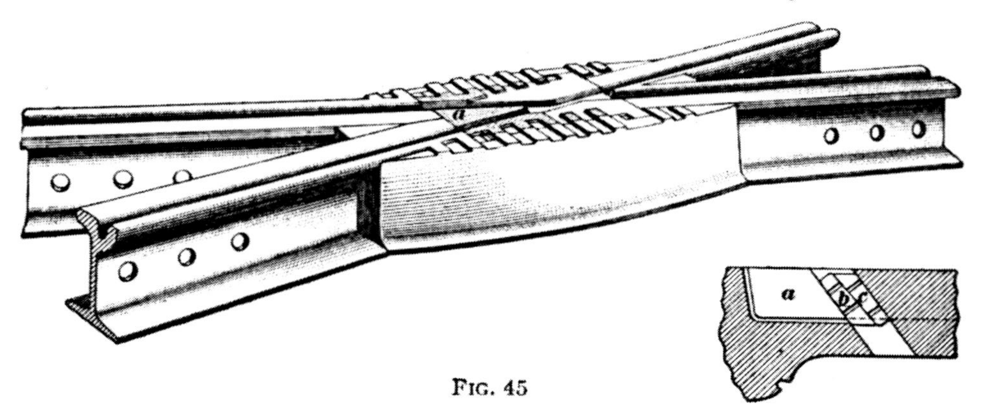

Fig. 45

is so constructed that the radii become shorter as the middle point of the curve is approached from either end. A compound curve is easier riding than a simple curve. Street-railway curves are usually designated by the radius, in feet, at the center. Long curves of light rail are sprung in, as a rule; that is, the rail is pried over with a bar and spiked into position, the paving being relied on to keep the track in place. The main objection to "springing in" a curve is that in course of time the ends of the rails may straighten out and make an angle at the joint; this difficulty is most likely to occur with heavy rails or on curves of short radii. The car trucks in rounding such a curve will change direction in jumps, instead of gradually, and impart to the car a disagreeable, jerky motion not to be found on a curve that is smooth and regular. On curves of heavy rails and moderate radius, a portable rail bender should be used, while shorter curves should be bent to a templet with a power bender. With ordinary **T** rails, curves having

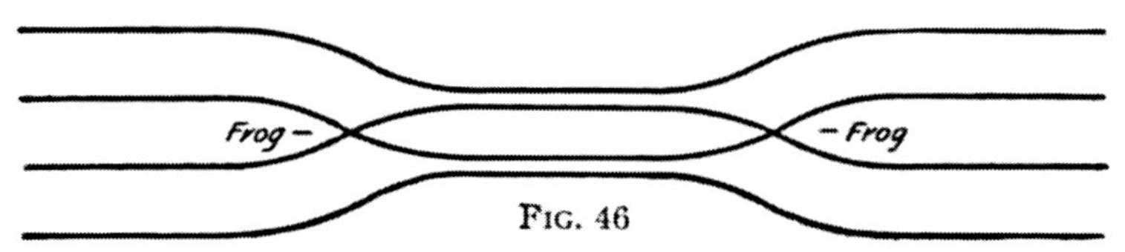

Fig. 46

a radius of 500 feet or over can be sprung in, but with girder rails or high **T** rails 800 to 1,000 feet is the smallest allowable radius.

54. Easement, transition, or **compound curves** are formed by combining curves of different radii, so that the entrance of the car into the curve shall be gradual, and a sudden shock avoided. The curve at the point where it branches from the straight part of the track, or a tangent, as it is called, is of long radius and the radii are gradually decreased until the radius of the center of the curve is reached. All steel companies who make electric-railway special work, have standard transition curves made up of a number of radii such that the length of arc of any one radius is not over 5 feet, and cars ride very smoothly around such curves.

55. Testing Clearances.—Single-track curves must be laid out so that cars will go around them freely without either end overhanging the corner of the sidewalk or striking any obstruction. On double-track curves, two cars should be able to pass each other without danger. The layout of a double-track curve therefore depends on the length and width of cars to pass on it, on the car overhang at the ends of single-truck cars and in the center of double-truck cars, on the distance between the track centers, the curvature, the elevation of the outside rail, and the length of the wheel base. Car fenders should be considered, since a fender increases the effective length of the car.

The best plan is to lay out on paper and to scale a plan of the proposed curve; then, by means of a pasteboard dummy that scales the dimensions of the outside lines of the car, the actual clearance at all points can be readily determined. The positions of the car wheels may be indicated by holes in the dummy through which the track can be seen, or transparent paper may be used, so that the dummy can be made to take the right path around the curve. For the safety of passengers, a clearance of at least 12 inches should be allowed on each side of the car if cars are to pass each other on curves. Special precautions are necessary where the center-pole method of line construction is used.

56. Rail Elevation on Curves.—Wheels tend to climb the outer rail in curves because the tendency for the car to move straight ahead is overcome by the pressure of the head of the outer rail against the flanges of the wheels. If the flange of a wheel is much worn or chipped or if there is considerable space between rails at a joint in the curve, the tendency for the wheel to climb is increased, because the flange may catch on the rail. Elevating the outer rail or lowering the inner one or dividing the total change between the two eliminates a

TABLE III

RAIL ELEVATIONS

Radius of Curve, Feet	Speed of Car	
	6 Miles per Hour	19 Miles per Hour
	Elevation of Rail, in Inches	
1,200	.078	.780
900	.156	1.287
600	.234	1.989
450	.312	2.613
300	.429	3.939

portion of the side thrust on the rail. Where cars round curves at high speed, elevation of the outer rail is necessary; these elevations for certain stated conditions may have the values indicated in Table III. On curves as installed in city streets where the car speeds are low, rail elevation is usually not essential.

MAINTENANCE OF TRACK

57. Smooth, level, even track is absolutely essential for easy riding cars and minimum cost of maintenance of superstructure, rolling stock, and track. Rough uneven track jolts passengers, causes the trolley to fly off and possibly injure itself or the overhead work, racks the rolling stock, batters the track into still worse shape, and may be the cause of serious accidents and loss of life.

Safe, agreeable, and efficient operation, therefore, demand the maintenance of good track. Substantially constructed new track, although higher in first cost, can be maintained at less cost than poorer track with cheaper construction. Proper maintenance requires *good lining, surfacing,* and *gauging;* only a trained eye can detect such defects unless they are very serious.

58. Lining the track refers to the elimination of horizontal kinks that tend to give cars a swaying motion; **surfacing** refers to the elimination of vertical kinks in the track that cause jouncing. Depressions in the track are caused by poor joints, soft spots in the roadbed, poor tamping of the ballast, or washouts under the ends of the ties. Poor surfacing increases the wear on rolling stock and rails and may cause accidents; when the wheel of a heavy, high-speed car drops into a depression, it delivers a blow that is liable to break the wheel, the axle, or the rail. The effect of such blows can be seen on old track with poorly maintained joints; on a double-track road, for example, the rail ends on the north side of the joints of the north-bound track and on the south side of the joints of the south-bound track are hammered into cup-shaped depressions, and are said to be **cupped.** On single track the rail ends on both sides of the joint become cupped. Furthermore, this continual pounding on low spots forces the rails laterally out of line and if this condition is long neglected it becomes very difficult to correct. Poorly lined or surfaced track also increases the resistance to train movement and consequently increases the energy required to operate cars. Measurement of the current required to start a car on different parts of supposedly level track will often show great differences.

59. Gauging refers to the maintenance of proper distance between rail heads and is very important especially on straight track. On curves, distances between rail heads a little more than the normal gauge if not beyond the limits of actual safety, are less objectionable, because the wheel flanges press against the outer rail, and if that is a true curve the running will be satisfactory. The rails on straight or curved track should not be close enough together to bind the wheel flanges.

60. Rail Breaks.—Rail breaks are most likely to occur in freezing weather, when the rails are most brittle and the roadbed most inflexible. Broken rails are extremely dangerous, especially in the outer rail of a curve, where the forward outer wheel flanges may catch in the break and cause derailment. Continuous thorough inspection is the only safeguard, and even this may not prevent an accident for a train may break a rail and be derailed thereby.

61. Weeds.—Weed growth on a track lubricates the rails and causes the driving wheels to slip. Weeds also conceal defects in the track and make approximate lining and surfacing by eye impracticable. Among the methods used to remove weeds are digging them out, cutting them off, killing them with high-voltage alternating current, and burning them by means of crude-oil burners extending from a special car; the last method is the cheapest and most convenient when much of this work is to be done.

ELECTRIC-RAILWAY CALCULATIONS

LINE CALCULATIONS

CONDITIONS AFFECTING THE SIZE OF FEEDERS

VARIABLE CONDITIONS

1. Line calculations, as here considered, deal with the determination of the sizes of conductors required between the station or substation and the cars to transmit the electric energy needed under specified conditions of operation.

No general rule can apply to the values to be used in all of the conductor calculations involved in the design of a given electric-railway system. The practice of dividing the line into insulated sections simplifies calculations, because each section may be considered as governed by its own load conditions. If these conditions could in any case be assumed with certainty, the problem for that section could be easily solved; but the solution would not apply to other sections, for it is unusual for two sections to have the same load conditions. Besides, a change of attractions along the line or a shift of suburban improvements may develop a gradual change in load conditions more serious than a daily or weekly shift. A certain line lay-out may meet all existing requirements, but subsequent changes in the requirements may unbalance the system. It may then be necessary to make new calculations and either change the amount of conductor and its disposition or install devices for maintaining the voltage, such as boosters and storage batteries. Conductors placed with good judgment may satisfactorily improve the voltage at a desired point, but in some cases the installation of devices for boosting the voltage at that point may result in more economical operation.

LIMITING CONDITIONS

2. Increasing the total cross-sectional area of the conductor system raises the voltage and saves energy by reducing the line loss and this saving can be approximately calculated. Knowing the cost of a unit of energy at the power station, the direct saving to be effected by a proposed increase of conductor may be obtained; then, knowing the cost of the added conductor, including installation, the interest on its total cost can be computed.

Improving the feeding system not only saves directly, by reducing the line losses, but it saves indirectly, because raising the voltage increases the car efficiency and speed and decreases the number of cars and men required to meet a certain timetable. Besides, the improvement in service attracts and creates travel by adding to the pleasure riding.

It pays to install more conductor if the cost of the energy saved in a year plus the added income each year exceeds the interest on the total cost of the added conductor. These statements may be expressed in the form of a limiting equation such as:

interest on the conductor cost = *value of energy saved* + *added income due to improved service*

The limiting value is that the value of the left-hand member of the equation should not exceed the value of the right-hand member.

FEEDER FORMULAS

3. Copper Conductors.—The resistance R of a conductor of commercial copper 1 foot long and 1 circular mil in cross-section is usually taken as 10.8 ohms. A conductor l feet in length and of 1 circular mil in cross-section has a resistance of $10.8\,l$, but if the cross-section of the conductor is a circular mils, the conductor resistance is, $R=\frac{10.8\,l}{a}$.

The resistance of any wire to the passage of direct current is equal to the drop in volts u through the length of the wire divided by the current I, or $R=\frac{u}{I}$.

As these two values of R are equal, $\frac{10.8\,l}{a}=\frac{u}{I}$;

or,

$$\text{circular mils } a=\frac{10.8\,l\,I}{u} \qquad (1)$$

Drop, in volts, in the feeder

$$u=\frac{10.8\,l\,I}{a} \qquad (2)$$

EXAMPLE 1.—A copper feeder 1,000 feet long must carry a current of 100 amperes, with a drop of not more than 20 volts. What must be the sectional area of the conductor, expressed in circular mils?

SOLUTION.—Substituting the values of $l=1{,}000$ ft., $I=100$ amp., and $u=20$ volts, in formula **1**,

$$a=\frac{10.8\times 1{,}000\times 100}{20}=54{,}000 \text{ cir. mils. Ans.}$$

EXAMPLE 2.—A copper feeder having a sectional area of 54,000 circular mils transmits a current of 100 amperes a distance of 1,000 feet. What is the drop, in volts?

SOLUTION.—Substituting the values of $l=1{,}000$ ft., $I=100$ amp., and $a=54{,}000$ cir. mils, in formula **2**,

$$u=\frac{10.8\times 1{,}000\times 100}{54{,}000}=20 \text{ volts. Ans.}$$

4. If the copper conductor alone is to be considered, the value 10.8 ohms per mil-foot is used and the drop u is that in the feeder alone. Formulas **1** and **2,** Art. **3,** may, however, be modified so as to take into consideration the approximate added drop in the track-return circuit by using the value 14 ohms per mil-foot for the combination of copper and steel-rail conductors and the value of u as the total drop in both the outgoing and the return conductors.

The size of the copper conductor is then,

$$\text{circular mils } a = \frac{14\, l\, I}{u} \qquad (1)$$

and the drop in the outgoing and the return conductors is,

$$u = \frac{14\, l\, I}{a} \qquad (2)$$

The formulas of this article should be used only when the track bonding is known to be in good condition. If the track is poorly bonded, the calculated results will not be reliable.

5. Aluminum Conductors.—Aluminum feeders are used to some extent in low-tension railway work. Comparison of conductors of different materials receives treatment in *Long-Distance Transmission of Electrical Energy*. The resistance per mil-foot of aluminum conductor is here taken as 17.17 ohms. The formulas relating to the outgoing aluminum feeder alone are,

$$a = \frac{17.17\, l\, I}{u} \qquad (1)$$

$$u = \frac{17.17\, l\, I}{a} \qquad (2)$$

EXAMPLE 1.—An aluminum feeder 1,000 feet long must carry a current of 100 amperes, with a drop of not more than 20 volts. What must be the sectional area of the conductor?

SOLUTION.—Substituting the values of $l = 1{,}000$ ft., $I = 100$ amp., and $u = 20$ volts, in formula **1,**

$$a = \frac{17.17 \times 1{,}000 \times 100}{20} = 85{,}850 \text{ cir. mils.} \quad \text{Ans.}$$

EXAMPLE 2.—An aluminum feeder having a sectional area of 85,850 circular mils transmits a current of 100 amperes a distance of 1,000 feet. What is the drop, in volts?

SOLUTION.—Substituting the values of $l=1{,}000$ ft., $I=100$ amp., and $a=85{,}850$ cir. mils, in formula 2,

$$u=\frac{17.17\times 1{,}000\times 100}{85{,}850}=20 \text{ volts. Ans.}$$

6. If the added drop in the return circuit is also to be considered, as explained in connection with copper feeders, the value 22 ohms per mil-foot for the combination of aluminum and steel-rail conductors gives approximate results when the rail bonds are in good condition. The formulas of Art. 5 become,

$$a=\frac{22\,l\,I}{u} \qquad (1)$$

$$u=\frac{22\,l\,I}{a} \qquad (2)$$

TRACK RESISTANCE

RAIL AND BOND DATA

7. Resistance of Steel Rails.—The resistance of steel rails varies greatly, as it depends on the composition of the steel. Track rails are selected for their wearing qualities; hence, hardness is essential, but the harder the rails the higher is the resistance. The service of conductor, or third, rails, however is not as severe as that of track rails, hence these rails are sometimes made of softer steel, in which cases they are better conductors than the track rails.

The relative resistance of steel track rails as compared to similar rails if made of copper is frequently taken as 12 to 1 or 13 to 1; the relative resistance of conductor rails made of soft steel to similar rails if made of copper is sometimes taken as 8 to 1. These ratios will be referred to as the *resistance ratios;* they do not apply to all rails, and the manufacturers should be consulted when greater accuracy is desired.

8. Comparison of Newly Bonded Joint and Solid Rails.—A comparison of a newly bonded rail joint and a corresponding length of solid rail indicates, in some cases, that the resistance of the joint is as low or sometimes lower than the resistance of the rail. After the rails have been in service, the jarring to which the joints are subjected may greatly increase the resistance from rail end to rail end.

A bond conductor of 0000 B. & S. wire has a sectional area of .1662 square inch, which, taking the resistance of rail steel as twelve times that of copper, is equivalent to $.1662 \times 12 = 1.994$ square inches of rail. Assuming that a bond 1 foot long connects two 60-pound rails, each rail having a sectional area

TABLE I

LENGTH OF RAIL EQUAL IN RESISTANCE TO 1-FOOT BONDS

Weight Rail Pounds per Yard	Rail Cross-Section Square Inches	Copper Section for Equal Conductance Square Inches	Kind of Bond							
			One 0000	Two 0000	One 000	Two 000	One 00	Two 00	One 0	Two 0
			Feet of Rail Equal in Resistance to Bond 1 Foot Long; Resistance Ratio 12 to 1							
60	6.0	.5	3.00	1.50	3.80	1.90	4.80	2.40	6.00	3.00
65	6.5	.54	3.26	1.63	4.10	2.05	5.20	2.60	6.52	3.26
70	7.0	.58	3.50	1.75	4.44	2.22	5.60	2.80	7.00	3.50
75	7.5	.63	3.76	1.88	4.74	2.37	6.00	3.00	7.52	3.76
80	8.0	.67	4.00	2.00	5.06	2.53	6.40	3.20	8.00	4.00
85	8.5	.71	4.26	2.13	5.38	2.69	6.80	3.40	8.52	4.26
90	9.0	.75	4.50	2.25	5.70	2.85	7.20	3.60	9.00	4.50
95	9.5	.79	4.76	2.38	6.00	3.00	7.60	3.80	9.52	4.76
100	10.0	.83	5.00	2.50	6.32	3.16	8.00	4.00	10.00	5.00

of 6 square inches, the resistance of the bond conductor will be equivalent to $6 \div 1.994 = 3$ feet of rail. The contact resistance between rails and bond terminals is not here considered.

The resistance of a rail joint is the resistance of the parallel paths consisting of the channel plates and the bond conductor and its contacts with the rail. The resistance of the paths through the channel plates, especially after the joint has been in service for some time, is so uncertain that they are not usually considered when the resistance of the joint is being calculated.

The cross-section of the bond conductor that is to be installed is usually proportioned to the size of rail used; thus, 90-pound rails should be bonded more heavily than 60-pound rails; the increase in bond conductance is preferably obtained by using two or more bonds in parallel for each joint.

Table I shows rail lengths equivalent in resistance to bonds of different cross-sections but all 1 foot long. The resistance of the bond contacts is not considered in this table. For example, if a joint with 70-pound rails is bonded with two No. 0000 bonds each 1 foot long, the parallel resistance of the two bonds is equivalent to that of 1.75 feet of rail, the resistance ratio being 12 to 1.

9. A rail is *bonded to full conductance* when the resistance of a given length of bonded rail is considered as being equal to the resistance of a continuous rail of the same length plus the total contact resistance of the bonds used on the rail. Rails on roads that operate under very heavy traffic conditions are sometimes bonded to full conductance, but usually it is unnecessary to install so much bond copper.

TABLE II

SECTIONAL AREA OF BONDS WITH RESISTANCE EQUIVALENT TO THAT OF RAILS

Weight of Rail Pounds per Yard	Resistance Ratio	
	12 to 1	13 to 1
	Circular Mils of Copper Equal to Steel of Resistance Ratio	
50	530,515	489,705
60	636,618	587,646
70	742,721	685,587
75	795,773	734,558
80	848,825	783,528
90	954,928	881,469
100	1,061,030	979,410

Table II shows the sectional area of copper conductors that have the same resistance per unit length as steel rails having resistance ratios of 12 to 1 and 13 to 1. Table III shows the resistance per mile of rails and tracks bonded to full conductance. Table IV shows the resistance per mile of conductor rails having a resistance ratio of 8 to 1.

TABLE III

RESISTANCE PER MILE OF RAIL AND TRACK BONDED TO FULL CONDUCTANCE, RESISTANCE RATIO 12 TO 1

Weight of Rail Pounds per Yard	One Rail, No Joints Ohm	One Rail, 176 Joints Ohm	Two Rails, 352 Joints Ohm	Four Rails, 704 Joints Ohm	Contact Resistance per Joint Ohm
60	.08765	.094954	.047477	.023739	.0000415
65	.08090	.087641	.043821	.021910	.0000383
70	.07513	.081378	.040689	.020345	.0000355
75	.07012	.075963	.037982	.018991	.0000332
80	.06573	.071204	.035602	.017801	.0000311
85	.06187	.067009	.033505	.016752	.0000292
90	.05843	.063270	.031635	.015818	.0000275
95	.05535	.059926	.029963	.014982	.0000260
100	.05259	.056955	.028476	.014239	.0000248

TABLE IV

RESISTANCE PER MILE OF CONDUCTOR RAIL BONDED TO FULL CONDUCTANCE, RESISTANCE RATIO 8 TO 1

Weight Rail Pounds per Yard	One Rail, No Joints Ohm	One Rail, 176 Joints Ohm	Two Rails, 352 Joints Ohm
60	.058433	.0632994	.0316497
65	.053933	.0584210	.0292105
70	.050080	.0542512	.0271256
75	.046746	.0506400	.0253200
80	.043820	.0474799	.0237399
85	.041246	.0446824	.0223412
90	.038953	.0421958	.0210979
95	.036900	.0398920	.0199460
100	.035060	.0379816	.0189908

RAIL-RESISTANCE FORMULAS

10. Track-Rail Resistance.—One square inch equals 1,273,000 circular mils, approximately. If 1 mil-foot of copper wire has a resistance of 10.8 ohms, a bar of copper 3 feet long and 1 square inch in sectional area will have a resistance of $\frac{10.8 \times 3}{1{,}273{,}000}$ ohm. If the resistance ratio is 12 to 1, the resistance of a bar of rail steel 1 yard long and 1 square inch in sectional area is $\frac{10.8 \times 3 \times 12}{1{,}273{,}000}$ ohm. The sectional area A in square inches of a rail may be determined approximately by dividing the weight per yard W_y, in pounds, by 10, or $A = \frac{W_y}{10}$. One yard of rail having a weight of W_y has a resistance approximately of $R_y = \frac{10.8 \times 3 \times 12}{1{,}273{,}000 \frac{W_y}{10}}$ ohm; therefore,

$$R_y = \frac{.003}{W_y} \text{ ohm} \qquad (1)$$

One thousand feet of rail having a weight per yard of W_y has a resistance $R_t = \frac{.003 \times \frac{1000}{3}}{W_y}$; therefore,

$$R_t = \frac{1}{W_y} \text{ ohm} \qquad (2)$$

One mile of rail having a weight per yard of W_y has a resistance $R_m = \frac{.003 \times \frac{5280}{3}}{W_y}$; therefore,

$$R_m = \frac{5.28}{W_y} \text{ ohm} \qquad (3)$$

EXAMPLE 1.—What is the resistance of 1 yard of rail-steel bar weighing 10 pounds a yard and having a resistance ratio of 12 to 1?

SOLUTION.—Substituting for W_y in formula **1**, its value 10, $R_y = .003 \div 10 = .0003$ ohm. Ans.

EXAMPLE 2.—What is the resistance of 1,000 feet of steel rail weighing 50 pounds a yard and having a resistance ratio of 12 to 1?

SOLUTION.—Substituting for W_y in formula 2 its value 50, $R_t = \frac{1}{50} = .02$ ohm. Ans.

EXAMPLE 3.—What is the resistance of 1 mile of steel rail weighing 100 pounds a yard having a resistance ratio of 12 to 1?

SOLUTION.—Substituting for W_y in formula 3 its value 100, $R_m = 5.28 \div 100 = .0528$ ohm. Ans.

11. Conductor-Rail Resistance.—The following formulas for determining the resistance of conductor rails of special steel are based on the resistance ratio of 8 to 1.

$$R_y = \frac{.002}{W_y} \text{ ohm} \qquad (1)$$

$$R_t = \frac{.666}{W_y} \text{ ohm} \qquad (2)$$

$$R_m = \frac{3.52}{W_y} \text{ ohm} \qquad (3)$$

12. Calculation of Track Resistance.—In the formulas of Arts. **10** and **11**, the resistance of a continuous rail is assumed. The formulas apply to bonded rails if they are bonded to at least full conductance. It should be understood that these formulas give results of only approximate accuracy and that these results do not necessarily check exactly with data, such as given in Tables III and IV.

A single-track road has two lines of rails in parallel; a double-track road, four lines of rails; and a four-track road, eight lines of rails; therefore, when considering the resistance of the portion of the return circuit formed by the rails, the resistance of one rail for the calculated distance should be divided by the number of rails in parallel. It is considered that the rails are bonded to full conductance and that the tracks are cross-bonded.

If the condition of the joints is not known, 100 or more of them should be measured for resistance in order to determine what resistance if any should be added to the resistance of the rails.

FEEDER PROBLEMS

DISTRIBUTION OF VOLTAGE IN THE TRANSMISSION CIRCUIT

13. It is necessary to provide sufficient conductor in the feeder portion and in the track-return portion of the complete circuit to maintain at the cars the voltage required for satisfactory operation. The weight of rail is determined by the traffic considerations and the resistance of the track circuit can then be estimated. The amount of current may be estimated from the number, weight, and speed of the cars liable to be in service at the same time on the section of road under consideration. The probable drop in the track circuit may then be calculated. On different roads the total drop in the transmission circuits usually varies from 10 to 20 per cent., where boosters are not used, and may be as high as 40 per cent. of the total generated voltage when a booster is in service.

The drop in the feeders, neglecting that in the mains and trolley wires, is equal to the total allowable drop in the transmission circuit minus the drop in the track portion of the circuit. The size of the feeders may be calculated if their length, the current, and the allowable drop, in volts, are known.

STATION AT ONE END WITH LOAD AT OTHER END

14. Fig. 1 shows the layout of a single-track road 5 miles long. The station is at the extreme left and the load consists of ten 10-ton cars requiring a total current of 200 amperes.

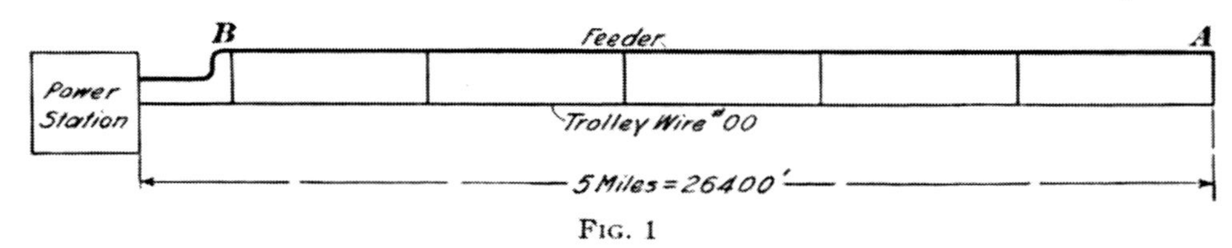

FIG. 1

The generator voltage is 500 and the voltage at the other end of the line when all cars are operating at that point must be 400; that is, the allowable drop is 100 volts. The track consists of 80-pound rails bonded to full conductance. The trolley wire is

No. 00 and is in parallel with the feeder, the two conductors being connected at intervals. What must be the size of the feeder $A\ B$ when the total load is at the end of the line remote from the station, a condition that rarely occurs, but which calls for maximum sectional area of feeder conductor for a given drop.

The resistance of the track is taken as .0356 ohm a mile, as indicated in the fifth line of the fourth column of Table III. The resistance of the 5-mile track-return circuit is $.0356\times 5 = .178$ ohm; the drop, in volts, is .178 ohm$\times$200 amperes $=35.6$ volts; and the drop, in volts, for the feeder and trolley in parallel is $100.0-35.6=64.4$ volts.

15. Size of Copper Feeders.—A copper feeder to carry 200 amperes 5 miles, or 26,400 feet, with a maximum drop of 64.4 volts may be calculated by substituting the values of l, I, and u in formula **1**, Art. **3**. The sectional area of the combined feeder and trolley is, $a=\frac{10.8\times 26{,}400\times 200}{64.4}=885{,}000$ circular mils.

The sectional area of the No. 00 trolley wire is approximately 133,000 circular mils, therefore the copper feeder must have a sectional area of $885{,}000-133{,}000=752{,}000$ circular mils. A 750,000 circular mil cable is the nearest standard size and this should be used.

16. Size of Aluminum Feeders.—The size of an aluminum feeder to be used in place of the copper feeder may be determined for the problem of Art. **14**, by first finding the part of the total current that the No. 00 copper trolley carries. Substitute in formula **1**, Art **3**, the values, $a=133{,}000$; $l=26{,}400$, and $u=64.4$; $133{,}000=\frac{10.8\times 26{,}400\times I}{64.4}$; from which,

$$I=\frac{133{,}000\times 64.4}{10.8\times 26{,}400}=30 \text{ amperes.}$$

If the trolley current is 30 amperes, the current in the aluminum feeder is $200-30=170$ amperes. Substituting the values of l, I, and u in formula **1**, Art. **5**, $a=\frac{17.17\times 26{,}400\times 170}{64.4}$ $=1{,}197{,}000$ circular mils, about. The nearest standard size of aluminum cable has a sectional area of 1,000,000 circular mils. If this were used the drop would be a little over that allowed in the problem.

The mil-foot constant for copper is taken as 10.8 and the constant for aluminum as $10.8 \times 1.59 = 17.17$; therefore, a simple method to determine the equivalent size of an aluminum feeder when the size of the copper feeder is known is to multiply the sectional area of the copper feeder by 1.59. The sectional area of the copper feeder of this problem is 752,000 circular mils and $752,000 \times 1.59 = 1,196,000$ circular mils, the sectional area of the aluminum feeder. The difference in the results obtained by the two methods is negligible.

STATION AT ONE END WITH LOAD EVENLY DISTRIBUTED

17. Under ordinary operating conditions, the cars are distributed with approximate regularity along the length of the road. The current in the feeder is greatest at the station end and gradually decreases to zero at the distant end. If the cars are evenly distributed, it may be assumed when making feeder calculations that one-half of the total current is transmitted the whole length of the feeder or that the total current is transmitted from the station to a point distant one-half the length of the feeder.

If an even distribution of the load of ten cars is assumed in the installation shown in Fig. 1 and the feeders calculated for 50 volts total drop when the total current is considered as transmitted one-half the total length of the road, the following results are obtained.

18. The value of l in formula **1**, Art. **3**, is now $26,400 \times \frac{1}{2} = 13,200$ feet; the value of I is the total current, 200 amperes; the resistance of $2\frac{1}{2}$ miles of single track laid with 80-pound rails is $2\frac{1}{2} \times .0356 = .089$ ohm and the drop due to the passage of 200 amperes is $.089 \times 200 = 17.8$, the value of u being $50 - 17.8 = 32.2$ volts. Substituting values in the formula, the sectional area of the combined feeder and trolley is, $a = \frac{10.8 \times 13,200 \times 200}{32.2} = 885,000$ circular mils.

Subtracting the sectional area of the trolley, the sectional area of the feeder is 885,000 − 133,000 = 752,000 circular mils for a copper feeder.

An aluminum feeder for the same service should have a sectional area of 752,000 × 1.59 = 1,196,000 circular mils.

19. The size of the feeder is the same as that required to supply a similar total load banked at the end of the system, but with a maximum allowable drop of 100 volts. With the same size of feeder and total load, a distributed load causes one-half of the drop that a banked-end load causes. If the drop is to be the same in both cases, the distributed load will require a feeder only one-half the size of that for a banked-end load.

If one-half of the current is considered as transmitted the whole length of the road, the value of $l = 26{,}400$ feet and the value of $I = 100$ amperes and the calculated result is identical with that of the other method.

STATION AT CENTER OF ROAD WITH LOAD BANKED AT ONE END

20. With the station at the center of the road and the total load banked at one end, the amount of conductor is much less than with the station at one end and the total load banked at the distant end. In the case of ten cars, requiring a total current of 200 amperes, at *A* or *B*, and the station at *S*, Fig. 2,

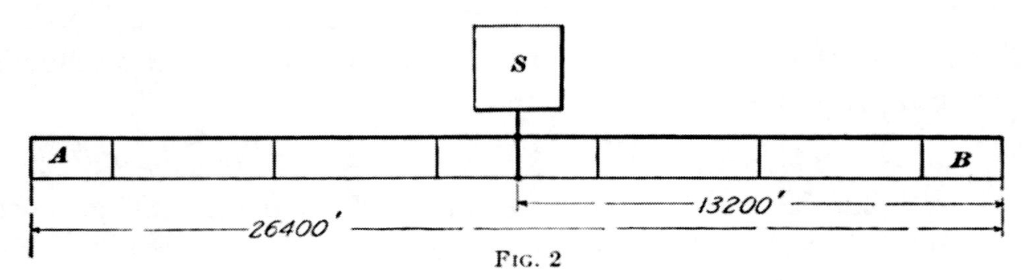

FIG. 2

with the generator voltage 500; the maximum allowable drop 100 volts, and the single track laid with 80-pound rails, the drop in the track circuit $2\frac{1}{2} \times .0356 \times 200 = 17.8$ volts. The value of u is $100 - 17.8 = 82.2$; the value of $I = 200$ amperes; and the value of $l = 13{,}200$ feet. Substituting these values in formula **1**, Art. **3**, $a = \frac{10.8 \times 13{,}200 \times 200}{82.2} = 347{,}000$ circular mils for copper feeder and No. 00 trolley wire. The sectional area of the copper feeder is $347{,}000 - 133{,}000 = 214{,}000$ circular mils; so a No. 0000 copper feeder would probably be used.

An aluminum feeder for the same service should have a sectional area of $214{,}000 \times 1.59 = 340{,}000$ circular mils.

The result of locating the station at the center instead of at the end of the road as in the problem of Art. **14** is to permit the use of a feeder approximately one-fourth the size.

21. Allowance for Rail Drop.—An allowance for rail drop is incorporated in the constants of the formulas of Arts. **4** and **6**. These constants are based on the assumption that the resistance of the rail return is not usually as high as the resistance of the overhead portion of the circuit nor less than .25 per cent. of it. The formulas are approximate for calculating feeders when the bonded joints are in good condition. If after installation the total drop is found to exceed that assumed, the bonding of the joints probably needs attention.

STATION AT CENTER WITH SECTIONALIZED LINE

22. Fig. 3 shows the general arrangement of feeders, mains, and trolley wires for a double-track road that is divided into seven sections by insulators *a*, *b*, *c*, *d*, *e*, and *f*. The number of cars in each section, the total current taken by each section, the length of the sections and of the feeders *1* to *7* which extend from station *g* are as indicated. The two trolley wires are of No. 00 copper; feeders No. *1*, *2*, *3*, *4*, and *5* are of copper and feeders *6* and *7* of aluminum. The feeders are connected to the centers of the sections and the mains and the trolley wires are tied together at frequent intervals. The rail-return circuit is assumed to be in such condition that the constants 14 ohms per mil-foot for the combined copper and track and 22 ohms for the combined aluminum and track may be used, as explained in Arts. **4** and **6**. The drops in the mains and in the trolley wires will not be considered in the feeder calculations. A total

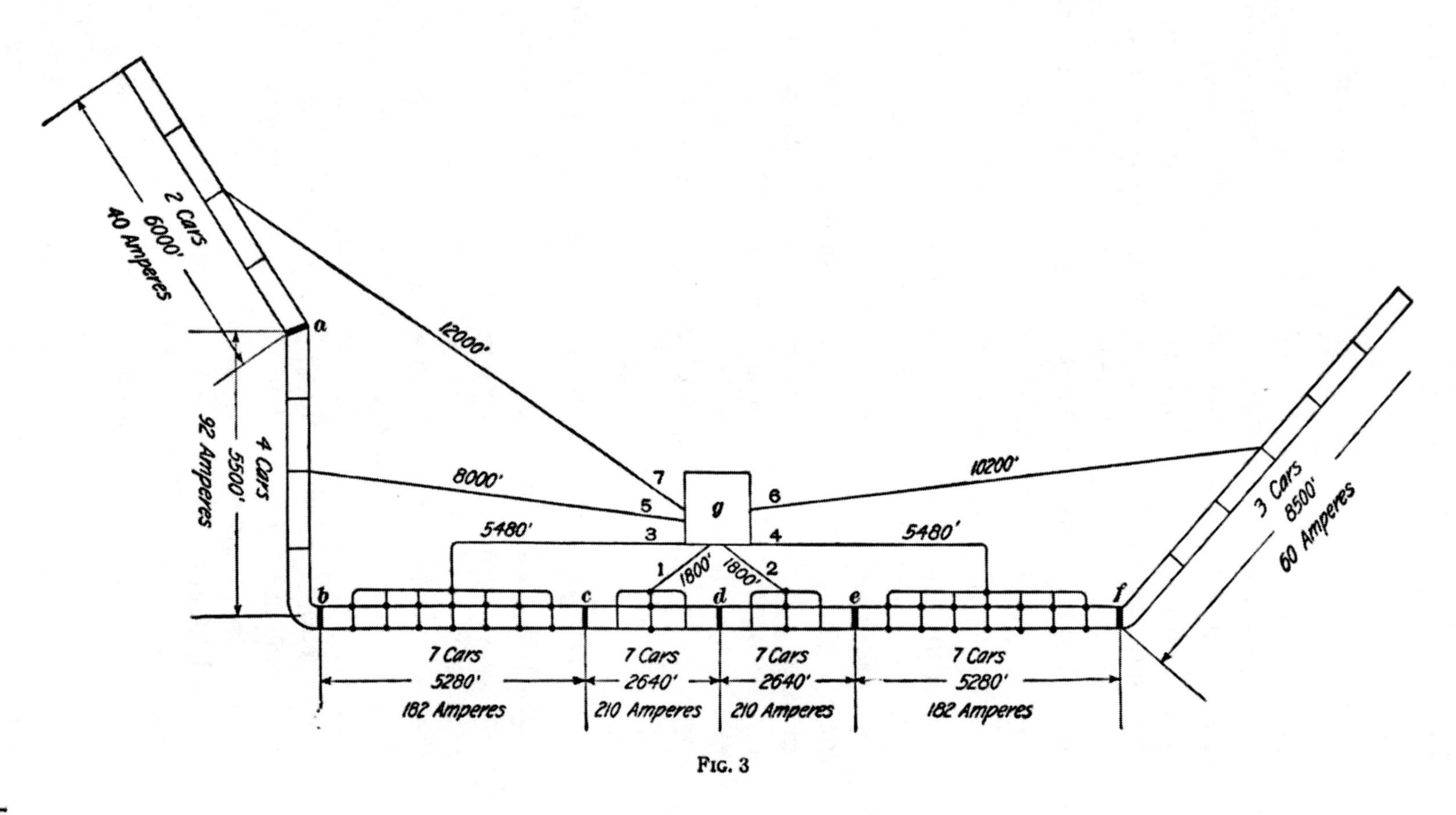
2 Cars
6000'
40 Amperes
a
12000'
4 Cars
5500'
92 Amperes
8000'
7
5
3
g
6
4
10200'
5480'
5480'
1
1800'
1800'
2
b
c
d
e
f
3 Cars
8500'
60 Amperes
7 Cars
5280'
182 Amperes
7 Cars
2640'
210 Amperes
7 Cars
2640'
210 Amperes
7 Cars
5280'
182 Amperes

FIG. 3

of thirty-seven cars are used to enable the road to operate a 2-minute time table. The bus-bar voltage is 600 and a line voltage of 550 is maintained at all sections. The permissible drop in the transmission circuit of each section is, therefore, 50 volts under the assumed conditions. The sections in which the traffic is congested take more current than the outlying sections because there are more cars on the central sections and they start oftener. The following data relates to the various sections:

Number of Feeder	Length of Feeder Feet	Material of Feeder	Number of Cars in Section	Number of Tons	Current Supplied Each Section Amperes	
					Per Ton	Per Feeder
1	1,800	Copper	7	70	3.0	210
2	1,800	Copper	7	70	3.0	210
3	5,480	Copper	7	70	2.6	182
4	5,480	Copper	7	70	2.6	182
5	8,000	Copper	4	40	2.3	92
6	10,200	Aluminum	3	30	2.0	60
7	12,000	Aluminum	2	20	2.0	40

23. Copper Feeders.—Formula **1**, Art. **4**, is used to determine the sectional area of the copper feeders. The values for the different feeders that are to be substituted in the formula may be taken from Fig. 3 or from the tabulated data of Art. **22**.

For feeders Nos. 1 and 2, $a = \frac{14 \times 1{,}800 \times 210}{50} = 106{,}000$ circular mils, about. The nearest standard conductor is No. 0. In this case, however, the carrying capacity of the conductor dictates the choice of at least a No. 00 conductor, as will be explained later.

For feeders No. 3 and 4, $a = \frac{14 \times 5{,}480 \times 182}{50} = 279{,}000$ circular mils. The nearest standard single conductor has a sectional area of 250,000 circular mils. Two No. 00 conductors having a combined sectional area of 266,200 circular mils would probably be used.

For feeders No. 5, $a=\frac{14\times 8{,}000\times 92}{50}=206{,}000$ circular mils.

The nearest standard conductor is No. 0000, which has a sectional area of 211,600 circular mils and would probably be used.

24. Aluminum Feeders.—Formula **1**, Art. **6**, is used to determine the cross-sectional area of the aluminum feeders.

For feeder No. 6, $a=\frac{22\times 10{,}200\times 60}{50}=269{,}000$ circular mils.

The nearest standard aluminum conductor has a sectional area of 250,000 circular mils and would probably be used.

For feeder No. 7, $a=\frac{22\times 12{,}000\times 40}{50}=211{,}000$ circular mils.

The nearest standard aluminum conductor is No. 0000, having a sectional area of 211,600 circular mils, and this would probably be used.

25. Current-Carrying Capacity of Feeders.—When calculating the size of a long feeder for a permissible drop in volts, the calculated result will usually indicate a size of feeder large enough to carry the necessary current safely. When the size of a short feeder is calculated, however, the result may indicate a size of feeder that is not large enough to carry the current safely. It is well, therefore, to check the calculated wire with a table showing the current-carrying capacity of conductors.

In the case of the feeders *1* and *2* of Fig. 3, the calculated feeder No. 0 is not large enough to carry safely a current of 210 amperes, therefore, a No. 00 or larger feeder should be used for each of these sections. The feeders indicated by the other calculations are of sufficient sectional area to carry safely the required current.

When checking the carrying capacity of an aluminum feeder, reference should be made to the carrying capacity of the equivalent copper conductor as indicated in *Long-Distance Transmission of Electrical Energy.* The carrying capacity of an aluminum feeder may be safely assumed equal to that of a copper conductor of equal length and resistance; in fact, the aluminum conductor being larger has greater radiating surface and should remain cooler than the copper conductor when carrying the same current.

Table V indicates approximately the current-carrying capacity of copper feeders with an allowance of about 25° F. rise in

TABLE V

CURRENT-CARRYING CAPACITY OF COPPER FEEDERS

Gauge Number	Approximate Sectional Area of Feeder Circular Mils	Capacity of Feeder Amperes
	500,000	509
	400,000	426
	350,000	388
	300,000	355
	250,000	319
0000	211,600	275
000	167,800	237
00	133,100	195
0	105,500	168
1	83,690	143

temperature above the surrounding air. The current-carrying capacity of a conductor does not increase in direct proportion to its sectional area because the cross-section increases as the square of the diameter and the radiating-surface area as the first power of the diameter.

26. Checking Feeder Calculations for Drop in Voltage.—The data of Tables VI and VII may be used to check the results of the feeder calculations. Take for instance feeder 7. The resistance of 1,000 feet of No. 0000 aluminum cable, Table VII, is .08 ohm approximately; 12,000 feet has a resistance of .96 ohm. The feeder drop with a current of 40 amperes

TABLE VI

PROPERTIES OF STRANDED COPPER CABLES

B. & S. Gauge Number	Diameter of Bare Cable Mils	Area of Bare Cable Circular Mils	Weight of Cable, in Pounds		Number of Feet per Pound	Resistance of Cable at 68° F. Ohms	
			Per 1,000 Feet	Per Mile		Per 1,000 Feet	Per Mile
	1,632	2,000,000	6,100	32,208	.164	.00518	.02733
	1,412	1,500,000	4,575	24,156	.219	.00690	.03644
	1,152	1,000,000	3,050	16,104	.328	.01035	.05466
	1,000	750,000	2,288	12,078	.437	.01380	.07290
	819	500,000	1,525	8,052	.655	.02070	.10930
	728	400,000	1,220	6,442	.819	.02590	.13660
	679	350,000	1,068	5,636	.936	.02960	.15620
	630	300,000	915	4,831	1.093	.03450	.18220
	590	250,000	762	4,026	1.312	.04140	.21860
0000	530	211,600	645	3,405	1.550	.04890	.25830
000	470	167,800	513	2,709	1.950	.06170	.32580
00	420	133,100	406	2,144	2.460	.07780	.41080
0	375	105,500	322	1,700	3.210	.09810	.51800
1	330	83,690	255	1,347	3.940	.12370	.65300
2	291	66,370	203	1,072	4.930	.15600	.82400
3	261	52,630	160	845	6.250	.19670	1.03900
4	231	41,740	127	671	7.870	.24800	1.30900

TABLE VII

PROPERTIES OF STRANDED ALUMINUM CABLES

B. & S. Gauge Number	Diameter of Bare Cable Mils	Area of Bare Cable Circular Mils	Weight of Cable, in Pounds			Resistance of Cable at 75° F. Ohms	
			Per 1,000 Feet	Per Mile	Feet per Pound	Per 1,000 Feet	Per Mile
	1,152	1,000,000	920.0	4,858	1.087	.01695	.0895
	996	750,000	690.0	3,645	1.450	.02260	.1193
	814	500,000	460.0	2,430	2.040	.03300	.1790
	725	400,000	368.0	1,944	2.720	.04240	.2240
	679	350,000	322.0	1,701	3.110	.04840	.2560
	621	300,000	276.0	1,458	3.620	.05650	.2980
	567	250,000	230.0	1,215	4.350	.06780	.3580
0000	522	211,600	195.0	1,028	5.730	.08000	.4230
000	464	167,800	154.4	816	6.480	.10100	.5330
00	414	133,100	122.4	647	8.160	.12700	.6730
0	368	105,500	97.1	513	10.300	.16000	.8470
1	328	83,690	77.0	407	13.000	.20200	1.0690
2	291	66,370	61.0	323	16.400	.25500	1.3500
3	261	52,630	48.5	256	20.600	.32200	1.7000
4	231	41,740	38.5	203	26.000	.40600	2.1440

is 38.4 volts. Assuming that 90-pound rails are used on the double track and that the distance along the track from the distant end of feeder *7* to the station, Fig. 3, is 3.1 miles, approximately, the resistance of 1 mile of double track with 90-pound rails, Table III, is .015818 ohm; 3.1 miles will have a resistance of .049 ohm. The drop for a current of 40 amperes is 1.96 volts. The drop in feeder and track return is 38.4 +1.96=40.36 volts. This indicates that the bonds must be in poor condition to bring the total drop to 50 volts when a No. 0000 feeder is used.

Copper feeder *5* may be checked in a similar manner. The resistance of 8,000 feet of No. 0000 copper conductor, Table VI, is .0489×8=.3912 ohm. The resistance of 2 miles of track is .015818×2=.031636 ohm. The total resistance of feeder and track is .3912+.031636=.422836 ohm. The drop in the feeder and track when carrying a current of 92 amperes is .422836×92 =38.9 volts. The rails may be bonded to considerably less than full conductance and yet the total drop will not exceed 50 volts when feeder *5* is of No. 0000 copper wire.

SINGLE-PHASE ALTERNATING-CURRENT ROAD

27. When alternating current is used for propulsion, the drop in voltage in the feeder and track return of a system is much greater than when the same value of direct current is used, because of the inductance of the circuit. It has been experimentally determined that the impedance of the overhead copper feeder and trolley wire with 25-cycle alternating current is 1.5 times their resistance to the passage of direct current; and that the impedance of the track return is 6.6 times its resistance with direct current. These are figures of approximate value only. Under conditions of similar values of current and drop in voltage, the feeder for alternating-current propulsion would therefore be much larger than that required for direct-current propulsion.

It is customary, however, to use higher voltages and correspondingly lower currents in alternating-current railway work. Alternating voltages as high as 11,000 are in use, and with the same per cent. drop as in direct-current work, the allowable drop, in volts, for the alternating-current feeders is high enough to permit the use of feeders of moderate size.

28. The calculation of feeders for a single-phase alternating-current road may be made with approximate accuracy as follows: The drop in the track return for the same value of direct current as the alternating-current load is calculated and this value multiplied by 6.6 gives the track-return drop for alternating current. The total alternating-current drop is found by multiplying the impressed line voltage at the station by the per-cent. drop that is to be allowed. Subtracting the alternating-current track return drop from the total drop gives the alternating-current drop to be allowed for the feeder. The direct-current drop for the feeder equals the alternating-current drop divided by 1.5. When the direct-current drop for the feeder is determined, the size of copper feeder may be calculated by formula **1,** Art **3,** and for an aluminum feeder by formula **1,** Art. **5.**

THIRD-RAIL ROAD

29. Conductor rails have considerable current-carrying capacity, and with roads on which the load is not heavy, the conductor rails serve to transmit the entire current, thus making

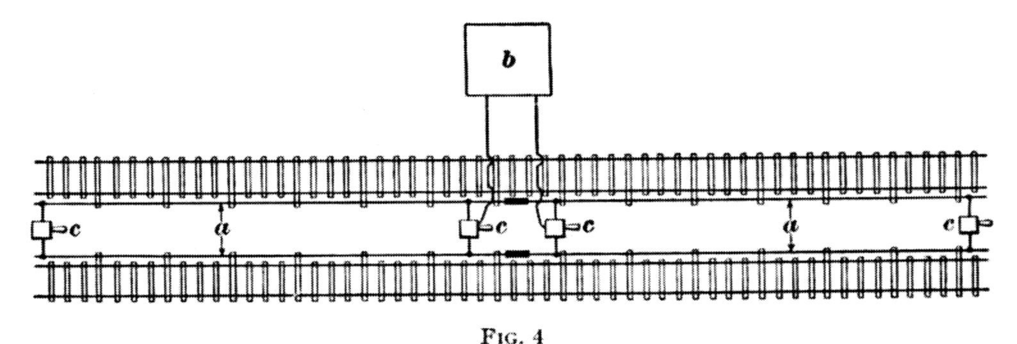

Fig. 4

unnecessary the use of copper feeders. On third-rail roads involving the operation of trains over long distances, however, copper feeders are usually installed.

Fig. 4 shows a double-track road supplied with current through two conductor rails *a* from a substation *b* at the middle of the system. The station voltage is 650 and the allowable drop is 100 volts when one-half of the total load is banked at either end of the road. The load consists of four 50-ton cars, which require an average current of 2.5 amperes per ton, the total current is $4 \times 50 \times 2.5 = 500$ amperes; so a load of 250 amperes is considered as being banked at each end of the road. The track is laid with 90-pound rails. The two conductor rails are discarded 70-pound track rails with a resistance ratio of 12 to 1. These rails are normally connected in parallel and have a total sectional area of 17,825,000 circular mils and in case of trouble may be divided into four sections by the switches *c*. The total distance that the two conductor rails may be extended from the substation and still keep within the allowed drop in voltage is to be determined.

30. The approximate distance l that the conductor rail or rails of a third-rail road may be extended from the station or substation may be calculated from the formula,

$$l = \frac{a\,u}{190\,I}$$

in which, a = total sectional area of conductor rails used to supply current to section of road under consideration;

u = allowable drop in these conductor rails and track return, in volts;

I = current in these conductor rails.

Substituting the values stated in the problem,

$$l = \frac{17{,}825{,}000 \times 100}{190 \times 250} = 37{,}526 \text{ feet} = 7.1 \text{ miles.}$$

The conductor rails may be extended to a distance of 7.1 miles from the substation in either direction, without the drop exceeding 100 volts.

31. If 70-pound conductor rails having a resistance ratio of 8 to 1 were substituted for the discarded track rails the allowable distance would be, $l = \frac{a\,u}{146\,I}$; or $l = \frac{17{,}825{,}000 \times 100}{146 \times 250} = 48{,}836$ feet = 9.25 miles.

32. The results of the calculation of Art. **31** may be checked as follows: The resistance of the two 70-pound conductor rails in parallel, bonded to full conductance, and having a resistance ratio 8 to 1 is, by Table IV, $9.25 \times .0271256$ ohm, and the drop is $9.25 \times .0271256 \times 250 = 62.73$ volts. The resistance of 9.25 miles of four lines of 90-pound track rail in parallel, bonded to full conductance, is, by Table III, $9.25 \times .015818$ ohm and the drop is $9.25 \times .015818 \times 250 = 36.58$ volts. The total drop is $62.73 + 36.58 = 99.31$ volts. This checks closely with the allowable drop of 100 volts.

IMPORTANCE OF LOW-VOLTAGE DROP

33. In order to maintain satisfactory service on an electric road, it is important that the voltage at any portion of the line be approximately the value for which the car motors were designed. Low line voltage may be due to low generator voltage, or to excessive drop in the line caused by overload, poor condition of the track bonds, or small feeders.

Low line voltage is likely to result in damage to the car apparatus because of the abuse to which it is subjected by the motorman in an endeavor to keep the car on schedule time. With low voltage the speed of the car when the controller is on its final running position is much lower than normal and motormen are apt to make up time by rapid acceleration with the result of causing burnt controller fingers, motor windings, and commutators.

Excessive drop also means a large amount of electric energy lost in the transmission circuit. The relative cost of generating electric energy and of installing increased conductor determines whether or not it is advisable to use additional feeders.

LINE TESTS

BOND TESTS

34. Test of the conductance of rail joints should be made at intervals to determine their electric condition. The usual method is to compare the resistance of a length of 3 feet of rail, including the bonded joint, with the resistance of 3 feet of solid rail. The resistance of the joint is expressed as equivalent to the resistance of a certain number of feet of solid rail of the same sizes as the jointed rail. The bond makers assume for newly made joints a rail equivalent as low as 2.5 feet and this may be realized with careful installation. The advantage of bonds with low rail equivalent is that a good rail-return path is assured for possibly several years without further attention.

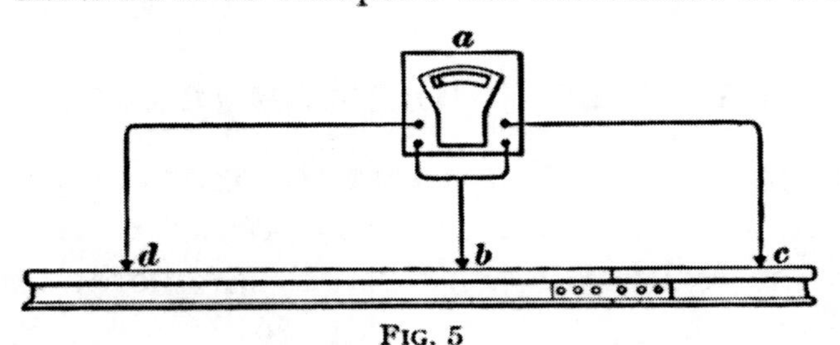

FIG. 5

The resistance of the rail-return portion of the circuit, especially on double-track roads, is but a small part of the total resistance through feeders and rails; therefore, the joints may uniformly deteriorate somewhat without materially affecting the drop, in volts, of the transmission circuit. In cases where the load is not very heavy less copper can be used for the bonds, resulting in a higher rail equivalent; but care should be taken that the joints are maintained in good condition.

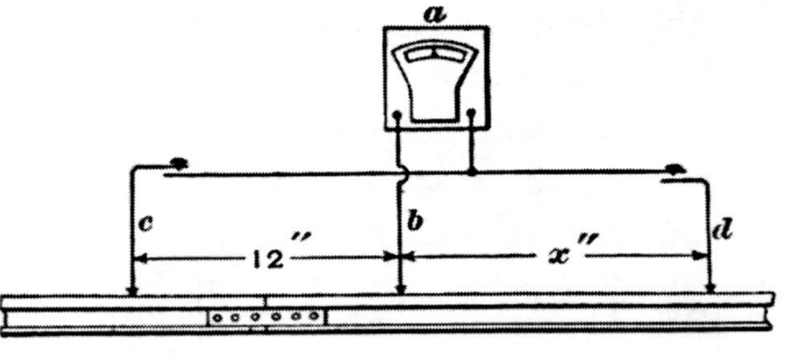

FIG. 6

A large number of tests of joints in actual service indicate that the rail equivalent of the bonded joint is seldom less than $6\frac{1}{2}$ feet and that an average value is about 12 feet.

Among the methods employed to test bonds are those employing a differential voltmeter, a Wheatstone bridge, and a bond testing car.

35. Differential Voltmeter Method.—Fig. 5 indicates the connections of a differential voltmeter *a* when used to test bonded joints. The voltmeter has two coils and is provided with three terminal leads *b*, *c*, and *d*. Terminals *b* and *c* of one coil are in contact with the rail equal distances from a joint between them and terminal *d* at some point farther along the rail. The current in the rail causes voltage drops between the contacts, and these voltages establish currents in the two coils. The two magnetic fluxes set up by the coils are in opposition and contact *d* is moved along the rail until the needle of the instrument rests at zero. This indicates that the drop in voltage between *b* and *c* equals that between *b* and *d*. The resistance of the length of rail *b c* including the joint is then equal to the resistance of the length of solid rail *b d*. The lengths of rails between terminals are read from the scale to which the rail contact points are attached.

36. Bridge Method With an Adjustable Contact. Fig. 6 indicates the connections of a millivoltmeter when used for testing a bonded joint. The millivoltmeter *a*, with a scale having the zero point at the center, is provided with three terminals, two of which *b* and *c* are a fixed distance apart and are placed equidistant from the joint between them. The other terminal *d* is moved along the rail until a balance of the bridge resistance is indicated by the instrument needle resting at zero. When the bridge is balanced, the number of feet of rail between *b* and *d* is the solid rail equivalent of the joint.

37. Bridge Method With Fixed Contacts.—In Fig. 7 is shown a bond-testing outfit operated on the Wheatstone bridge principle with three fixed contacts *a*, *b*, and *c* that make good connections with the rails when the supporting rod is pressed down by the operator's foot. The plate *p* is placed over the center of the joint. There is 3 feet of solid rail between *a* and *b* and 3 feet of rail including the joint between *b* and *c*. The bridge is balanced by adjusting the resistance sections *d*, Fig. 8, by means of switch *s* mounted on the test box carried by the operator. The telephone receiver *e* worn over the operator's ear and connected in series with a mechanical circuit-interrupting device *f* then makes minimum sound when the interrupter is operated. If the balanced position of switch *s*

Fig. 7

is as shown, the resistance of *g h* is to that of *a b* as the resistance of *h k* is to that of *b c*. The contact points of switch *s* are numbered to give the resistance of the bonded joint expressed in terms of 3-foot rail lengths. If the balance point is on contact No. 1 the resistance of the section of rail including the joint is equal to 3 feet of solid rail; a balance on contact No. 1.5 means that the resistance of the joint section is equal to $3 \times 1.5 = 4.5$ feet of solid rail. Joints indicating a high resistance are usually marked for repairs.

In some forms of bond testers, the telephone receiver is replaced by a galvanometer. A small switch button on the instrument controlling the resistance sections is turned until the galvanometer needle rests at zero. The rail equivalent is then read from a scale at the point at which a needle attached to the rotating button rests.

38. Autographic Bond-Testing Car.—Fig. 9 (*a*) indicates the more important connections of the Herrick bond-testing apparatus as installed on a test car for the purpose of making a record, on a moving chart, of the comparative condition of the bonded joints as the car passes over the track.

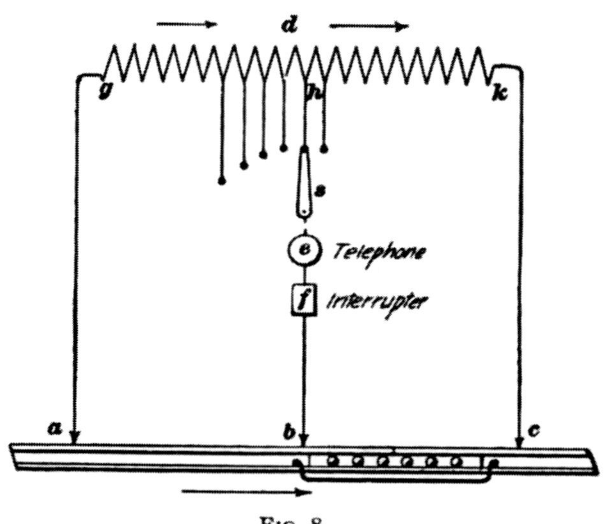

Fig. 8

In order to assure sufficient current in the rail for readable deflections on the instruments, the generator of a motor generator set *a* is connected through a reverse switch and truck wheels to the rails. The reverse switch serves to make the direction of the local current agree with that of the rail-return current from other cars on the section.

The drop in voltage between the brush terminals *b* and *c* that bear on the rail treads is measured by a sensitive millivoltmeter *d*. The needle and scale of this voltmeter are shown enlarged just above *d*. A portion of the needle *e* of the millivoltmeter *d* is of metal and when deflected moves over, but without touching, a metal sector *f* and several metal segments *g*. Each scale segment is connected to a corresponding metal point in a row of terminals *h* at the recording device.

A paper chart, the movement of which is caused by the car mechanism, passes between the row of terminals *h* and a metal

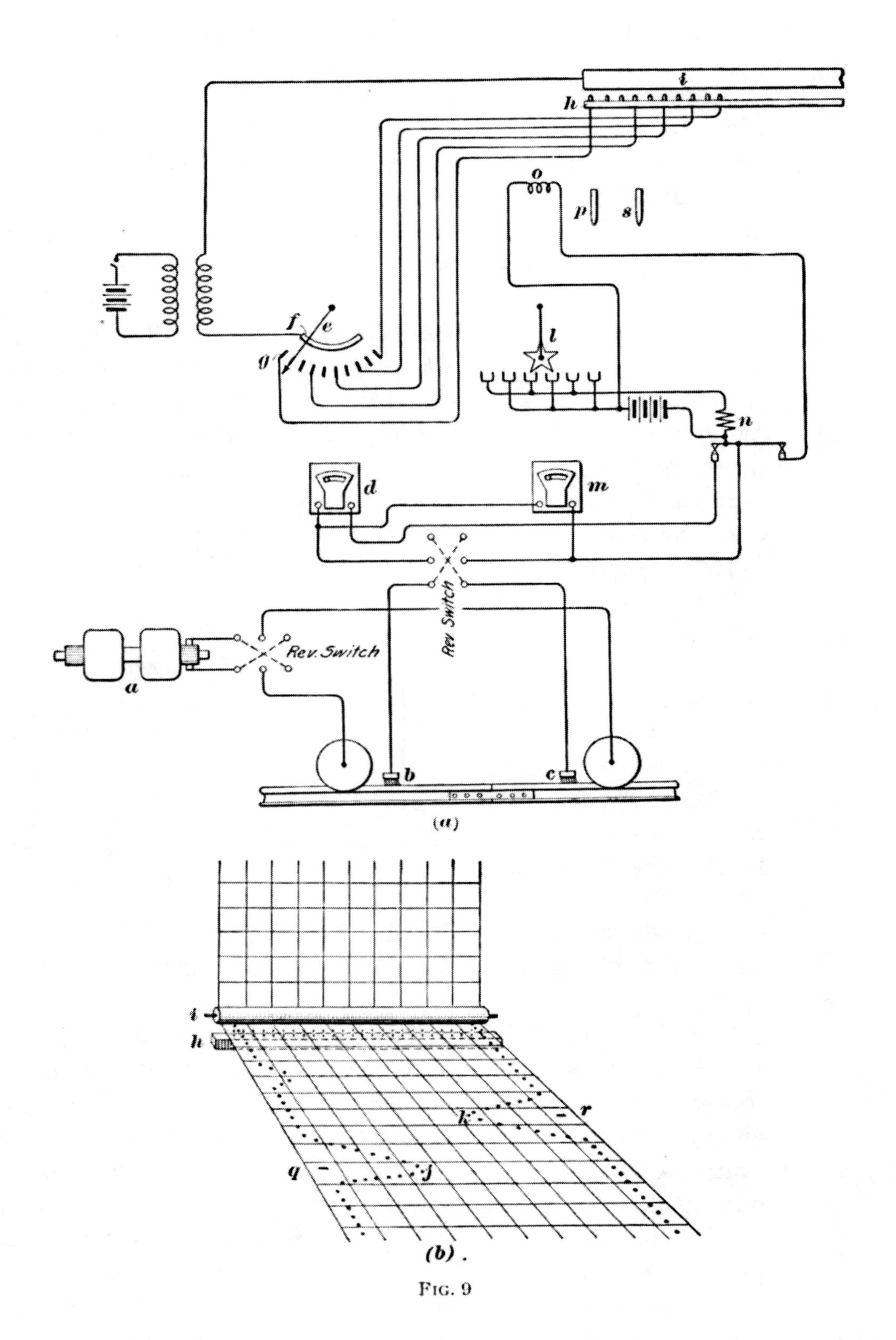

Fig. 9

roll *i* as indicated in the detail sketch, view (*b*). A small induction coil operated by a battery and interrupter has one secondary terminal connected to sector *f*, view (*a*), and the other terminal to roll *i*. The secondary circuit includes spark gaps between *f* and *e*, between *e* and the particular segment *g* over which the needle may be moving at any instant; and between the point on *h* corresponding to the active segment *g* and the roll *i*. When the interrupter operates, sparks jump through the three gaps and the chart will be perforated.

The position of the hole depends on the deflection of the needle *e* which in turn depends on the drop in volts between *b* and *c*. Bonds in good condition are indicated by holes near the margin of the chart and faulty bonds by holes near the middle of the chart, as at *j* and *k*, view (*b*).

The apparatus for tests on one rail is indicated; there is, however, duplicate apparatus that makes a record for the other rail. The faulty bond indicated at *j* is on one rail of the track and *k* indicates a faulty bond on the other rail.

39. If the bonded joint is in unusually bad condition, the needle *e*, Fig. 9 (*a*), swings far to the right, making a spark hole nearer the middle of the chart than normal. The needle *l* of a less sensitive millivoltmeter *m* which is connected normally in parallel with millivoltmeter *d* is moved sufficiently to cause a star wheel to dip into mercury cups. A local-battery circuit is then completed including relay *n*. This relay operates a switch that cuts out of the circuit millivoltmeter *d*, thus protecting it from overload damage, and at the same time closing a circuit through coil *o*, which causes a pen *p* to make a small mark, like *q* or *r*, view (*b*), on the margin of the chart at the point where a faulty bond is also indicated by the spark holes. As soon as the faulty bond is passed the needle *l* drops toward zero, and millivoltmeter *d* is again cut into circuit.

A pen *s*, view (*a*), near the middle of the chart, is moved by a magnet controlled by a push button operated by the tester so as to make a mark on the chart as the line poles are passed by the car. At the same time the number of the pole is printed on the chart by means of a numbering stamp. In some cases a small quantity of blue powder is automatically squirted on the track at the location of faulty bonds. After a test run, the chart should be carefully inspected and the faulty bonds repaired.

Bond tests are of most service when made periodically so that successive results of tests can be compared and deteriorations noted.

FEEDER AND TRACK-RETURN TESTS

40. After a road has been in operation for a considerable period, the drop in voltage on certain sections may be much greater than on other sections operating under similar conditions. Tests may be made to determine whether the fault is in the feeder or the track-return portions of the circuit.

Fig. 10 indicates the connections for a test on the transmission circuit. The generator at the station is shown at *a*;

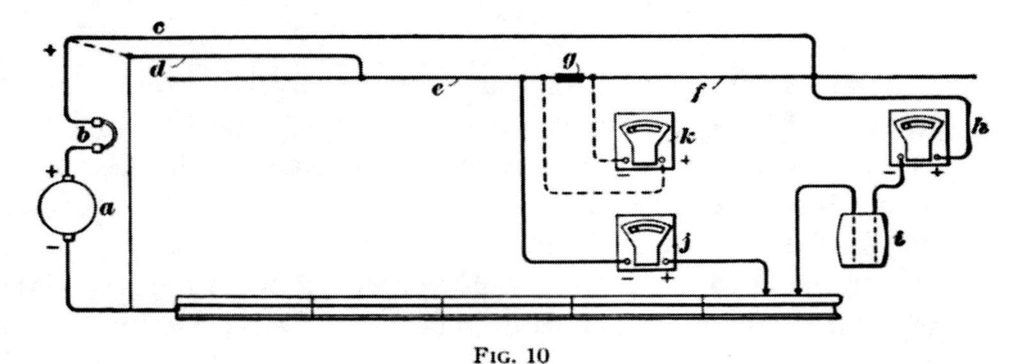

FIG. 10

a circuit-breaker at *b*; a feeder to the section under test at *c*; an adjacent feeder, used as a voltmeter lead, at *d*, this feeder being temporarily grounded at its station end; trolley sections at *e* and *f*; a section insulator at *g*; an ammeter at *h*; a water rheostat at *i*; a voltmeter at *j* in position for measuring the drop in voltage of the track return; and the same voltmeter at *k* connected for measuring the drop in voltage in the feeder.

The test is usually made at night when all car load can be removed from the section. The breaker *b* is closed and the water rheostat is adjusted to get readable deflections of the voltmeter. Care should be taken that none of the load current passes through any part of the conductors used as extension leads of the voltmeter, as *d*, *e*, and *f*. The voltmeter connections to the rail should be made near the point of the load-current connection, but should be a separate contact. The ammeter indicates the total current in the circuit; the voltmeter *j* indicates the drop in voltage in the track return; and the voltmeter reading divided by the ammeter reading is equal to the resistance of the rail-return portion of the circuit.

To determine the drop in voltage in the feeder, the voltmeter *k* should be connected as indicated by the dotted lines and a connection made between the left ends of *c* and *d*, the station ground connection of *d* being removed. This voltmeter reading divided by the ammeter reading equals the resistance of the feeder *c*.

The test determines the resistance of the two portions of the transmission circuit; therefore, the portion responsible for the excessive drop is determined. A test of this kind on the track is useful in determining the total resistance of a length of track including rails and bonded joints.

GENERAL ENGINEERING FEATURES

RAIL CALCULATIONS

41. Relation of Rail Weight to Sectional Area.—The relation of rail weight to sectional area (see Art. **10**) is expressed by the formula,

$$W_y = 10\,A$$

in which W_y = weight of rail, in pounds per yard;
A = sectional area, in square inches.

EXAMPLE 1.—A rail weighs 95 pounds per yard; what is its sectional area, in square inches?

SOLUTION.—Substituting for W_y its value 95 in the formula, $95 = 10\,A$, $A = 95 \div 10 = 9.5$ sq. in. Ans.

EXAMPLE 2.—A rail has a sectional area of 8 square inches; what is its weight per yard?

SOLUTION.—Substituting for A its value 8 in the formula, $W_y = 10 \times 8 = 80$ lb. per yd. Ans.

42. Weight of Rails Per Mile.—The weight of rails per mile W_m of single track, expressed in long tons, is equal to the weight per yard W_y of the selected rail multiplied by $\frac{11}{7}$, or

$$W_m = \tfrac{11}{7} W_y \qquad (1)$$

The formula for double-track construction is,

$$W_m = \tfrac{22}{7} W_y \qquad (2)$$

EXAMPLE.—A single track is laid with rails weighing 100 pounds per yard; what is the weight per mile of track, expressed in long tons?

SOLUTION.—Substituting, in formula 1, the value of 100 for W_y, $W_m = \frac{11}{7} \times 100 = 157$ tons. Ans.

CURVE CALCULATIONS

43. Methods of Designating Curvature.—The method of stating the curvature of simple curves on interurban tracks is to state the number of degrees subtended by a chord 100 feet in length, the ends of which touch the center line of the track. A *chord* of a circle is any straight line between two points in the circumference of the circle; it subtends the angle between radii drawn from its ends. Thus, in Fig. 11, $a\,b$ is a chord that subtends the angle $a\,c\,b$. A line perpendicular to the chord at its middle point and ending in the circumference of the circle, as $d\,e$, is the *middle ordinate* of the chord.

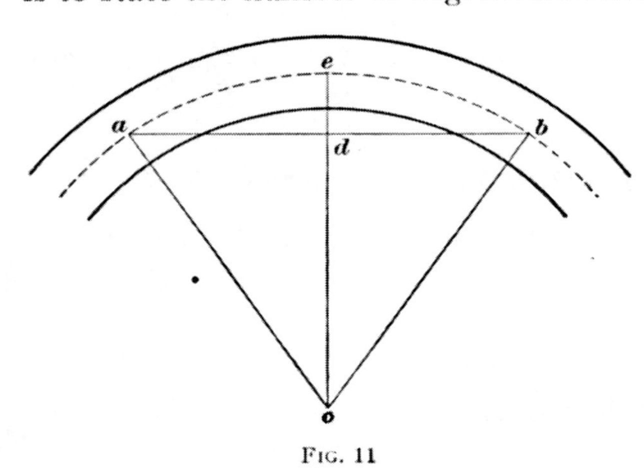

FIG. 11

When considering track curvature, it is usual to refer to the center curve of the track, which is equally distant from either track rail. The difference in curvature between the center curve and either rail curve is very small for track curves having a long radius, but is much greater for track curves of small radius.

For city work, it is customary to designate the track curvature of simple curves by the length, in feet, of the track radius of the arc formed by the center line of the track at the curve. This radius is indicated by $c\,e$ in Fig. 11.

The radius of the inner-rail curve is found by subtracting one-half the track-gauge distance plus the width of the rail head from the track-curve radius; and the radius of the outer-rail curve is found by adding one-half the track-gauge distance to the track-curve radius. The track-curve radius equals the radius of the outer-rail curve minus one-half the track gauge.

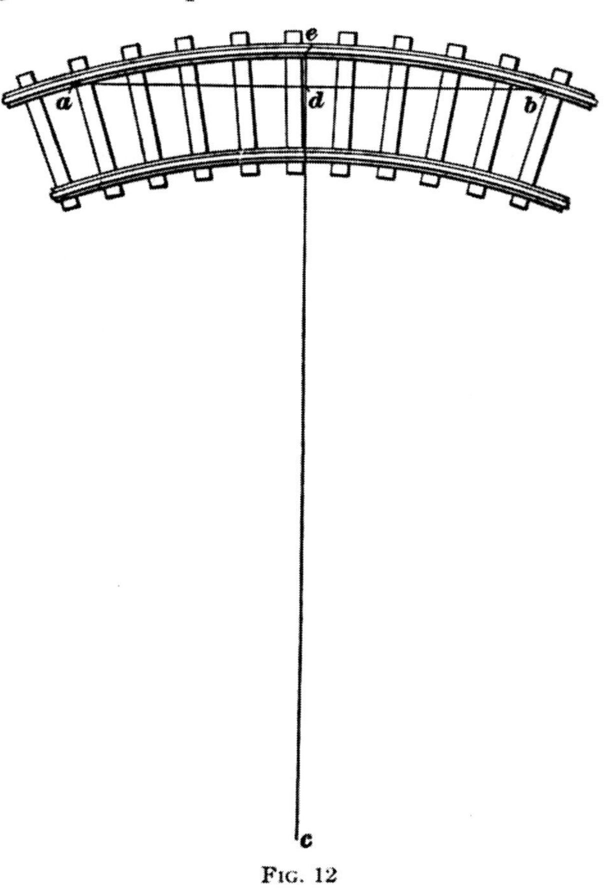

Fig. 12

44. Assume that the outer-rail curvature of a simple railway curve is as shown in Fig. 12. Stretch a tape line from a point a, touching the rail head on the inner side of the curve, to a point b touching the rail head at another point and note the length, in feet, of chord $a\,b$. With the tape held taut, measure the length, in feet and decimal of a foot, of the middle ordinate $d\,e$. From these two measurements the radius of curvature $c\,e$ of the outer rail can be calculated, using the formula

$$r = \frac{a^2 + b^2}{2\,a}$$

in which r = radius of outer rail, in feet;
a = middle ordinate, in feet;
b = half the length of chord, in feet.

EXAMPLE.—On a simple curve for city work a 20-foot chord is laid out on the outer rail, and the middle ordinate is 18 inches. The width of the rail head is 2.25 inches = .19 foot, and the track gauge is 4 feet $8\frac{1}{2}$ inches = 4.708 feet. What is the radius of: (*a*) the curvature of the outer rail? (*b*) the track curvature? (*c*) the curvature of the inner rail?

SOLUTION.—(*a*) Substituting the value of a = 18 in. = 1.5 ft., and b = 10 ft. in the formula, $r=\dfrac{1.5^2+10^2}{2\times 1.5}=34.08$ ft. Ans.

(*b*) The radius of the track curvature is $34.08-\dfrac{4.708}{2}=31.73$ ft. Ans.

(*c*) The radius of the curvature of the inner rail is

$$31.73-\left(\frac{4.708}{2}+.19\right)=29.19 \text{ ft. Ans.}$$

45. Bending Rail to a Given Radius.—Rail bending for short curves for city streets is usually done at the rolling mill. Rail benders are often used on the installation work for bending rail for longer curves. When bending a rail for a simple curve of any given radius r, the middle ordinate of any arbitrarily chosen chord on the selected curve should be known; then the curvature of the rail can be tested during the process of bending, for the middle ordinate for a given chord is the same for any position of the chord on a simple curve.

If the radius and the length of the chord are known the middle ordinate can be calculated by the formula

$$a=r-\sqrt{r^2-b^2}$$

in which the letters have the same meaning as in the formula of Art. **44.**

EXAMPLE.—It is desired to predetermine the length of a middle ordinate of a 20-foot chord used to check a simple outer-rail curve, the radius of which is 34.08 feet.

SOLUTION.—Substituting the values of r and b in the formula,

$$a=34.08-\sqrt{34.08^2-10^2}=34.08-32.58=1.5 \text{ ft. Ans.}$$

46. Construction of Simple Curves.—Simple curves can be laid out with a tape line, as shown in Figs. 13 and 14.

A line *a b* is laid out perpendicular to rails at the end of one section of the tangent, or straight portion of the track, and a corresponding perpendicular line *c d* at the end of the other section of the tangent track. A steel square held against the rails aids in laying out the perpendicular lines. The point *e* at which the lines *a b* and *c d* cross is the center of the rail and track curves. The line *a e* is the radius of the outer-rail curve and the line *e f* the radius of the inner-rail curve. The radius *a e* is used to locate points on the curvature for the outer rail, and

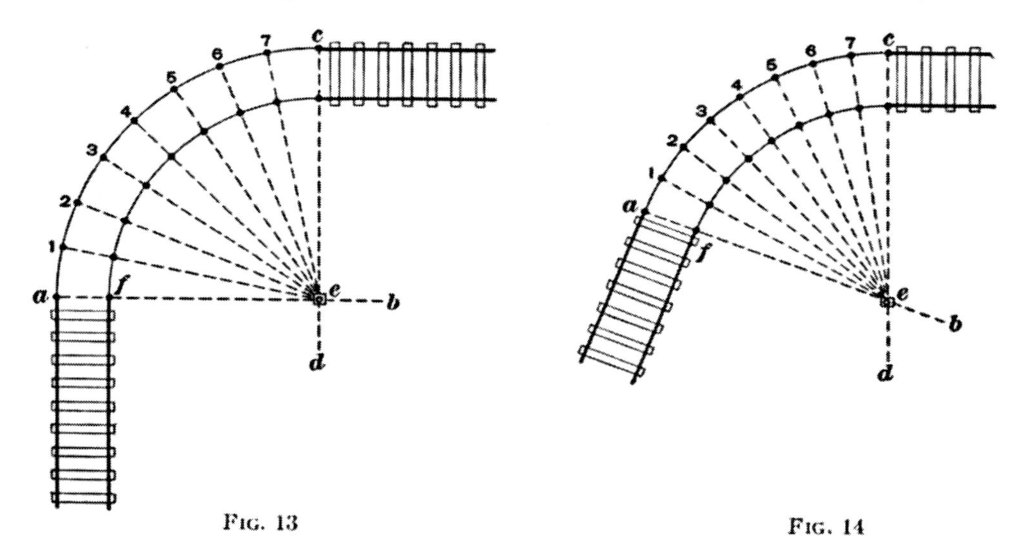

FIG. 13 FIG. 14

these are marked with stakes. Points on the curvature of the inner rail are also located in a similar manner by using the radius *e f*.

Templates of wood may be prepared to which the rails are bent by the rail bender, or a rail may be bent a little at a time and laid on the ties to see whether its curve conforms to the location stakes. As rail ends are hard to bend to a true curve, sections at the ends of the bent rails are sometimes cut out and new holes drilled for the rail joints. The improvement in curvature often warrants the extra labor.

GRADE CALCULATIONS

47. Per Cent. Grade.—The per cent. of a grade is usually considered in railway work as being numerically equal to the number of feet that a car is raised when traveling over 100 feet of track; thus, if a car is raised 6 feet in passing over a distance of 100 feet of track, the grade is 6 per cent. In general, the per cent. grade g is

$$g = \frac{r \times 100}{d},$$

in which r is the number of feet the car is raised in passing over a track d feet in length.

EXAMPLE.—On a certain grade a car is raised 4.3 feet in traveling over 344 feet of track; what is the per cent. of grade?

SOLUTION.—Substituting the values of $r = 4.3$ and $d = 344$ in the formula, the grade is $g = \frac{4.3 \times 100}{344} = 1.25$ per cent. Ans.

48. Grade Rise.—If the per cent. of grade g is known, the number of feet r that a car is raised when passing over a distance d can be calculated by the formula

$$r = \frac{g\,d}{100}$$

in which g is taken as a whole number.

EXAMPLE.—A car runs over 344 feet of track that has a grade of 1.25 per cent.; through what distance is the car raised?

SOLUTION.—Substituting the values of $g = 1.25$ and $d = 344$ in the formula, $r = \frac{1.25 \times 344}{100} = 4.3$ ft. Ans.

TRACK TESTS

49. Curvature Tests.—Fig. 15 indicates the **straightedge method** of testing a simple curve for accuracy of curvature. This consists in holding a straightedge $a\,b$, provided with an adjustable perpendicular projection $d\,e$ at its middle point, against the head of either the inside or the outer rail.

This test is made on the inside of the rail curve. If points *a*, *e*, and *b* touch the rail head simultaneously when the straightedge is applied to one part of the curve, they should touch at all parts of the curve; if they do not, the curve is not true. This method is useful on curves sufficiently short for a 6-foot to 10-foot straightedge to show a middle ordinate long enough for differences to be easily noticed. A steel tape and rule may be used in place of the straightedge for curves of longer radius.

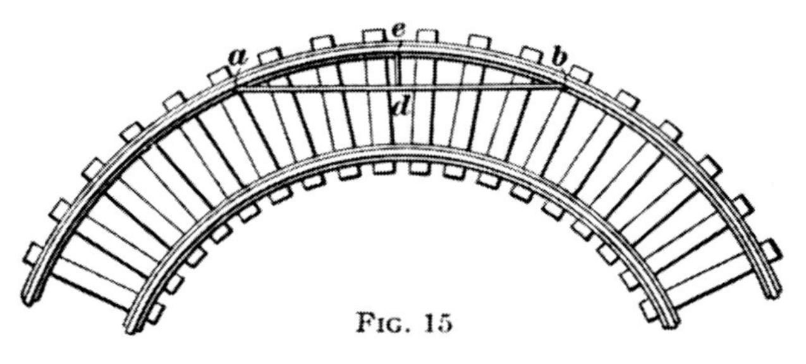

Fig. 15

50. A test by the **sighting method** is indicated in Fig. 16. It may be used where the radius of curvature is too great for even a tape line to be used. From a point *a* at a joint on the outer rail sight across to a point *b* on the same rail that is so located that the sighting line is tangent to the gauge line of the inner rail at some point *d*. The middle ordinate is now the track gauge, usually 4 feet $8\frac{1}{2}$ inches. The number of rails between *a* and *b* should be counted and this number should be approximately the same when the test is made on any portion of the simple curve.

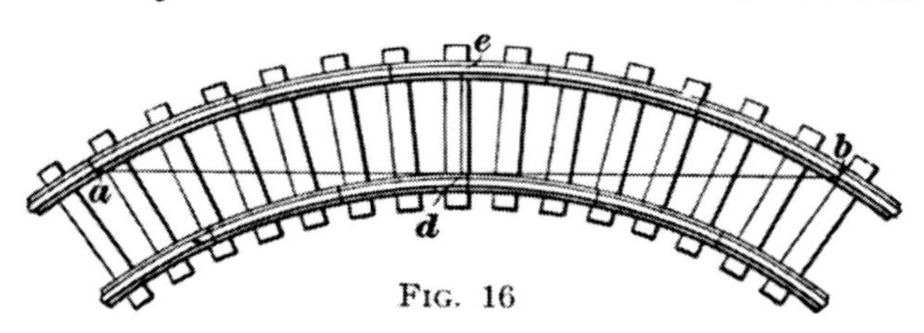

Fig. 16

51. Grade Test.—The per cent. grade of a track may be tested as indicated in Fig. 17. A level *a* is placed on a long straightedge *b* and the straightedge adjusted so as to be level; a rule or tape is used to measure the distances *c d* at right angles to *b* and the track distance *d e*. The per cent. of grade can then be calculated as explained in Art. **47**, where *r* is the distance *c d*, Fig. 17, and *d* is the distance *d e*.

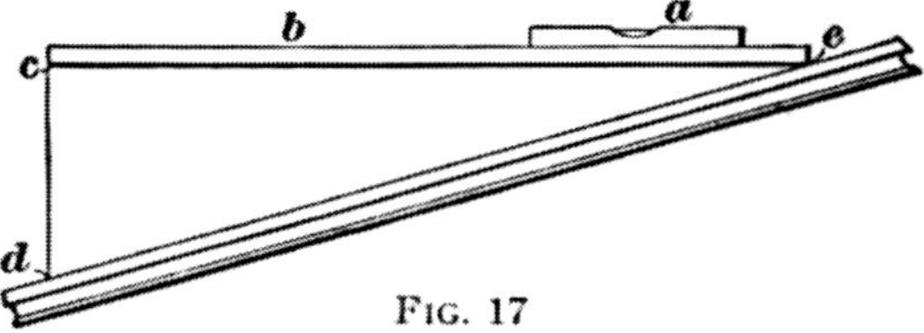

Fig. 17

ELECTROLYSIS

52. Elementary Principles.—The form of electrolysis here considered relates to the gradual eating away of metal objects buried in the ground, such as gas and water pipes, which action is likely to occur when the pipes act as conductors for portions of the return current of a railway system.

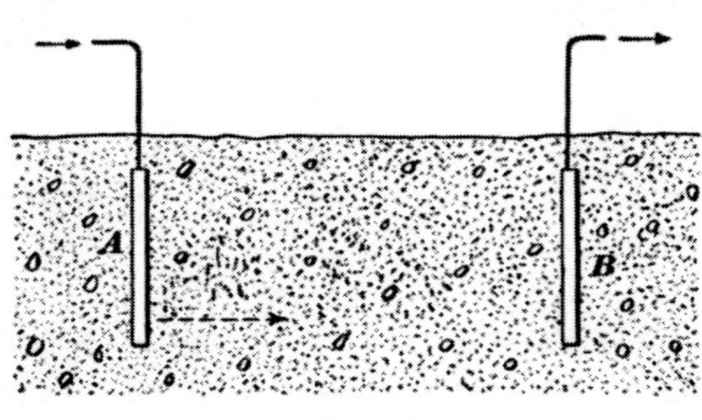

Fig. 18

If two iron plates A and B, Fig. 18, are buried in the ground and so connected that electricity flows from A through the earth to B, plate A will be eaten away or pitted while plate B will not be damaged. This is practically the same electro-chemical effect that takes place in electroplating, where metal is taken from a plate, or anode, and deposited on the article to be plated. The point to notice is that wherever electricity flows *from* a metal conductor into damp earth, the conductor is eaten away, provided that the drop in voltage between the conductor and the adjacent earth is sufficient to effect the chemical decompositions; but where electricity flows

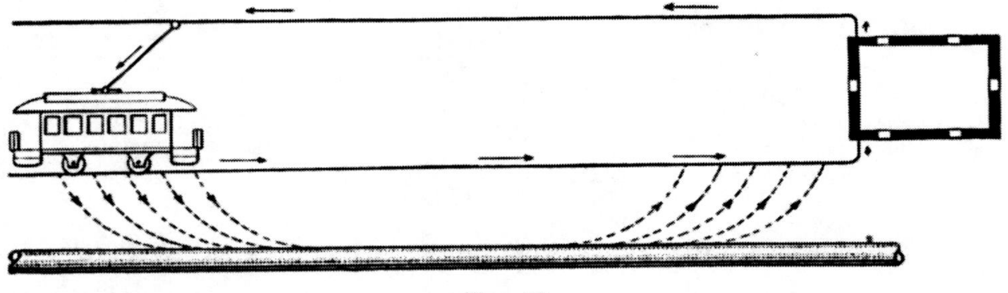

Fig. 19

from the earth *into* the conductor, the latter is not damaged. The rate at which the metal will be eaten depends on the strength of the current.

53. Electrolysis Due to Railway Currents.—Fig. 19 indicates in a simple manner how electrolysis may be caused by the flow of stray electricity of an electric-railway system. The normal path of the current, if the rail-return section is in good condition, is indicated by the full-line arrows. In case of poor bonding, some of the return electricity may flow from the rail to an adjacent pipe and from the pipe to the ground near the station, as indicated by the dotted arrows. The pipe is likely to be damaged at the point near the station where the electricity flows from the pipe to the earth.

Fig. 20 indicates a modification of the simple case shown in Fig. 19. The stray electricity leaves the rail *a*, Fig. 20, enters the pipe *b*, and flows through this pipe until a better path is found in the lead covering of the cable *c*. Electricity flows

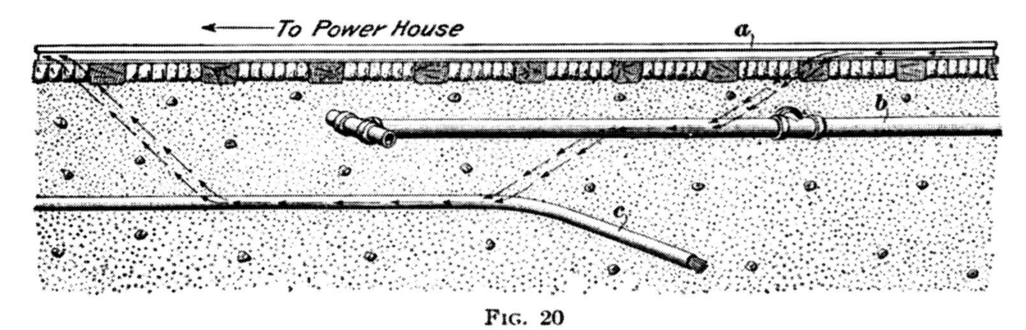

Fig. 20

along *c* until the track again forms a better path, when it flows back to the rail. The path of the stray electricity is indicated by the arrows. Electrolytic action may occur at the points where the electricity flows from the rail, the pipe, and the cable.

54. Prevention of Electrolysis.—A large system of piping forms a conducting network of very low resistance in parallel with the track, hence it is very difficult to prevent part of the current from leaving the track and entering the pipe. However, if proper steps are taken, the bad effects of electrolysis can be largely avoided; the following are the main points that experience has shown should be observed:

1. The trolley wire should be made the positive side of the system.

2. The track should be thoroughly bonded and the bonds maintained in good condition.

3. Any metallic connections that may exist between piping or cable systems and the track should be located and removed.

4. Return feeders should be run out from the station and connected to those pipes that carry the greater part of the current; thus, the current in the pipes will be drained off without passing from the pipes to ground.

5. Where service pipes, cables, or underground conductors pass under tracks or through other regions where they are exposed to electrolytic action, they can often be protected by covering them with glazed tile or by placing them in a trough filled with asphalt.

6. If in any part of a system the rail return carries an excessive current, return feeders should be run so as to relieve the rail of part of the current and prevent an excessive drop in voltage along the rail. The greater the drop in voltage in the rails, the greater is the tendency for the current to pass off to neighboring pipes.

The remedy given under 3 is important. Very often accidental connections exist between the rails and pipe so that current can pass directly to the piping system. This is specially the case where pipes run across iron bridges that also carry railway tracks. Before attempting to drain off the current from a piping system, all metallic connections between track and pipe must be removed. Where pipes pass across iron bridges, the best plan is to insulate the pipe from the bridge, or if this is impossible, insulate the pipe by the insertion of insulating joints at either end of the bridge.

Remedy 4 is very commonly practiced and gives good results if properly applied. The return feeders should be attached to the pipes that carry the most current and, as a rule, the current so returned to the power house will not be more than 5 or 6 per cent. of the total current; if it exceeds this amount it is probable that there is a metallic connection somewhere between the track and pipes. Pipes crossing under street-car tracks are subject to electrolysis, and when laid or repaired are easily covered with tile or run in a box.

RAILWAY MOTORS

PRELIMINARY CONSIDERATIONS

OPERATING REQUIREMENTS

1. The motors commonly used in America for electric-railway work are of the series-wound type because of their strong starting torque. The larger number of railway motors are intended for operation with direct current, but by a modification in design, series-wound commutator motors are adapted for operation with either direct current or single-phase alternating current. A car equipped with these motors and with the necessary switching apparatus may pass directly from a road operating with one kind of current to a road operating with the other kind.

Railway motors must withstand harder service, both mechanically and electrically, than most motors in stationary applications. Motors for electric locomotives are sometimes mounted within the cab, but for electric cars are generally placed on the trucks, where they are exposed to dirt and dampness. Because of limited space allowance, the motor must be built very compactly and its capacity relative to its size is high. The frame is usually of cast steel; but the frame halves of some motors are formed of steel plates pressed into the desired shapes. The frame must protect the active parts from water and dirt and yet allow ventilation of armature and field coils. The heat set up by the current in the windings may be removed from the motor frame by proper ventilation, so as to prevent rapid deterioration of the insulating material.

GEAR REDUCTION

2. The armatures of large motors for electric locomotives are in some cases installed directly on the axles of the trucks; in other cases the motors are mounted in the cabs and connected by driving rods and connecting-rods to the locomotive drivers. In such cases the armatures rotate at the same rate as the drivers.

The railway motor of moderate size is usually provided with a pinion on the armature shaft that engages with a gear mounted on the axle of the car. In order to protect the pinion and gear from dirt, to lessen wear, and to diminish the noise of operation, they are enclosed in a gear-case that contains a quantity of heavy oil.

The armature revolves at a comparatively high speed and the car axle at the lower speed desired for the service in which the car is engaged. The reduction in speed depends on the relative number of teeth in the pinion and gear; the less the number in the pinion, as compared with those in the gear, the greater will be the reduction in speed of the car axle from that of the motor armature.

3. The **gear ratio** of an equipment will here be understood as the ratio of the number of teeth in the gear to the number in the pinion. While this is the usual way of expressing gear ratio, it is sometimes given as the ratio of the number of teeth in the pinion to the number in the gear. The pinion has, in nearly every case, fewer teeth than the gear, so that there is little cause for confusion no matter which way the ratio is stated. If, then, a motor has 14 teeth in the pinion and 68 in the gear, the gear ratio is $68 : 14 = 4.86 : 1$ and the motor armature rotates 4.86 times as fast as the axle.

Table I gives the speed of car axles, in revolutions per minute, for different car speeds and diameters of wheels. By multiplying the revolutions given in the table by the gear ratio in any given case, the speed of the motor armature is obtained.

EXAMPLE.—A car is mounted on 33-inch wheels and runs at a speed of 20 miles an hour; how many revolutions a minute do the motor armatures make if there are 65 teeth in each axle gear and 15 in each pinion?

Solution.—The speed of the car axle, Table I, for 33-in. wheels and a speed of 20 mi. an hr., is 203.7 rev. a min. The gear has 65 teeth and the pinion 15, hence the gear ratio is 65 : 15. The speed of the armature is, therefore, $203.7 \times \frac{65}{15} = 883$ rev. a min., approximately. Ans.

TABLE I

REVOLUTIONS OF CAR AXLE CORRESPONDING TO VARIOUS CAR SPEEDS

Speed of Car Miles per Hour	Speed of Car Feet per Minute	Speed of Car Axles in Revolutious per Minute						
		30-Inch Wheels	31-Inch Wheels	32-Inch Wheels	33-Inch Wheels	34-Inch Wheels	35-Inch Wheels	36-Inch Wheels
6	528	67.2	65.0	63.0	61.1	59.3	57.6	56.1
8	704	89.6	86.7	84.0	81.5	79.1	76.8	74.7
10	880	112.0	108.4	105.0	101.8	98.9	96.1	93.4
12	1,056	134.4	130.0	126.0	122.2	118.6	115.2	112.1
14	1,232	156.9	151.7	147.0	142.6	138.4	134.5	130.7
16	1,408	179.2	173.4	168.0	163.0	158.2	153.6	149.4
18	1,584	201.7	195.1	189.0	183.4	178.0	172.9	168.1
20	1,760	224.0	216.8	210.0	203.7	197.8	192.1	186.8
22	1,936	246.5	238.4	231.0	224.1	217.5	211.3	205.5
24	2,112	268.8	260.0	252.0	244.4	237.3	230.4	224.2
26	2,288	291.3	281.8	273.0	264.8	257.1	249.7	242.9
28	2,464	313.8	303.4	294.0	285.2	276.8	268.9	261.4
30	2,640	336.1	325.1	315.0	305.6	296.6	288.2	280.2
32	2,816	358.4	346.8	336.0	326.0	316.4	307.4	298.8
34	2,992	380.9	368.4	357.0	346.3	336.2	326.6	317.6
36	3,168	403.4	390.2	378.0	366.7	356.0	345.8	336.2
38	3,344	425.8	411.8	399.0	387.1	375.7	365.0	354.9
40	3,520	448.0	433.6	420.0	407.4	395.6	384.2	373.6
42	3,696	470.6	455.2	441.0	427.8	415.3	403.5	392.3
44	3,872	493.0	476.8	462.0	448.1	435.1	422.6	411.0
46	4,048	515.4	498.5	483.0	468.5	454.8	441.9	429.7
48	4,224	537.6	520.0	504.0	488.8	474.6	461.1	448.4
50	4,400	560.2	541.8	525.0	509.2	494.4	480.3	467.0

MOTOR RATING

4. The rating of a railway motor is commonly based on the mechanical output that the motor can maintain at the car axle for a stated time without causing the rise in temperature in any part of the motor to exceed certain limits. Another rating is the continuous current input at a stated voltage that a motor can take from the circuit without exceeding certain temperature limitations.

Because of the variable conditions under which railway motors operate, no brief and definite rating that applies to all cases can be assigned to these motors. The American Institute of Electrical Engineers has adopted rules for tests to determine the rating of railway motors, which to a considerable extent have also been adopted by the manufacturing companies. These rules are modified from time to time and persons interested should consult the latest revision.

DIRECT-CURRENT MOTORS

DESCRIPTION OF MOTOR

5. Fig. 1 shows a direct-current railway motor of modern type. The motor is provided with commutating poles and arranged for forced-air ventilation. The enclosed frame *a* is of cast steel in one part and known as the *box type* to distinguish it from the frame that is in two parts and known as the *split type*. The screened intake for the air-cooling system is shown at *b*; the suspension lugs, which serve to hold the motor to the truck, at *c*; the iron gear-case, which is bolted to the motor frame and encloses the pinion on the armature shaft and the gear on the axle of the truck, at *d*; the cover over the opening provided for the inspection of the brush rigging and the commutator, at *e*; the pinion-end bracket, which is unbolted when the armature with the pinion on the shaft is to be withdrawn from the frame, at *f*, view (*a*); the commutator-end bracket, at *g*, view (*b*); the axle bearings, at *h*, and the armature and field-coil leads, at *i*.

6. The armature, one main pole piece with its coil, and one commutating pole piece with its coil, are shown in Fig. 2. The armature core shown has 29 slots in which are placed

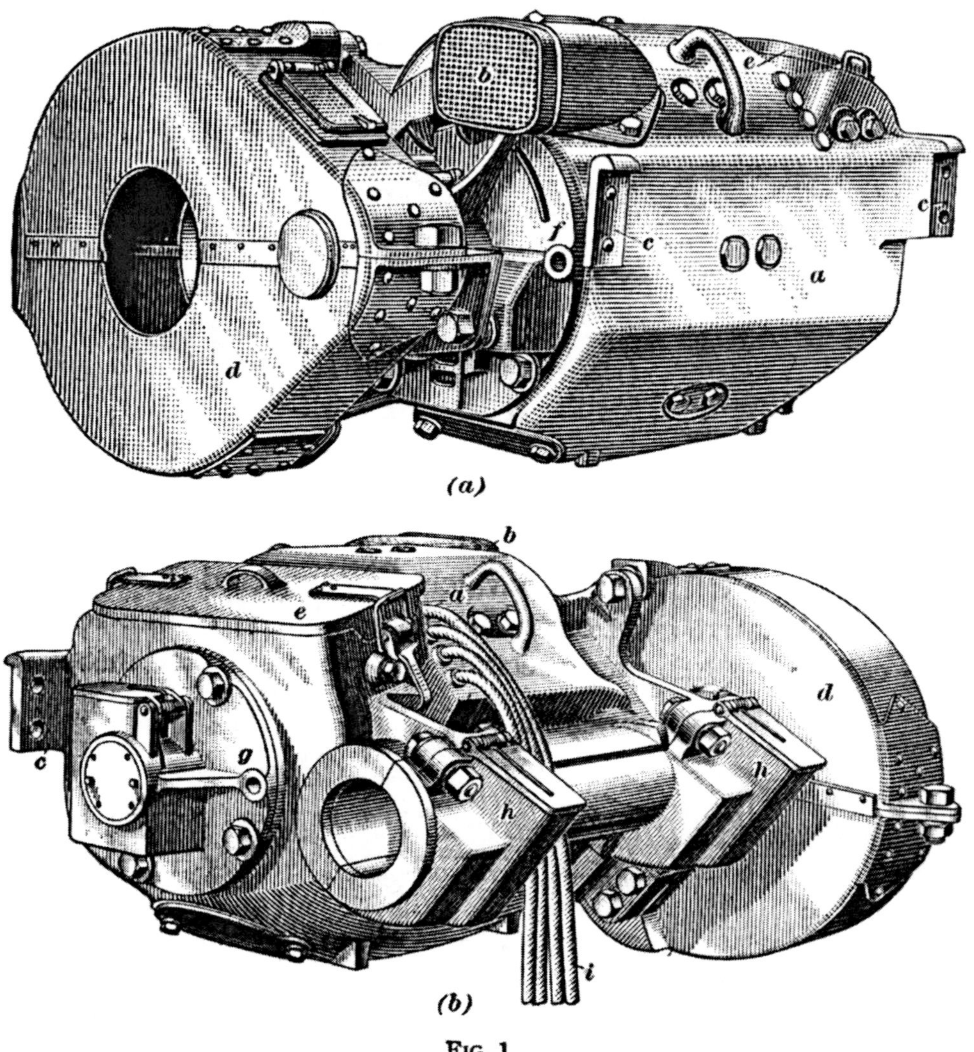

Fig. 1

29 groups of coils, each group containing 4 coils, making the total number of coils 116. The main pole pieces are laminated and the commutating pole pieces are of drop-steel forgings. All of the pole pieces are bolted to the frame of the motor. The field coils are wound with asbestos-insulated copper strap,

impregnated with insulating compound, and their surfaces are protected by layers of japanned webbing. The core laminations of this armature are assembled and keyed directly on the

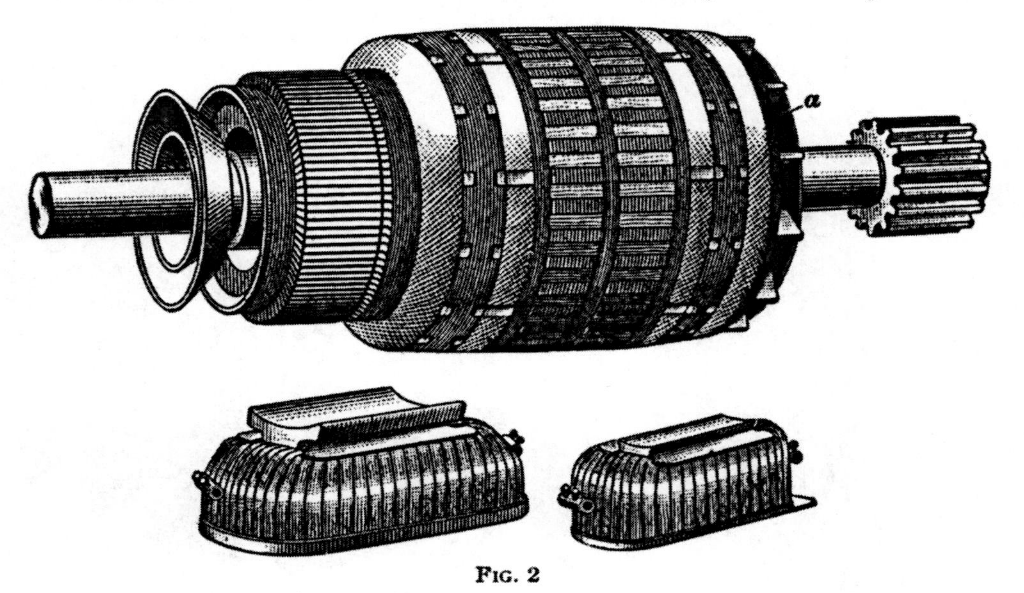

Fig. 2

shaft on which the commutator spider is also pressed. Ventilating holes are provided lengthwise through the core. The commutator has 115 bars, to which are connected the 230 leads of the 115 active coils of the armature, 1 coil of the total of 116 coils being left unconnected. The commutator micas are

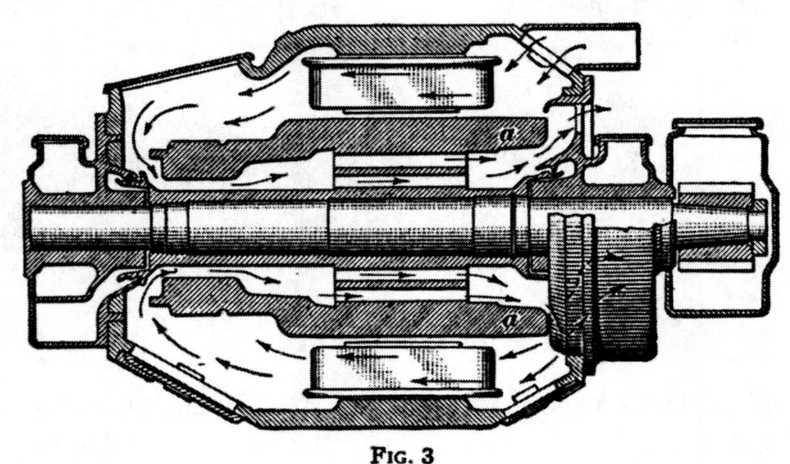

Fig. 3

grooved, or undercut, to a depth of $\frac{3}{64}$ inch. This custom, which aids in the prevention of sparking, is quite generally followed. All of the motor bearings are oil-lubricated.

7. A centrifugal fan *a*, Fig. 2, at the pinion end of the armature core revolves with the armature and sets up a circulation of air in the direction indicated in Fig. 3. The fan is located at the end of the armature core marked *a*, Fig. 3. The air enters through the screened opening *b*, Fig. 1 (*a*), passes over the field coils and the outside surface of the armature, Fig. 3, then through the commutator and lengthwise holes in the armature core out through the centrifugal fan and openings in the pinion-end bracket of the frame. A constant stream of cool air thus enters the frame, circulates around the heated parts, and absorbs and carries away heat. The motor, therefore, runs cooler for a given load than it would if the fan were omitted.

FEATURES OF CONSTRUCTION

8. The construction of the modern railway motor is the result of many years of experience of designers and operators of this type of apparatus. Some of the details of construction

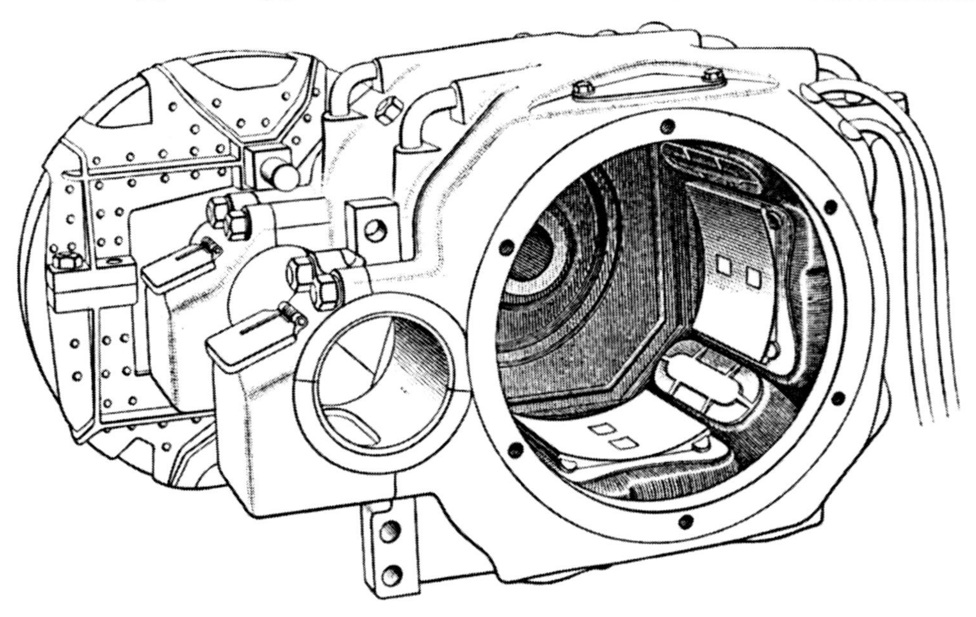

Fig. 4

have been developed especially to meet the difficult requirements of railway service, while others apply to motors intended for either railway or stationary service. Some of the more important of these features are here treated, but they do not necessarily belong to the same type of motor or to the motors made by one company.

9. Commutating Poles.—The relative positions of the main pole pieces and the commutating pole pieces are indicated in Fig. 4. There is usually the same number of main pole pieces as commutating pole pieces and each of the latter is located midway between a pair of the former. The object of the commutating poles is to improve commutation for either direction of rotation of the armature; the theory of this action has been explained in a previous Section.

Many of the electric troubles of car motors are due to sparking between the brushes and the commutator. The motor

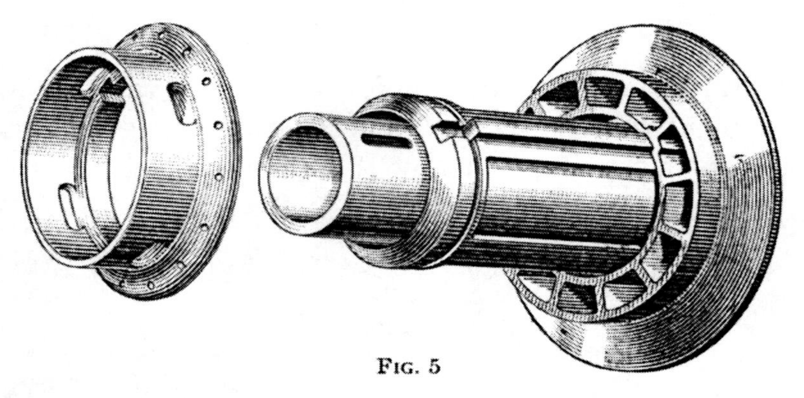

Fig. 5

load is variable and this causes variations in the armature reaction. Unless the effects of these variations are neutralized by the flux of the commutating poles, the positions of the neutral spaces will be changed, thus tending to produce sparking. The motor in most car applications must revolve in either direction; therefore, the brushes are usually set at the no-load neutral point. The application of commutating poles to railway motors thus causes good commutation under all ordinary load conditions and with fixed positions of the brushes.

10. Armature Spider.—In some railway motors, the armature-core punchings are assembled on a spider that also serves to support the commutator. The armature shaft is forced into the spider under heavy pressure. With this construction, a shaft damaged in operation can be pressed out and repaired or a new one substituted without disturbing the armature winding and connections. Ventilating space can also

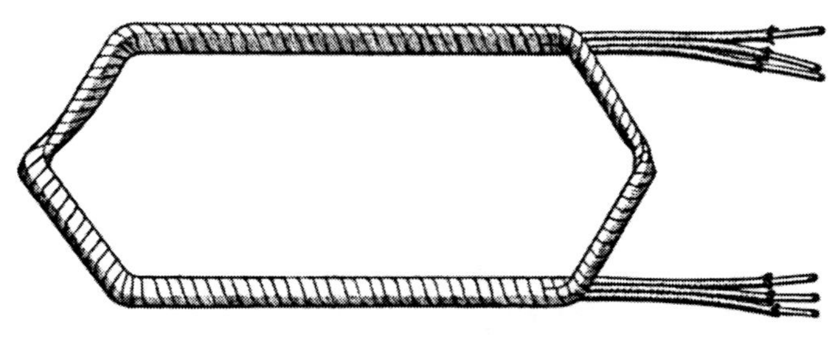

Fig. 6

be easily provided between the spider ribs on which the core is assembled.

Fig. 5 indicates one form of armature spider. The core laminations are held between the fixed shoulder on the spider and the removable end ring.

11. Armature Coils.—Fig. 6 shows a group of three individual coils assembled into a single armored cell. Some coils are formed with round wire and others with copper strap. Cotton, mica, linen tape, and insulating compound are used to insulate the coils from one another in the cells and the cells from the surfaces of the armature slots. The coils are so placed and joined together as to form a *series*- or *wave-connected* winding. This arrangement requires only one positive and one negative set of brushes, which may be placed at points on the commutator convenient for inspection.

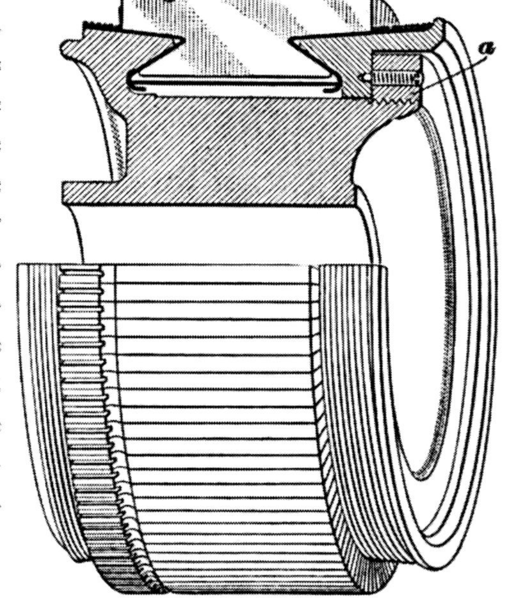

Fig. 7

12. Commutator.—Fig. 7 shows one form of commutator used on railway motors. The bars are made of hard-drawn copper and insulated from one another and the commutator shell by mica strips and cones. The shell is in two parts, which are forced by heavy pressure into the **V**-shaped notches in the ends of the commutator bars, after which the ring nut *a* is set up tight against the shell and locked by small screws, one of which is shown. The bars are thus firmly clamped together. In this case the shell is mounted on a spider designed for forcing on the armature shaft; in other cases, the commutator shell is pressed on an extension of the armature spider.

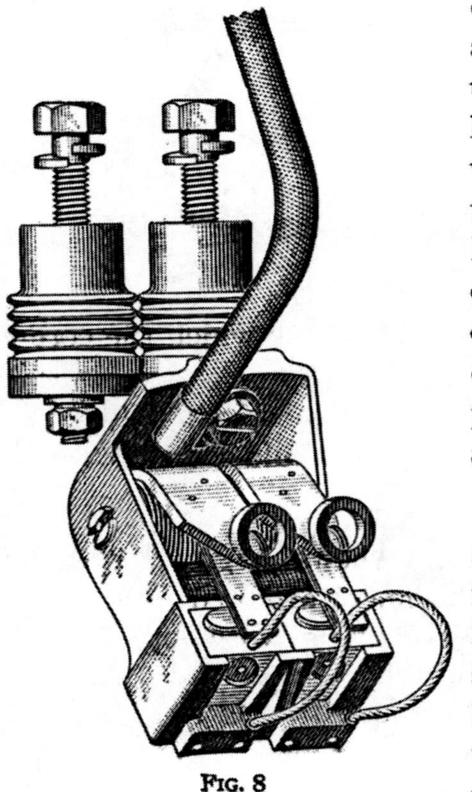

Fig. 8

13. Brush Holders.—Fig. 8 shows a brush holder arranged for two carbon brushes. Each holder is supported by two steel studs cemented into porcelain bushings. The studs pass through holes in the motor frame and are bolted firmly in position. The brushes are pressed against the commutator by adjustable springs. Flexible copper shunts between the carbon brushes and the holders relieve the springs from carrying large currents.

14. Pinions.—Pinions are usually mounted on a tapered seat formed at the end of the armature shaft, as indicated in Fig. 9. With this construction, the pinion can be more readily removed than when mounted on a straight seat, as in early practice.

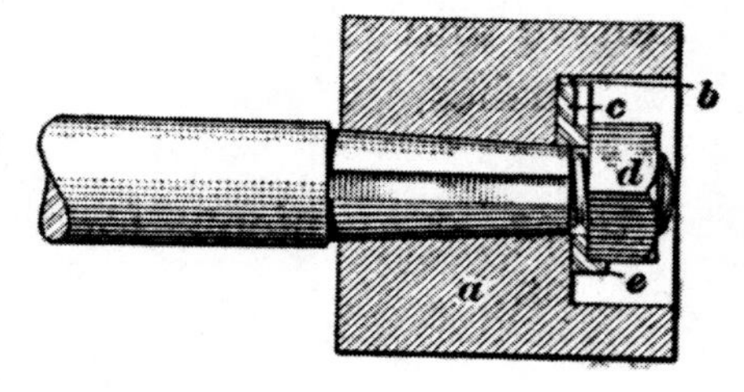

Fig. 9

In some cases the pinion nut, Fig. 9, is placed in a counterbore at the end of the pinion, in order to save space in the gear-case. A method of locking the nut is also indicated. The pinion *a*, mounted on its tapered seat, has a small groove *b* cut at the bottom of the counterbore. A portion of the steel washer *c* is hammered into this groove. The nut *d* is run on and screwed home and one edge of the washer is turned up against the nut

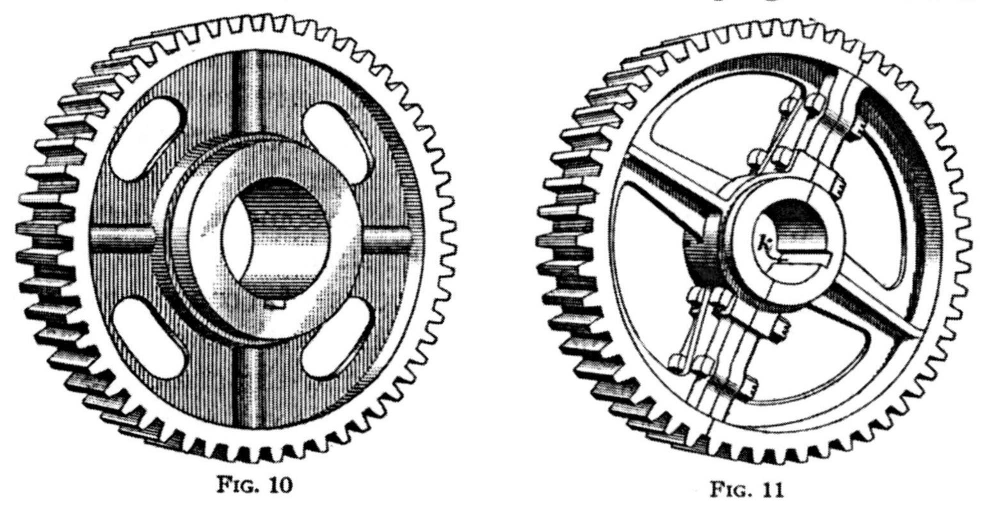

FIG. 10 FIG. 11

with the aid of a thin tool that will lift the edge of the washer as indicated at *e*. The washer prevents the nut from turning, the pinion prevents the washer from turning, and the pinion key prevents the pinion from turning on the shaft.

15. Gears.—The type of solid axle gears shown in Fig. 10 is put on the axle under heavy pressure; further protection against slip and turning on the axle is furnished by a key. The style of gears shown in Fig. 11 are made in two parts that may be clamped together around a shaft. A keyway is shown at *k*. The life of gears is increased by special hardening treatments of the steel and by the use of special lubrication.

16. Lubrication of Bearings. Both the axle and the armature bearings of the earlier motors were lubricated with grease. In such cases the bearing must heat and melt the grease before lubrication starts. Oil lubrication of railway motors is now, however, quite generally employed.

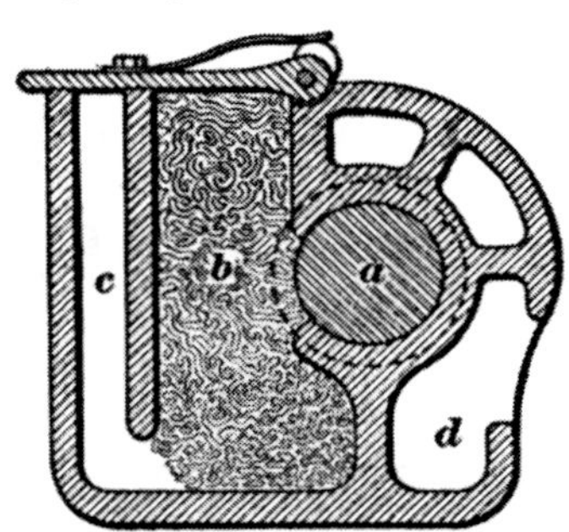

FIG. 12

Fig. 12 indicates one method of applying oil to the armature bearings. Axle-bearing lubrication is effected in practically the same manner. The armature shaft *a* is exposed directly to the oil-soaked waste stored in chamber *b*. The oil is poured into the bearing through the separate chamber *c*. The waste in chamber *b* absorbs the oil and delivers it to the surface of the shaft. After the oil passes through the bearing it is collected in a drip chamber *d* and, after filtration, may be used again.

Oil guards mounted on the armature shaft close to the bearings prevent oil from reaching the commutator or the armature windings. The oil from the guards is sometimes deflected to the drip chamber or thrown to the street.

In some motors, the bearing sleeve is of bronze overlaid with a thin surface of Babbitt. If the Babbitt melts, the shaft will be supported by the bronze sleeve and the armature will thus be prevented from striking the pole pieces.

ALTERNATING-CURRENT MOTORS

DESCRIPTION OF MOTOR

17. Most of the alternating-current motors used in America are of the series-wound commutator type that may be used

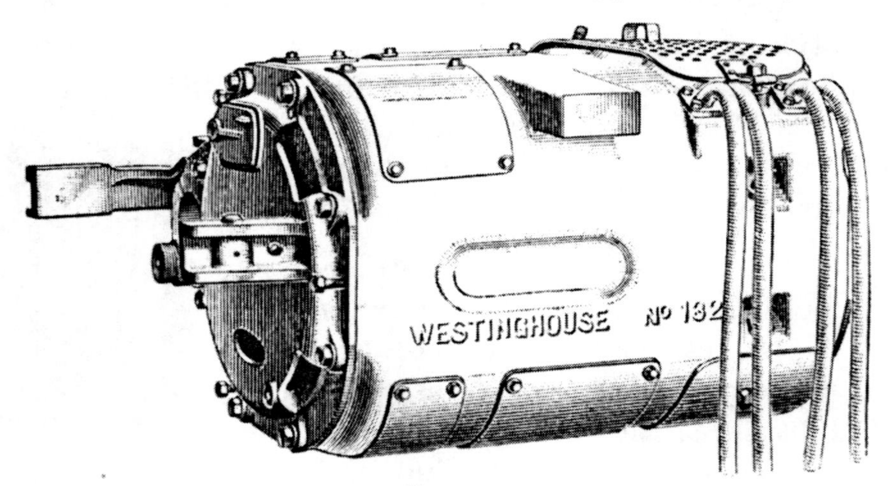

FIG. 13

with either direct current or alternating current. Fig. 13 shows the exterior of a 100-horsepower motor of this type provided with a box frame and detachable end brackets for the removal of the armature. The laminated field core is so constructed as to form four projecting pole pieces, on which are mounted

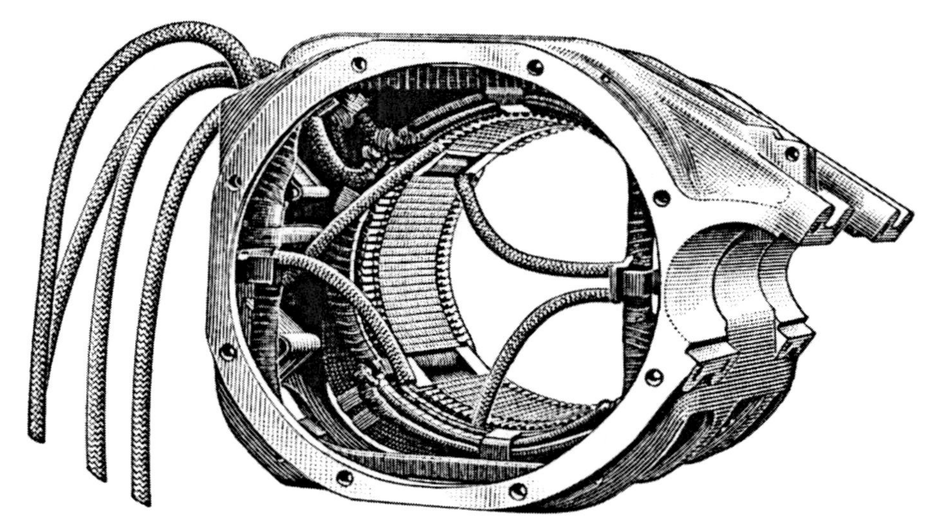

Fig. 14

the four main field coils, each consisting of a few turns of heavy copper strap.

A **compensation winding,** consisting of copper bars connected at the ends by copper straps so as to form a continuous

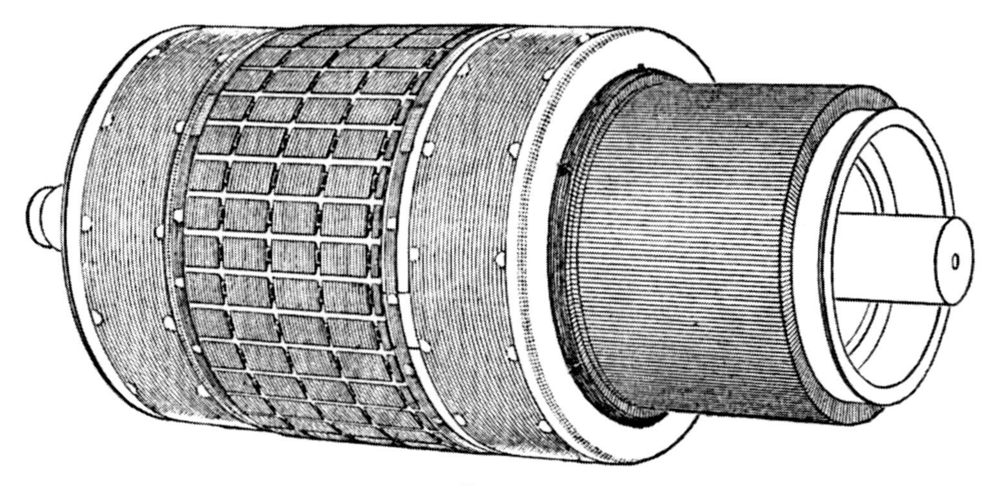

Fig. 15

winding, is distributed in the slots in the pole pieces, as shown in Fig. 14, in order to obtain a distribution of magnetomotive force similar to that of the armature conductors.

The compensating winding of a motor intended for either direct- or alternating-current circuits is connected in series with the armature and the main-field winding. In a motor used exclusively on alternating-current circuits, the compensating winding is short-circuited, either as a whole or by sections, and is separate from the armature circuit. The compensating winding thus becomes the secondary of a transformer of which the armature forms the primary.

Fig. 15 shows an armature used with this type of motor. The armature is parallel-wound and requires four sets of brushes. The motors are usually wound for 250 volts and 25 cycles.

FEATURES OF OPERATION

18. Reversal of Current.—When an alternating current is established in the field and armature windings of a series motor, the current reverses through each winding at the same time. The polarities of the pole pieces and of the poles formed on the armature core, therefore, reverse at practically the same

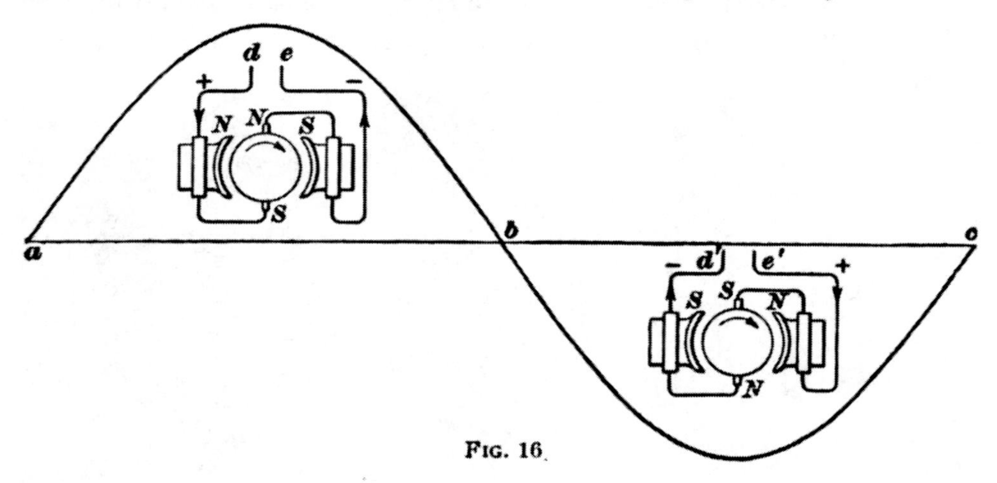

Fig. 16

instant and continuous rotation in a given direction is thus maintained. In Fig. 16, curve *a b c* represents one cycle of an alternating current. In the portion *a b*, the direction of current is assumed to be from motor lead *d* to lead *e*, resulting in the polarity and rotation shown. At *b* the direction of current is reversed. Lead *e'* is now positive and lead *d'* negative, the polarities of field poles and armature poles are thus reversed, but the direction of rotation remains unchanged, as indicated.

19. Self-Induction of Windings.—When operating with alternating current an electromotive force of self-induction is set up in the windings. In the case of the main-field windings, this electromotive force is reduced to its lowest value by decreasing the turns of the exciting field coils to the least number capable of providing the necessary magnetic flux. The less the number of turns that are acted on by the passage of the alternating magnetic flux through them, the smaller becomes the electromotive force of self-induction.

The mutual induction between the conductors of the compensating winding and the adjacent conductors of the armature windings greatly reduces the effect of self-induction in both windings. The compensating action is similar, whether the compensating coils are in series with the armature and main field coils or whether they form a separate circuit or circuits excited by the transformer action of the armature coils.

Besides improving the power factor of the motor, the current in the compensating winding so neutralizes the armature reaction that proper commutating conditions may be secured with fewer turns on the field winding, and thus a weaker field than would otherwise be possible.

20. Effect of Transformer Action on the Armature. When the armature is rotating in the magnetic field established by alternating current in the main field coils, a counter electromotive force is generated in the armature coils in the same manner as in a direct-current armature. There is also set up in the armature coils an electromotive force due to the alternating magnetic flux passing through the pole pieces and armature. The latter action is practically similar to that of a transformer. The first, or mechanically generated, electromotive force is proportional to the speed; the second, or electrically generated, electromotive force is proportional to the frequency of the alternating current. The first electromotive force is the useful one, as it is the counter electromotive force of the motor.

In half of the coils between any two adjacent brushes, the electromotive force generated by transformer action is opposite in direction to the electromotive force generated by transformer action in the remaining coils of the group, for one-half of the coils are affected by the flux emanating from a north pole, while the other coils are simultaneously affected by the flux entering a south pole. The effects of transformer action are thus practically neutralized and the counter electromotive force is affected but little, if any, by the transformer action, provided the brushes are properly placed.

21. When an armature coil is short-circuited by a brush during commutation, it forms a closed secondary circuit, the primary of which is the adjacent main field coil. The ohmic resistance of the coil itself is low, and to prevent the establishment of a large current, the armature coils have but one turn each, which limits the electromotive force generated in the closed coils, the frequency of the system is low, and the leads connecting the junction of adjacent coils to the commutator bars in some motors have sufficient resistance to limit the current to a safe value. The leads are active only when the coils to which they are connected are being commutated.

In railway motors of ordinary size, the resistance leads are usually tapped to the armature winding at the rear end of the core and brought to the commutator bars through the same slots as the main winding. In some very large motors, the leads are tapped to the armature winding at the front end of the core, carried to the rear end, doubled back to the front, and then connected to the commutator.

MOTOR CHARACTERISTICS

22. When the suitability of a motor for a given class of service is to be determined, information regarding its performance is a necessary factor. These data are obtained from tests made by the manufacturers, and are usually presented in the form of curves, known as **motor characteristics,** or **performance curves.** These curves commonly show the relation between values of the current, speed, tractive effort, horsepower, and efficiency. As the heat of the windings limits the output of a motor, and this heat is due to the current, the other properties considered are usually referred to the values of current taken by the motor under various operating conditions. In the curves, the current values are usually laid out on the axis of abscissas and the other values on the axis of ordinates.

The performance curves shown in Fig. 17 relate to data for a 100-horsepower single-phase motor when operating with direct current and a terminal voltage of 150. For low speeds the motors operate four in series across a 600-volt circuit. The curves are drawn for a gear ratio of 63 to 20 and for 36-inch driving wheels.

23. Speed.—If the voltage is constant, only one armature speed corresponds to each value of motor current. For example, the car speed, Fig. 17, for a current of 400 amperes is 13 miles an hour. As the motor is series-wound, the armature speed decreases as the load increases, the car speed corresponding to a current of 600 amperes being 10 miles an hour.

An increase in line voltage increases the car speed in almost direct proportion, especially for lower and moderate speeds. If the speed of a car is 11 miles an hour at 500 volts, the car speed at 600 volts is $11\times\frac{6}{5}=13.2$ miles an hour, approximately. At higher speeds accurate results are not obtained with this proportion because of the greatly increased effect of air resistance.

24. Tractive Effort.—The armature conductors are moved by the interaction of the current in them with the flux of the pole pieces. The force applied to the conductors is

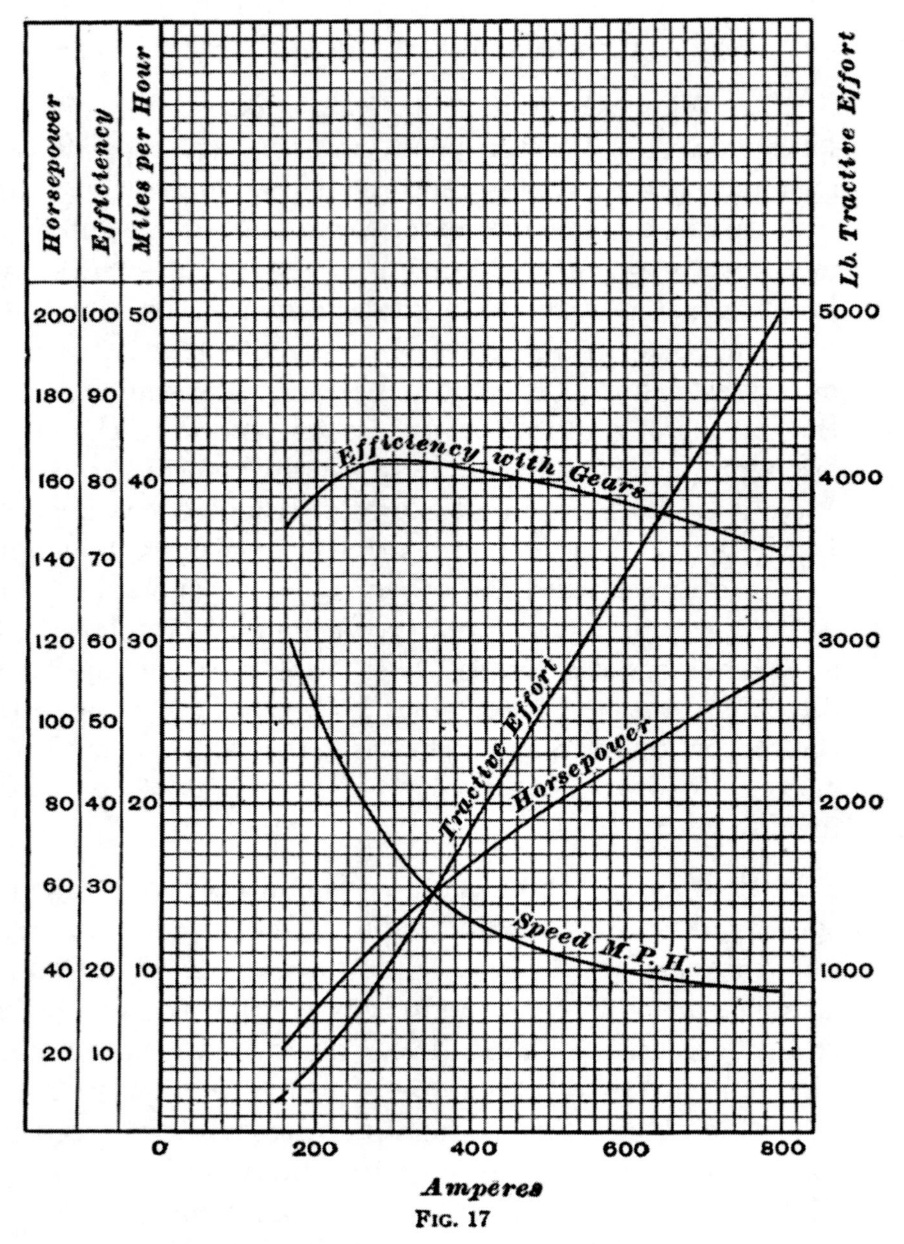

FIG. 17

transmitted through the pinion and gear to the car axle and thence to the car wheels fixed to the axle.

The force exerted at the tread of both of the wheels on the car axle is known as the tractive effort of the motor and is usually expressed in pounds. Any change in the gear ratio, wheel diameter, or current affects the value of the tractive effort. When the motor is taking a current of 400 amperes, the tractive effort is 1,850 pounds, as indicated in Fig. 17.

25. Horsepower.—The nominal rating of a railway motor is considered as the mechanical output at the car axle. The power at the car axle is less than that at the armature shaft due to losses in gears, bearings, etc.

The car speed in feet a second for a given current multiplied by the corresponding tractive effort in pounds and divided by 550, the number of foot-pounds a second equivalent to 1 horsepower, gives a point on the horsepower curve, Fig. 17. For example, the speed at 400 amperes is 13 miles an hour, or $\frac{13\times 5{,}280}{60\times 60}$ feet a second, and the tractive effort is 1,850 pounds. The power is $\frac{13\times 5{,}280}{3{,}600}\times\frac{1{,}850}{550}=64.1$ horsepower, approximately, as indicated in Fig. 17. For a current of 600 amperes, the horsepower is 91, approximately.

26. Efficiency.—The commercial efficiency is the ratio of the useful output to the total input. The efficiency of the motor when provided with gears is indicated in Fig. 17. For example, at 400 amperes input the efficiency is 81 per cent., approximately. The maximum efficiency is 82.5 per cent. at a current input of 300 amperes.

27. Comparison and Modification of Values.—By selecting a point on one curve, corresponding values on the other curves may be found by determining the place where a vertical line through the selected point cuts the other curves and then reading the values of these points by means of the data lines on the margins.

The characteristics, as given, usually apply to one motor, so that if the property sought involves all the motors on a car, the data obtained from the characteristic curve must be modified accordingly. For example, the tractive effort available for propelling a car equipped with four motors would be the tractive effort of one motor multiplied by four. Should the total current required to propel a car at a certain speed be desired, the total tractive effort being known, the tractive effort of one motor is found by dividing the total value by the number of motors and the corresponding current of one motor as ascertained from a characteristic curve such as Fig. 17, is then multiplied by the number of motors.

MOTOR INSTALLATION AND MAINTENANCE

INSTALLATION

28. Nose Suspension.—The term **motor suspension** applies to the manner of supporting the motor on the car axle and truck. Fig. 18 shows an arrangement called a *nose suspension*.

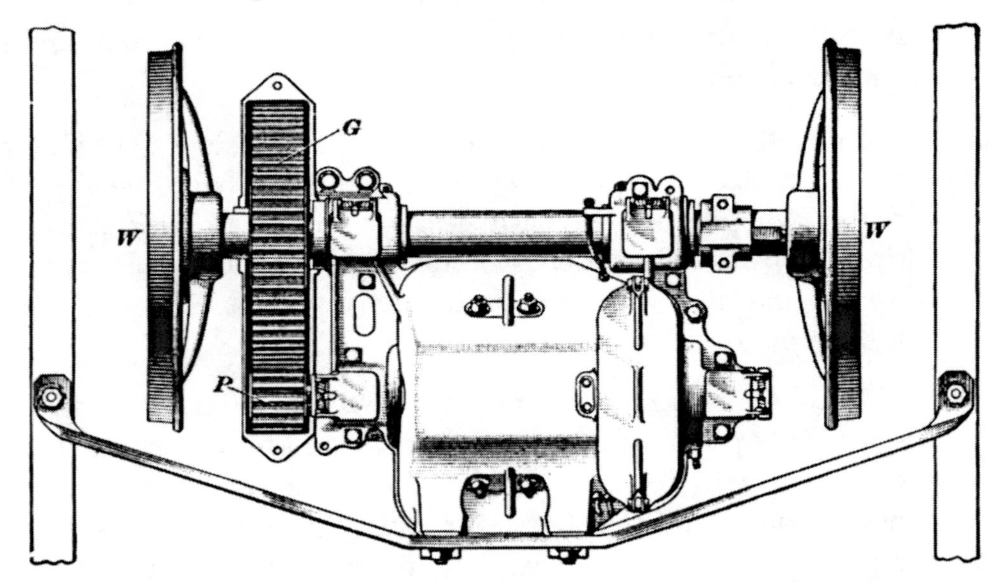

FIG. 18

One end of the motor is supported by bearings on the car axle and the other end on a suspension bar connected across the frame of the truck. The pinion *P* engages the gear *G* on the car axle on which the driving wheels *W* are also mounted.

Fig. 19 shows in more detail a type of nose suspension quite generally adopted as standard on account of its simplicity. A link *a* connects the motor to the suspension bar *b* and springs *c* and *c'* cushion the motor impact in either direction on an angle iron *d* riveted to the truck, thereby relieving the truck of direct shock.

29. The motor should be so installed on the axle that when the car is on a straight track, the motor rests neutrally in the axle space between the gear on one side and the axle collar on the other. The use of a crowbar to make parts fit, on either the motor or any portion of the suspension, should never be permitted, as such a course causes excessive wear of axle-bearing flanges. All suspension parts must be kept tight;

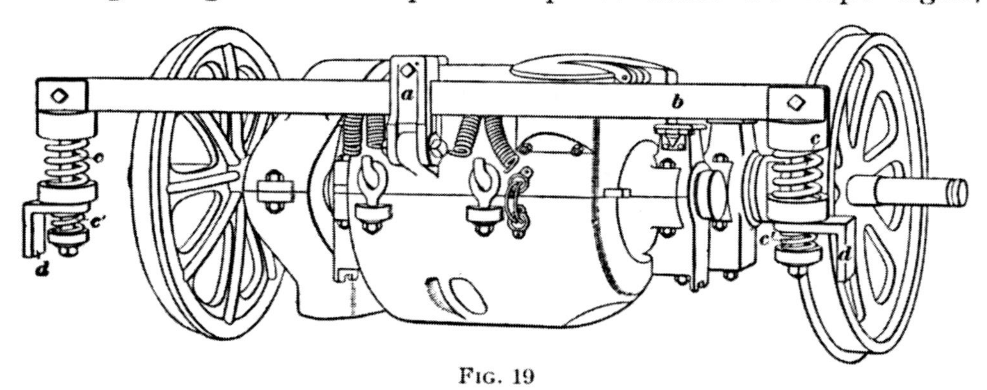

Fig. 19

otherwise, failure of the rigging may let the motor down when the car is in operation. When running over rough track, the motor jumps up and down, bringing both springs into action. In starting, where the motors are hung front to front or back to back, the top suspension spring of one motor and the bottom spring of the other are compressed. This continual jarring and movement tends to loosen suspension parts, which soon produce a ringing, rattling noise. The suspension springs should be neither too weak nor too strong; if too strong they will lack flexibility, and if too weak they will close every time the car gets a jolt; in either case the motor is subjected to vibration liable to jounce the brushes from the commutator and cause flash-overs.

30. Location of Motor.—On either a two-motor, two-axle, or a four-motor, four-axle car, a motor is geared to each axle and the motors should usually be hung between the axles of each truck. The motors are then near the truck centers and have little motion relative to the car body. Short leads may therefore be used, with little chance of grounding them. The stresses on truck parts are also less than when the motors are hung outside the axles, because there is less leverage.

When starting a car or when the motors are used to stop it, the pinions tend to crawl around their respective gears in opposite directions and the resulting stresses are greater with the motors outside of the axles than when they are hung between them.

When two motors are to be hung on an eight-wheel car, most master mechanics prefer to put both motors on the same truck, because then but one heavy truck is required and the first cost of trucks and the dead-weight to be hauled, are minimized. Furthermore, the brake rigging is simplified, car-floor trap doors are needed at but one end, and car wiring is made cheaper and simpler. Finally, where pit room is limited or pits are short, both motors may be reached while only one end of the car is over the pit.

MAINTENANCE

31. On most well-ordered roads, two kinds of car inspection are practiced: *quick inspection*, which is made every day or night or on alternate days or nights; and *overhauling inspection*, which is the thorough inspection attending overhaulings made at intervals, based usually on car mileage.

QUICK INSPECTION

32. Quick inspection is limited to weaknesses that may be observed without disassembly of parts. It includes inspection of air-gap clearances on motors having bearings of the grease-fed type, and inspection of cotter pins and nuts on important parts such as suspension rigging, brake rigging, motor frames, and gear-cases. As failure of a suspension or brake rigging is apt to be a very serious matter, these parts should be inspected as often as practicable. On high-speed cars, wheels should be inspected for cracks and chipped flanges at least once a day. The general adoption of steel instead of cast iron for car wheels has greatly decreased the trouble from broken wheels, and only the hammer test for cracks is required. The motor leads should be gone over, to eliminate contacts with the motor frame, suspension rigging, or other metal parts, and to detect loose connections or actual open circuits. The blowing of a circuit-breaker will of course give notice of an actual ground; but the object of the quick inspection is to find the abraded motor lead before it actually gives trouble.

An open circuit in a motor lead of a two-motor car, becomes evident from the slow action of the car during operation; but numerous cases are on record where four-motor cars have operated for weeks on three motors, because a lead of one motor had pulled out of the connecting sleeve. Cars in this condition have been turned in repeatedly for slow speed attributed to the brakes binding. Good contact of the motor ground wires should be insured at each inspection. Commutator covers should be lifted and inspection made for short brushes, tight brushes, broken brushes, tension fingers resting on the holder, instead of on the brush; burnt, broken, or weak tension springs, broken brush shunts, loose holders, holders resting on or too far from commutator, holders rubbing head of commutator, brushes riding over end of commutator, burnt armature head, burnt or loose string band, accumulation of carbon dust on brush holders, or on string band, or its equivalent, evidence of oil on commutator, rough commutator, black commutator, and red commutator indicating poor commutation or excessive heating. These simple causes of prospective trouble make a long list; but the eye of an experienced inspector can detect any of them at a glance. Any unhealthy motor condition, as a rule, is evidenced by the condition of the brushes and of the commutator; if the cause of their bad appearance is not evident and easily removed, the car should be marked to be held in. On motors in which only one brush per holder is used, most of the small troubles just enumerated are sufficient to cause a flash-over and the operation of a car circuit-breaker.

But on motors in which two or more brushes to a holder are used, the only evidence of irregularity may be a difference in the appearance of two zones of the commutator because one brush carries too much current and another carries little or no current.

33. On motors with grease-lubricated bearings, quick inspection should of course include the grease boxes; on all motors, the inspector should see that no grease- or oil-box covers are missing. Any accumulation of mud around the grease covers should be prevented by frequent use of a stiff wire brush, especially on armature grease-box covers. As the passing of the hand along the grids of a grid rheostat will readily reveal any broken units, this quick test should be made; broken grids often run for weeks without being reported and they may be the cause of blistered controller fingers, sparking brushes, or both.

OVERHAULING INSPECTION

34. Frame.—At the time of overhauling, the motors are entirely disassembled and all parts laid out where they can be inspected thoroughly. The frame is scraped, wiped inside and out, and carefully inspected for cracks, especially around the armature and axle journals. The entire inside of the frame, except the pole-piece seats, is painted with a compound whose glaze prevents the adherence of moisture. Formerly, cracked frames were thrown into the scrap; but by the use of any one of several different processes, they can now often be repaired.

Missing grease-box lugs, springs, and covers are replaced; hinges, gaskets, and linings of inspection covers and lids are renewed; all bearing chambers are thoroughly flushed with light oil, defective motor-lead bushings are replaced, nuts and bolts are put into working order, and missing cotter pins are replaced. In fact, the frame is put as nearly as possible into its original condition. The pole pieces are thoroughly scraped to give perfect seating, and the tapped holes are cleaned with oil.

35. Gear-Case and Gears.—Gear-cases may have cracked. bent, or have broken lugs, and if made of sheet metal the cases may be bent, from contact with raised paving stones or other obstructions. Bent cases should be straightened and broken lugs replaced. Defective felt dust guards around the axle hole should be repaired or replaced, and loose rivets replaced, or tightened, if necessary, care being taken not to bend the case. A new gear-case should be cleared of all internal burrs and any gear-case should be cleaned before installation. No dependence is to be placed on the old lubricant, as it may contain nuts or rivets that may be drawn up between gear and pinion and thus cause bending of the armature shaft. If the gear and pinion are in good condition, no further disturbance is necessary than to see that the nuts are tight and well secured. If, on account of broken or worn teeth, the gear must be replaced by a new gear, a new pinion should be installed and the replaced pinion reserved for use with a worn gear; the use of a worn pinion with a new gear shortens the life of the gear. A pinion that is to be used again should never be removed with a cold chisel, because the resultant flattening of the ends of the teeth is liable to bend the next armature shaft with which it is used. By the use of pinion pullers, which are usually available, a pinion may be forced from the shaft without injury.

The small clearance between a gear and its case should be equally distributed on both sides of the gear. As the gear is carried on the axle and the gear-case on the motor frame, the end play, due to axle-collar or bearing-flange wear, unless discovered and repaired, will cause the gear to rub and wear through the case.

When tightening a split gear, the bolts nearest the axle should be tightened first, so that the outside bolts may act as a lock. All bolts should be firmly tightened. In order that the motors may properly divide the load, all motors on the same car should have the same gear ratios and the gears and pinions should be properly mated, otherwise the wear on both will be excessive.

36. Pole Pieces and Field Coils.—Field coils should be examined for mechanical injuries and their resistance compared with that of a similar new coil known to be standard. Results will be misleading, however, unless the suspected coils are subjected to mechanical pressure similar to that due to the pole pieces or shields by which they are secured against vibration in the motor. The fastening devices, as well as the heat of the motor draw the turns of the coil together, and to obtain reliable results from resistance tests, these conditions must be reproduced. If any of the coils are abraded or are oil- or water-soaked, they should be sent to the coil room to be dried and recovered and new coils substituted. Before the old coils are replaced in the motor, they should be given, preferably by dipping, a coat of insulating paint that will take a gloss, the leads should be trimmed, tinned, and taped, if necessary, and the coils then set aside to dry. The coil insulation shrinks in service and some form of washer must be used to take up the clearance, for if the field coils are allowed to shake around in their fastenings, grounds or loose-connection troubles will result. If the coils are a loose fit on the pole pieces, they should be made tight by winding on dipped webbing; they should then be forced on the pole pieces with a rawhide mallet and a block of wood. The pole piece and coil are then placed in the motor and the bolts inserted and tightened, the rawhide mallet being used to tap the pole piece. One of the bolts is then removed and the clearance between the pole piece and its seat is tested with a piece of steel wire bent at right angles on the end and ground thin. If the knife edge cannot be forced in between the pole piece and its seat, a good seating may be assumed, in which case the bolt is replaced and screwed home; the other bolt is then taken out and the test is repeated.

37. A commutating pole piece should be so installed that the distances from the straight edges of the pole to the adjacent main pole pieces throughout their lengths should not vary more than $\frac{1}{16}$ inch. To test this adjustment make a hard-wood block of the proper thickness to bear against a main pole on one side and a commutating pole on the other.

38. The correct polarity of main and commutating poles should always be insured by test before reinstalling the armature or closing the motor frame. The main field coils alternate in polarity and so also do the commutating field coils. For a given polarity of main field coils and brush holders, the direction of rotation of the armature is best found by experiment.

In the case of a motor, the relative polarity of main and commutating poles is such that in the direction of rotation, an armature conductor must pass from a main pole of given polarity to a commutating pole of the same polarity and on to another main pole of opposite polarity. The absolute polarities

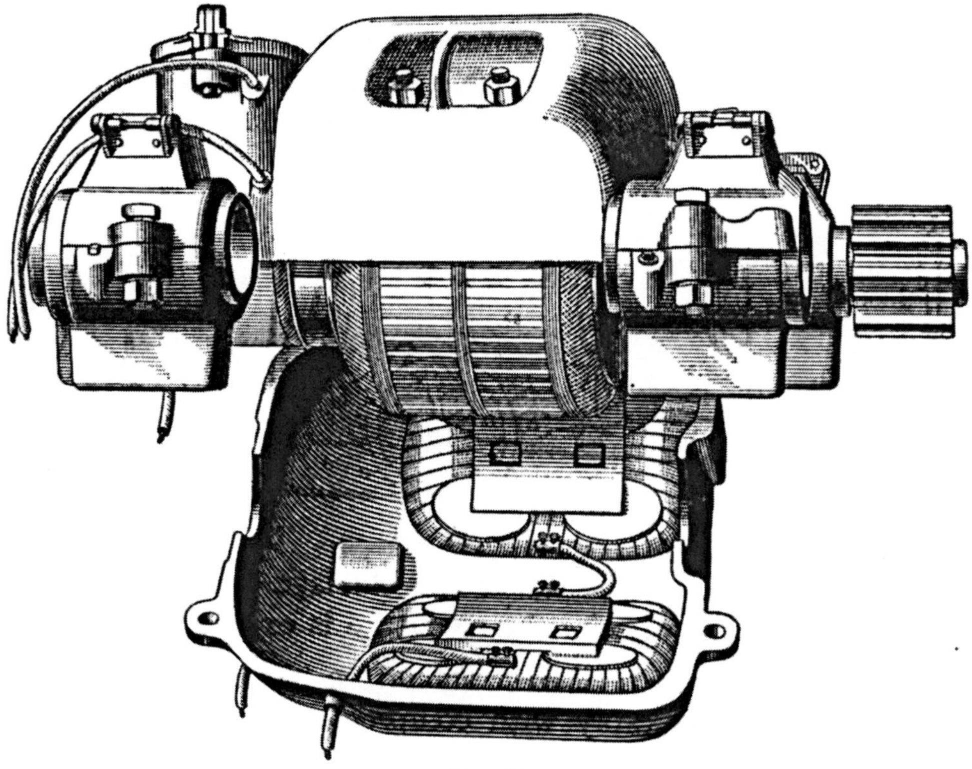

Fig. 20

of main and commutating poles may be tested with a compass, or their relative polarities with an iron nail. If a nail is loosely held between two poles of like polarity, it will turn to a position at right angles to a line between the poles. If the adjacent poles are of opposite polarity, the nail will then turn to a position parallel to the same line. Whenever the main and commutating poles are of the wrong polarity with regard to each other, the motor will spark badly and the commutating-coil terminals must be reversed.

39. Armature.—Some motors are provided with split frames, whose lower portions, with or without the armature, may be swung down, as indicated in Fig. 20. The arrangement permits an inspection of the interior of the frame without the removal of the motor from the truck. In other cases, the lower frame is entirely unbolted from the upper and is lowered into the car-barn pit by means of a portable hydraulic jack.

If the motor has a box frame, the armature is removed by unbolting one of the end brackets and sliding the armature through the opening. The motor is taken from the truck and placed on a stand provided with a sliding carriage *a*, Fig. 21; the armature is held between center *b* and *c*. The loosened

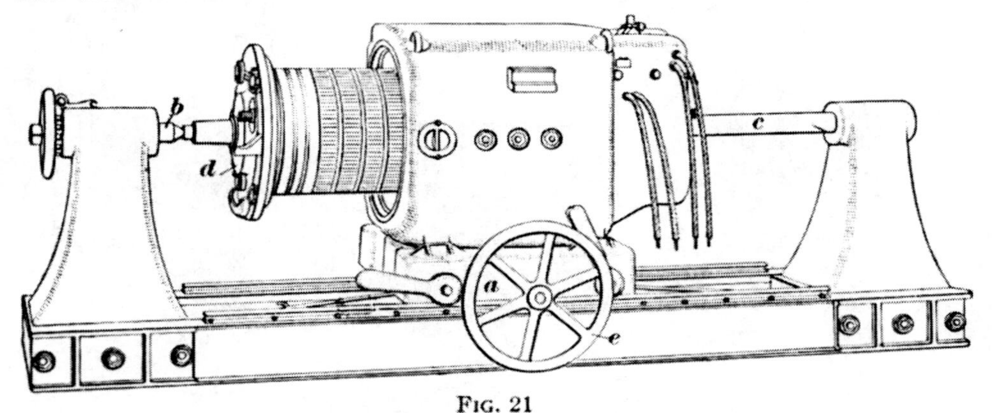

Fig. 21

bracket is shown at *d*. When the carriage is moved to the right by means of wheel *e* the armature is exposed.

40. Accumulations of dust should be blown from all ventilating ducts, preferably with compressed air. All moisture of condensation must be exhausted from the air hose before applying air to the armature. If practicable, the cleaning should be done under a hood piped to the atmosphere and exhausted with an air siphon, so that the dust ejected will not settle on everything in the vicinity. The armature should be swung in a lathe and tested for bent shaft; while there, the commutator may be turned, a new string band run on, and a new hood installed, if these repairs are required. The distance between the bearing surfaces must be gauged as a check against excessive end play, which should not exceed $\frac{3}{32}$ inch, even if it is necessary to put on new thrust collars.

Some shops straighten bent shafts, but as a rule, it is much better practice to press in new shafts, as a straightened shaft tends to resume its bent condition, and this condition may cause brush troubles and flash-overs. A record should be kept of any motor whose armature shaft has been straightened. When a shaft is renewed, the new shaft should be given the old one's number with a letter affixed and the old shaft should be scrapped, otherwise confusion in records will arise.

If the pinion nut thread is so badly damaged that it cannot be reclaimed, the shaft may be spliced with a piece upon which to cut a new thread, but the better way for all except small motors in very light service is to renew the shaft. The same course should be followed where it is necessary to cut more than two pinion keyways on account of the keyways having become hammered out.

If the armature connections show evidence of looseness, either the armature core or the commutator may be loose; if the core is loose, the chances are that it has worn its fastenings so that it should be reassembled. If the commutator is loose, the commutator bolts should be tightened while the commutator is hot, after which the commutator should be turned. All commutators should be rounded off at the end to minimize breaking out of the mica. After turning or grooving, all burrs reaching across the mica between bars must be picked out. Grooving is generally carried to a depth of about $\frac{3}{64}$ inch; after grooving, the edges of the bars must be smoothed with a scrapper to decrease the milling action on the brushes.

A better scheme than the common practice of laying extra armatures down on a rack, is to stand them up in holes bored for the purpose. Less floor space is required and the armatures are entirely covered with less trouble. Canvas bags slipped over them form a simple and effective protection. If oil gets on the commutator at any time, it should be thoroughly washed off and out of the grooves with gasoline. The flash-overs and breaker blowing caused by oil on a commutator will probably be attributed to almost everything else before the real cause of the trouble is located. Armatures should not be rolled on the floor, where they may come in contact with projecting nail heads, but should be rested on felt or similar substance, and should be carried from place to place, not rolled.

41. Brush Holders.—The type of brush holder that is mounted on studs fastened to the motor frame is free from many defects that characterized the old wood-insulated holders. Considerable attention, however, is required by the independent holders, to avoid wrong adjustments of holders and of tension springs, loose holders, loose brushes, loose leads, broken shunts, excessive armature end play, excessive accumulations of copper dust, etc. The last-named cause of trouble is minimized by the increased length of the leakage path, due to the corrugated insulators. Irrespective of the type of holder used, the brushes must be held the proper distance apart. On motors of the two-pole type, the set of the holders should be such that the distance from the center of contact of a brush to the center of contact of a brush in the opposite holder is one-half the circumference of the commutator; on a four-pole motor, the distance from center to center of brushes should be one-fourth the circumference of the commutator; and on a six-pole motor, one-sixth the commutator circumference. Instead of taking the stated fractions of the commutator circumference, one-half, one-fourth, or one-sixth of the total number of commutator bars are generally considered as the counts between brushes, and the bars between corresponding brush edges, forward or rear, are connected instead of between brush centers. A four-pole motor with a 215-bar commutator, for example, should have $215 \div 4 = 53\frac{3}{4}$ bars between the forward edges of brushes in adjacent holders. With the correct count between adjacent holders assumed, the location of the holders as a whole should be such that the armature speed is the same in both directions for a given load. The importance of having the count correct, as well as everything else pertaining to brushes and holders, cannot be emphasized too strongly. Most motor troubles can be traced, directly or indirectly, to brush or brush-holder irregularities.

42. Bearings.—Excessive wear or eccentric boring of bearings allows the armature core to strike the pole pieces;

if not repaired, this condition will soon destroy the armature. One of the first symptoms of core rubbing is the repeated blowing of the car circuit-breaker, due to the increased load imposed by the mechanical friction between the core and poles; the motor will heat, will use oil much faster than other similar motors, and will show a decided tendency to throw oil.

The air gap of a newly installed armature should be tested with a gauge to see that it is uniform all round. After two months of operation, wear is usually noticeable in the armature bearings, and they are usually changed at the time of overhauling, whether they really need it or not. In course of time, also, the armature shaft wears smaller so that standard bearings do not fit perfectly. In some shops several sizes of standard bearings, and armature shafts turned to these standard sizes are kept in stock, a shaft worn below the smallest standard size is either further turned and the bearing surface renewed with a sleeve or it is replaced with a new one. Other shops prefer to have a lathe at each place where overhauling is done, so that bearings kept in the rough may be bored to exact size.

43. Lubrication.—Good lubrication is absolutely essential to successful and economical operation, and oil has given better results at lower costs than did grease in older types of motors. In well-managed systems a careful record is kept of the oil consumed by each car, and if one car consumes much more oil than another, an investigation is made and the cause located and removed. The oil should be used economically; the usual way is to apply with a measure a specified quantity of oil to each bearing for a specified mileage. With careful lubrication, $\frac{1}{16}$ inch of good Babbitt will last 3 months or more; but the bearings are usually replaced oftener.

44. Armature and axle bearings should be as tight as possible without binding. An armature can usually be turned by hand, when a tendency to bind can be easily detected. A positive test for the fit of the axle-shaft bearings is to jack up the car wheels so that they can run freely, also the motor so as to remove friction from the car-journal bearings, and then operate the motor with an ammeter in series; the bearings should be as tight as may be without increasing the current above a value previously established by experiment.

On many roads maintenance of axle bearings is not given sufficient attention. Under ideal conditions, a rolling contact exists between pinion and gear in operation, and to get action of this sort the proper distance between gear and pinion centers must be accurately maintained. Wear in the axle bearings changes this distance and gear friction increases in a measure as bearing wear increases. One of the symptoms of such neglect is that pinion and gear-teeth do not wear evenly; the teeth wear thinner on one end than on the other.

45. Connections.—The leads should be brought out of all motors in standard order, and a standard order of connection to the car cable should be observed. Much time can thereby be saved in overhauling and in testing incident to trouble work. A wireman can then identify a positive brush or field lead at a glance, and no time will be lost in so connecting the motors that all tend to move their cars in the same direction for a given position of the reverse switches. Another advantage of standard connections is that a left-hand armature, as armatures of reversed polarity are generally called, can be detected as soon as it is installed in a motor because it will turn in the wrong direction.

46. Starting Resistance.—Sparking of brushes is sometimes due to the starting resistor being so sectionalized that current is applied to the motors in too great impulses. The resistor may have been wrongly divided in the first place, a part of it may have become short-circuited, or a parallel section of it may have become open-circuited. When overhauling the motors the starting resistor should be carefully inspected for loose, burnt-out, broken, or buckled units and for defective jumpers.

47. Circuit-Breakers.—Tests should also be made of the car circuit-breakers and of their adjustment to blow at a value consistent with the motor rating. Reliable circuit-breakers will save time and expense by revealing slight troubles that would soon become serious, if neglected.

ELECTRIC-CAR EQUIPMENT

MOTOR-CIRCUIT AND AUXILIARY APPARATUS

TRUNK CONNECTIONS

TRUNK WIRE

1. Trunk connections relate to the wires and devices traversed by the motor current from the current collector of the car to the controller, and are also sometimes considered as including the ground wires of the motors.

2. The **trunk wire** is the feed-wire from the collector; a portion of it carries the total current required for propulsion, heating, lighting, and signaling purposes.

Fig. 1 shows the general arrangement of the trunk wire for a small single-truck car. The current path from the trolley wheel to the controller trolley wire *T T* is indicated by the order of the numbers. The roof wires are protected by canvas tacked to the roof. Where the wire passes through the car bonnets, the passage is made water-tight by the use of lead or of special compounds made for the purpose. The vertical wires, or *risers*, pass through the bulkheads of closed cars or through grooves in the corner posts of open cars. The risers are often installed in iron conduits. In any case, the wires must not come in contact with nails, screws, or the sliding door. When the conduits are used bell mouths are employed at each end of the pipes to prevent abrasion of the wires and the upper end of the conduits are bent over and filled with insulation to prevent the entrance of water.

3. Fig. 2 shows an arrangement of the trunk wire for a car that operates on both third-rail and single-overhead trolley systems. The two trolley stands, located near the ends of the car, may or may not be connected; in either case a roof wire runs from each stand to one side of a single-pole, double-throw

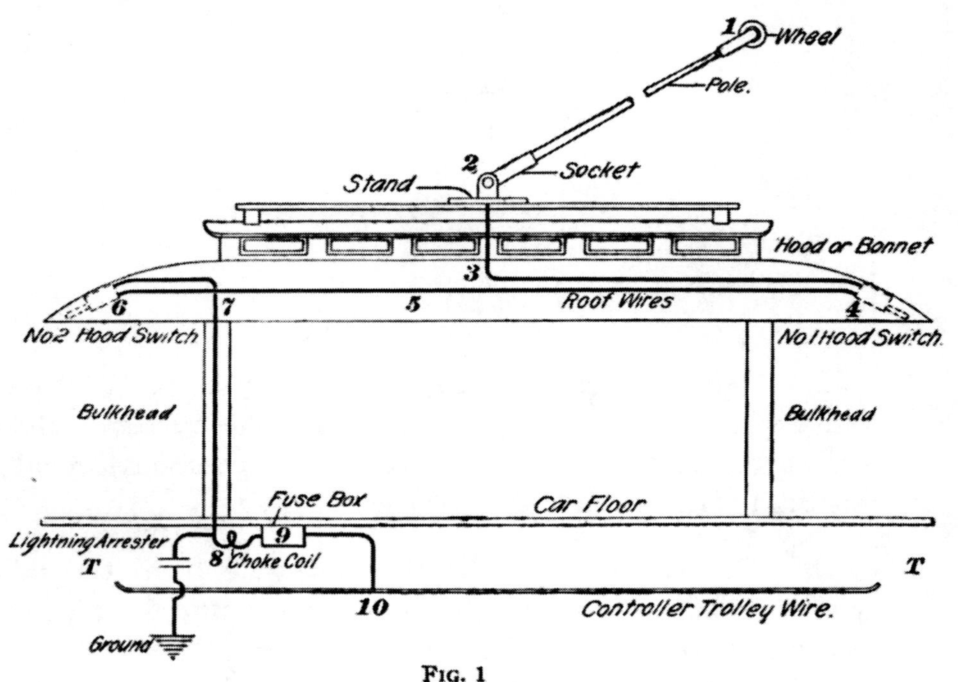

FIG. 1

switch, to the other side of which a riser from the third-rail trunk wiring connects. The change-over switches are located in the vestibules or cabs.

All of the contact shoes are connected by taps to a spine wire installed on the under side of the car floor. With this arrangement, the third rail may be placed on either side of the car and some of the shoes will be active.

The handle posts of the change-over switches are connected through circuit-breakers, located either in the vestibule or under the car, to the controller posts T. A circuit-breaker or enclosed fuse is sometimes connected to the spine wire at the point where it joins to the riser. In other cases, a shoe fuse is installed on the shoe beam above each contact shoe. By means of the change-over switch, the controller is connected to either the trolley stands or the contact shoes as desired and the collecting device not in service is entirely cut off from the live circuit.

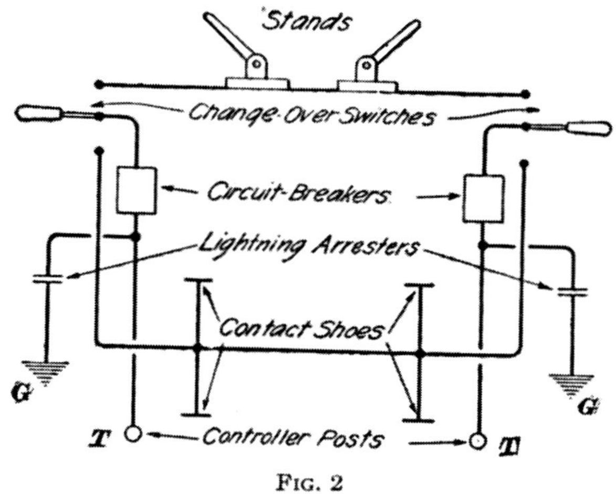

FIG. 2

4. Fig. 3 shows the arrangement of the trunk wire on a car operated on a conduit system. Besides the hood switches in the car bonnets two circuit-breakers, called *ground switches*, and two fuses (not shown) are installed under the car.

On conduit cars, extra facilities must be provided for the interruption of the motor circuit, because, while nominally operating on a metallic-return system, one side of the circuit is generally grounded. Circuit-breaking devices as close to the conductor rails as possible are installed to prevent damage to the car apparatus by accidental grounds.

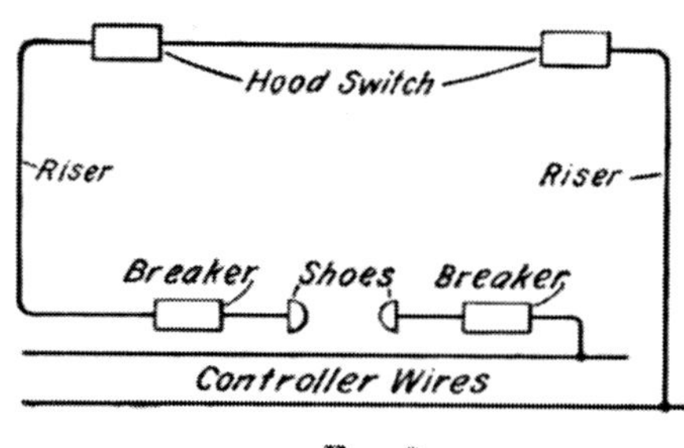

FIG. 3

5. Where motor cars are operated in trains that include intervening trail cars, the spine wires of the trail cars are joined with those of the motor cars by means of couplers; otherwise, when there is sleet on the third rail or on the trolley and different cars alternate in getting good contact, the resultant jerking will be liable to pull the train in two.

TROLLEY FITTINGS

6. The **trolley** is the device on the roof of the car for collecting current from the overhead conductor, called the *trolley wire*. The grooved *trolley wheel* rolls against the under side of the wire and makes electric contact with it; the wheel rotates on an axle in the *fork*, or *trolley harp*, riveted to the upper end of the trolley pole. The lower end of the pole is clamped to the upper portion of the *trolley stand*.

Short cars are usually equipped with only one trolley; some long cars have two trolleys either of which can be used. The

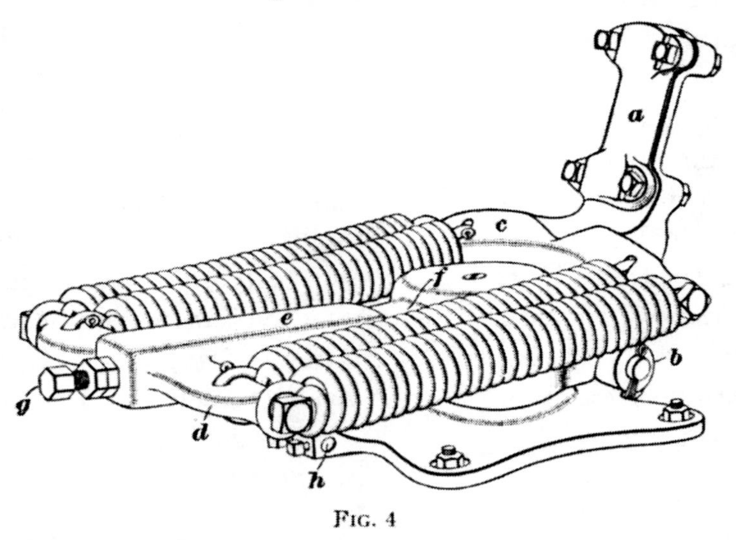

Fig. 4

wheel follows the trolley wire better when trailing over the rear end of the long car than when the trolley is placed at the center of the car.

7. The **trolley stand,** one type of which is shown in Fig. 4, consists of two parts, or members, the *spring base* carrying the socket clamp *a* for the trolley pole and the *trolley foot*, which is bolted to the car roof. The base is pivoted to the foot and can be swung around on a roller bearing to suit either direction of car movement. The socket *a* is also pivoted at *b* and has an arm *c*, to which four springs are connected. The other ends of these springs are joined to an arm *d*, which is part of a slide *e*. The slide is mounted on a projection *f* of the trolley base, and the tension of the springs, which force the trolley wheel against the wire, may be adjusted by the adjusting screw *g* at the end of the slide. The car-roof wire is connected to the trolley foot at *h*. On some stands a spring cushion stop is provided near the

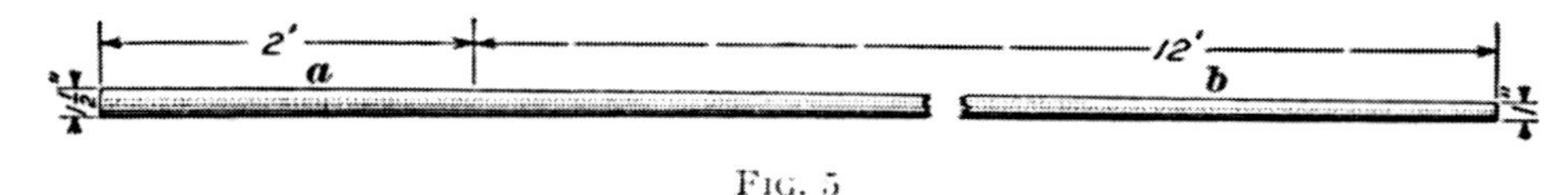

Fig. 5

lower part of the base to reduce the shock to the base and car, in case the trolley wheel leaves the wire.

8. The **trolley pole,** Fig. 5, is usually from 12 to 15 feet long, is about $1\frac{1}{2}$ inches in diameter for about 2 feet from the large end *a*, and then tapers through part *b* to a diameter of about 1 inch. Most poles are of hard-drawn steel tubing and offer great resistance to bending. Slight bends are generally straightened by using a post with a hole in it as a vise and then straightening the pole by hand pressure. Bad bends may be straightened by hammering, but in any case the pole should not be heated.

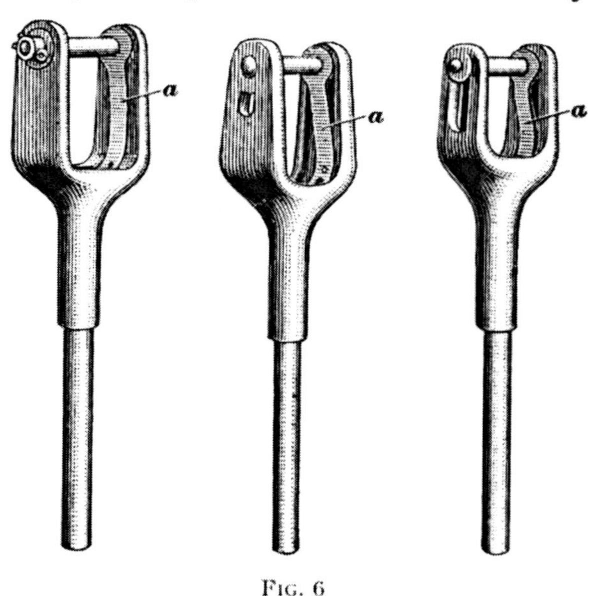

Fig. 6

9. Trolley harps, Fig. 6, are made of iron or brass and are shaped so they will not easily catch on the overhead work if the wheel leaves the wire. One end of the copper spring *a* is attached to the harp and the other end bears against the side of the wheel. The springs relieve the bearings by conducting a portion of the current from the wheel to the harp.

10. Trolley wheels, one of which is shown in Fig. 7, are made with great care to get the proper degree of hardness; if too soft they wear away too rapidly, and if too hard they injure the trolley wire. Much study has been devoted to the proper metal composition for the wheel and to the most suitable shape

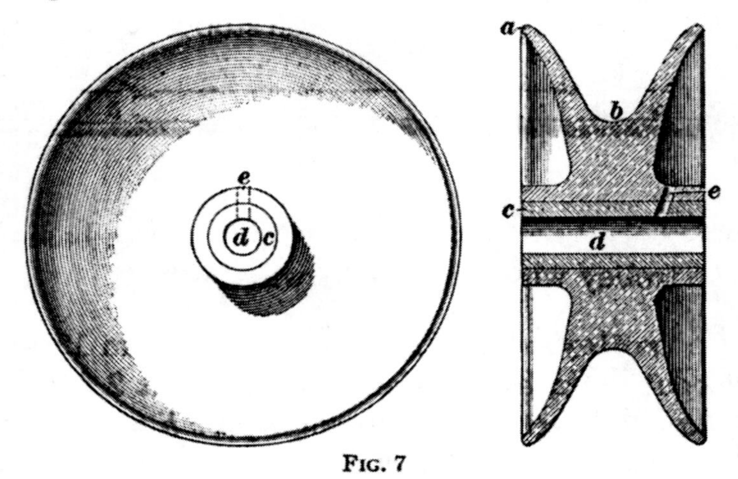

Fig. 7

for the flanges *a* and for the groove *b*. A bushing is provided at *c* and the axle, which passes through hole *d*, is oiled through hole *e*. Wear in the bushing causes the wheel to rattle. To

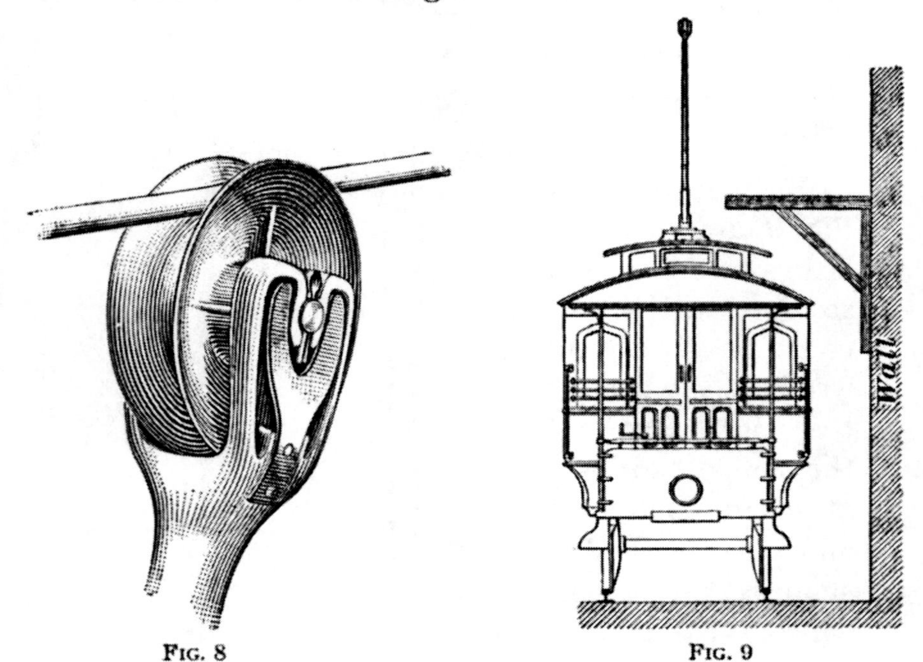

Fig. 8 Fig. 9

keep an ordinary bushing in good order, it should be oiled every 15 or 20 miles.

Fig. 8 shows an assembled harp and wheel with the wheel engaging the trolley wire. Fig. 9 shows a platform that may be used for the systematic inspection and oiling of wheels and other trolley parts.

11. The pressure of the wheel against the wire usually ranges from 20 to 40 pounds, depending on the tautness and alinement of the trolley wire, the speed of the car, and the condition of the roadbed. The compression required on the trolley-stand springs depends on the desired pressure of the wheel against the wire, the weight, fittings, and length of the pole. A spring balance attached to the wheel may be used to test the wheel pressure against the wire; the reading of the balance is taken when the pull exerted on the wheel through the balance is just sufficient to cause the wheel to leave the trolley wire. When the car is in operation, if the pressure is too weak, the wheel may leave the wire; and if too strong, the wear of the wire, the wheel, and the wheel bearings will be excessive. The wheel pressure is regulated by adjusting the tension of the trolley-stand springs.

Fig. 10

TROLLEY ACCESSORIES

12. Trolley catchers are devices intended to prevent damage to overhead wiring, in case the trolley wheel leaves the wire. The catcher is usually installed on the dash iron of the car and can be carried from one end of the car to the other as desired. A rope connects the upper end of the trolley pole to the catcher. Should the wheel leave the wire, the mechanism of the catcher will act and hold the pole after the wheel has risen a very short distance. The pole in this position is less liable to cause damage than if it were allowed to assume a vertical position.

13. A trolley retriever, Fig. 10, is a device somewhat similar to the trolley catcher, but in case the wheel leaves the trolley wire, the retriever pulls the pole down until the wheel is 3 or 4 feet below the wire and, therefore, clear of all interference with other overhead wires. A portion of the retriever

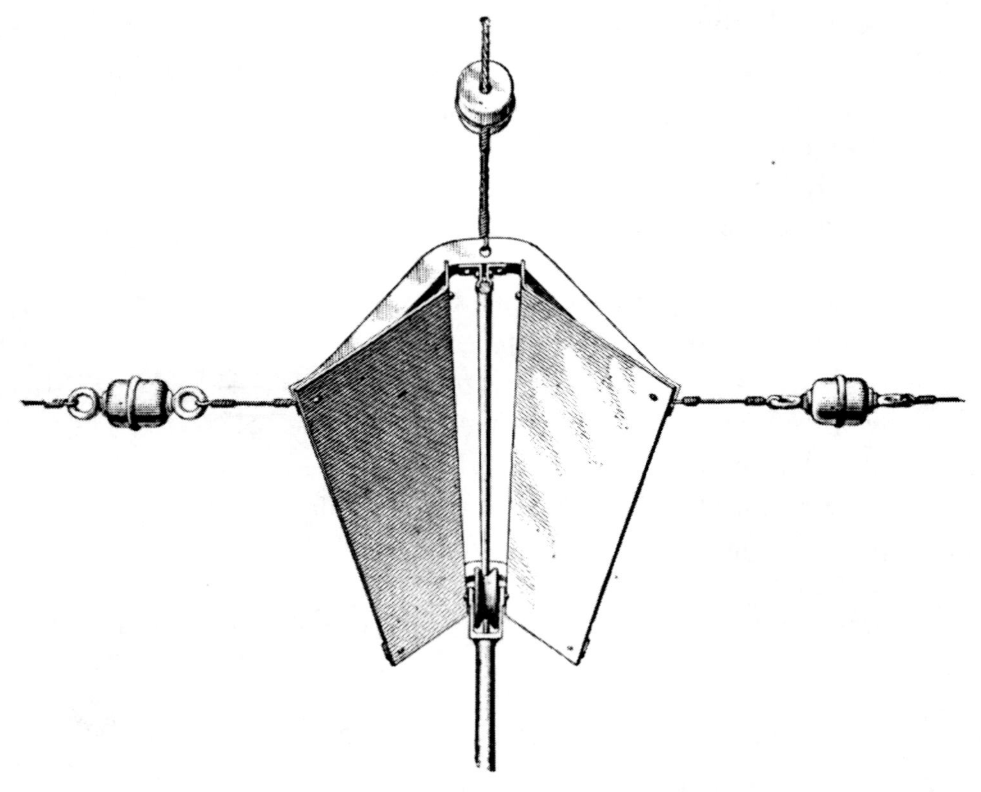

FIG. 11

mechanism allows the rope to wind on or unwind from a reel as the trolley wheel rises or falls while following the wire. When the wheel leaves the wire, the rope is wound on the reel and the pole is lowered by the action of a strong spring that is suddenly set in action within the retriever.

14. **Trolley guards,** Fig. 11, are sometimes installed on the trolley wires at the crossings of electric and steam roads, under bridges, and at the approach of tunnels. In case the trolley wheel leaves the wire, the guard will lead it back into its proper position on the wire and the car will not lose current in what may be a dangerous situation. The guard is held in place by special suspension wires.

CIRCUIT-BREAKERS

15. Broadly considered, a **circuit-breaker** is a device for readily opening and closing a circuit, either manually or automatically. As generally understood, a circuit-breaker, usually referred to simply as a *breaker*, is a switch that opens automatically when dangerous current conditions occur in the circuit of which it is a part. The broader definition includes both switches and automatic breakers and both will be discussed under this head.

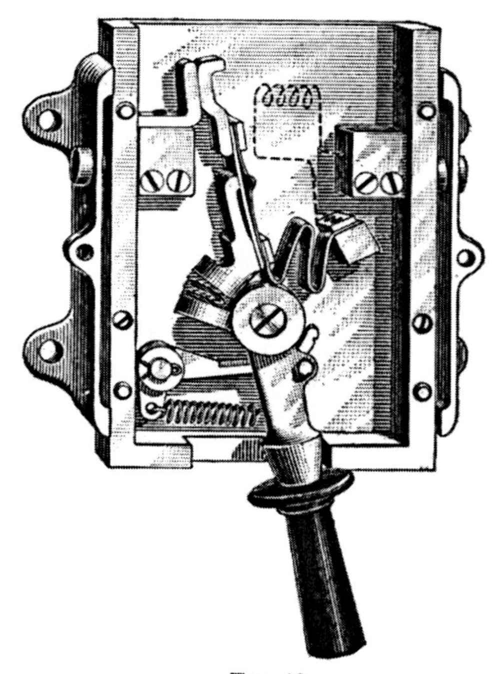

FIG. 12

The types of breakers used on cars to open the motor circuit depend on the nature of the operating system, to a certain extent on the size of the car, and on whether the control is manual or automatic. When operated by hand, the device is usually installed in the car hood, within easy reach of the motorman; hence the name, *hood switch*. The manually operated switches in the two car hoods are usually connected in series, as in Fig. 1.

Automatic breakers are usually placed in independent circuits, as shown in Fig. 2, to enable the motorman to close the circuit quickly after a breaker has opened. If the two breakers were in series and the rear one opened, the circuit could only be closed after considerable delay.

16. Breakers Tripped by Hand.—Fig. 12 indicates one part of a manually operated, quick-break main switch for an equipment not exceeding 200 horsepower in capacity. The part shown is for attachment to the roof over the motorman's head. A magnetic blow-out coil, indicated by dotted lines, and located in the lower portion of the box, is connected between the right-hand switch terminal and the block to which the switch arm is electrically joined.

Fig. 13 shows a type of quick-break switch used in motor circuits of multiple-unit cars to open the circuit leading to the car apparatus when inspections or tests are being made. The switch is intended to carry safely the full-load current, but not to be used for opening the circuit except, when in an emergency,

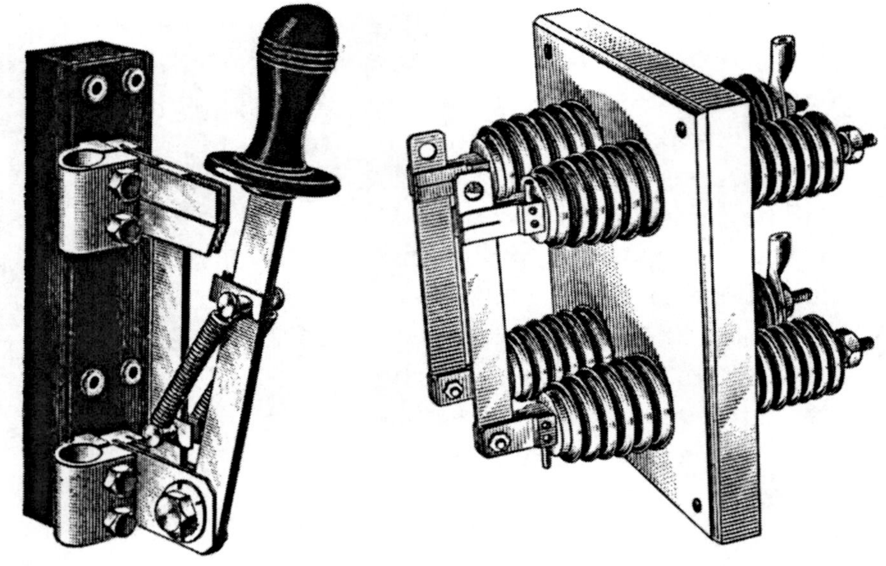

FIG. 13 FIG. 14

the automatic breaker in the same circuit fails to operate. When the switch is opened, the upper blade first leaves the switch jaws and puts in tension the two springs connecting the upper and lower blades. When the tension becomes sufficiently strong, the lower blade leaves the switch jaws very suddenly and opens the circuit.

Fig. 14 shows a disconnecting switch used for disconnecting high-tension car wiring during inspections and tests. This switch is not intended for opening loaded circuits. It is operated by means of wooden poles provided with hooks that may be inserted in the holes in the switch blades.

17. Breakers Automatically Operated.—When a loaded high-tension car circuit is to be opened, an oil switch is often used. Fig. 15 shows a type of oil switch used on a multiple-unit car operating on a single-phase, alternating-current system. A magnet *a*, when energized, operates through a system of levers to close switch contacts *b* installed in a case

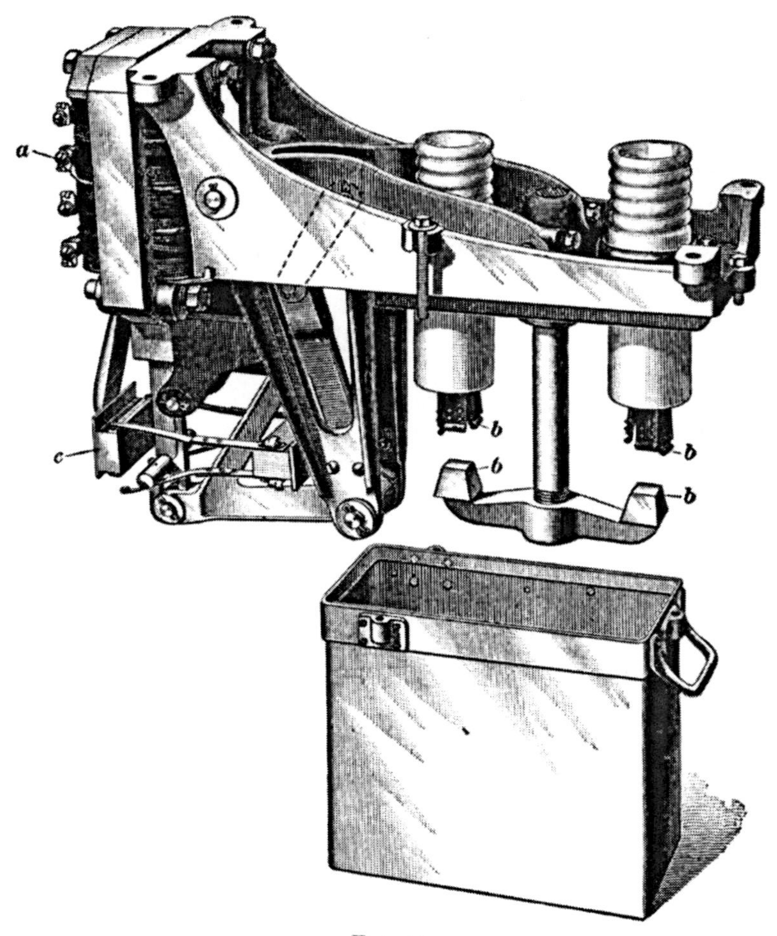

Fig. 15

containing insulating oil. The lower contacts *b* are mounted on a crosshead supported on a wooden rod. If current through the operating coil ceases, the switch opens by gravity.

In the open position of the switch, as shown, a pin on an extension of the armature depresses two spring contacts away from the contact piece *c* and in the closed position the same pin raises two other springs from the same contact piece. The springs and contact piece constitute what is known as *interlock switches*, the purpose of which is to make the closing or opening of certain control circuits dependent on the position of some main switch. They are used in some methods of speed control of multiple-unit trains. This switch may be arranged to be operated by either closing or opening the circuit of the operating coil *a* by means of a hand-operated switch, or automatically by means of a relay that may be installed in some control circuit.

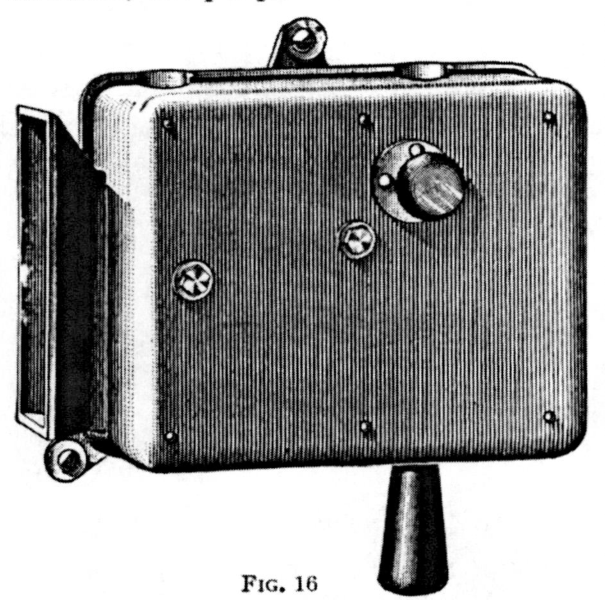

Fig. 16

The insulating oil must be non-explosive and non-freezable in the coldest weather encountered.

18. Figs. 16 and 17 show a circuit-breaker much used in cars taking not over 200 amperes. This breaker opens automatically on overload and can be opened manually by pressing the button shown outside the case in Fig. 16. It is closed by moving the handle, shown projecting below the case, to the right. The terminals *h* and *i*, Fig. 17, serve to connect the circuit, to the magnet coil *v*. If the current in the

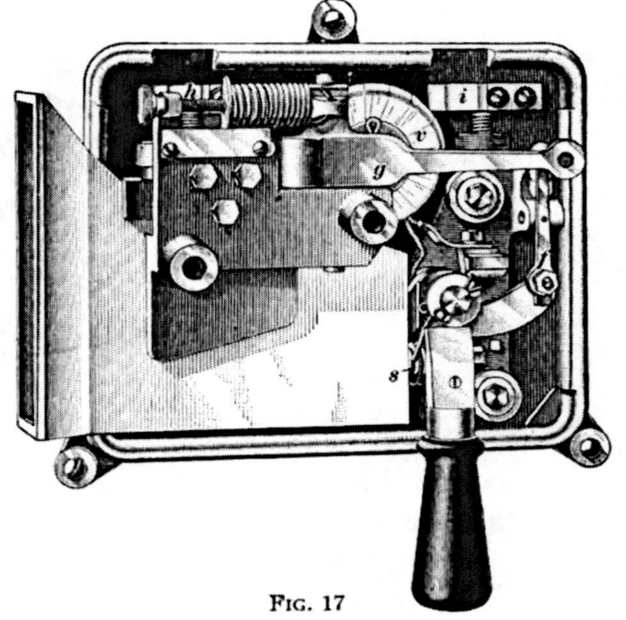

Fig. 17

circuit exceeds the value for which the breaker is adjusted, or if the button outside the case is depressed, the armature *g* is brought against the releasing lever *f*; the switch contacts located in the blow-out flue shown at the left, then snap open. The compression spring *s* which is located between the top part of the handle and the switch arm, serves to hold the contacts firmly together when the switch is closed. When a loaded circuit is opened by this breaker, hot gases are expelled through the blow-out flue.

FUSES

19. In some installations, a **fuse** is used in conjunction with a breaker as a further element of safety, should the breaker fail to operate.

Fig. 18 shows an **enclosed fuse** and its enclosing asbestos-lined iron box. The fuse is connected in circuit by means of the

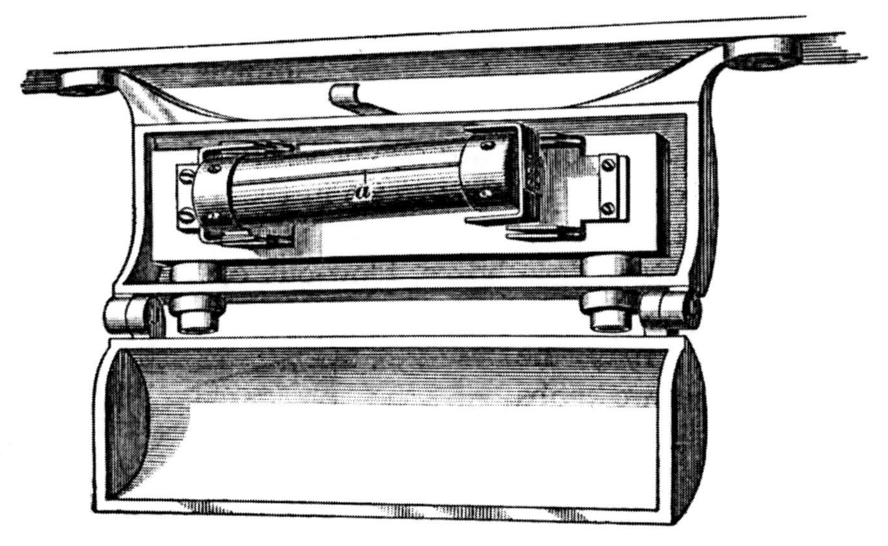

Fig. 18

switch clips shown near the ends of the fuse cylinder. When the fuse has blown, the *telltale* wire *a* also burns out and indicates that the main fuse has operated. A new fuse can be easily placed in the clips.

20. Fig. 19 shows the interior of a type of **magnetic blow-out fuse box**, the hinged cover being dropped. The circuit wires enter the box through holes in the cover and are fastened to the fuse terminal blocks by the nuts shown. The fuse consists of a copper ribbon stretched between the blocks. The cross-section of the fuse is decreased at the center by a hole punched in the fuse. The central part of the fuse thus has the least current-carrying capacity, and the fuse will melt at that point. The arc takes place some distance from either terminal, and the terminals are less liable to damage than if the fuse melted at a point near one of them. In the fuse box shown, the current in the fuse sets up in the iron plates on the box and cover the magnetic flux that helps to extinguish the arc. In high-tension fuse boxes, blow-out coils are provided. A new fuse is placed in the clamps, which may be tightened by the two knobs shown at the sides of the box.

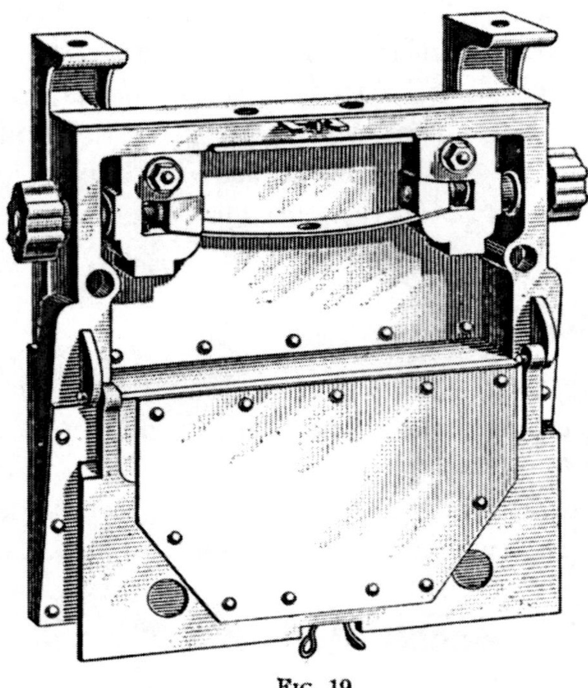

Fig. 19

21. Fig. 20 shows a type of **expulsion fuse** sometimes used in high-tension car circuits. The fuse is inserted in a tube made of insulating material. Terminal pieces on the tubes fit in switch clips mounted on porcelain insulators. When a fuse blows, the sudden accumulation of gas in the chamber at the

Fig. 20

base of the fuse expels the metal vapor and blows out the arc. The tube may be removed from the clips and a new fuse wire inserted.

MOTOR RHEOSTATS

22. Motor rheostats consist of resistors with taps leading to the controller. The resistance in the motor circuit can

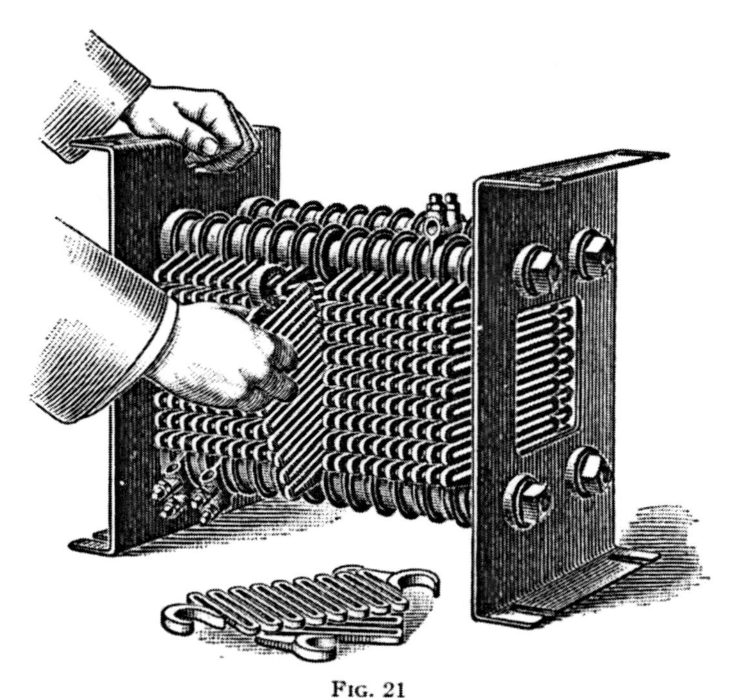

Fig. 21

thereby be cut out in steps as the car speed increases and the motors thus allowed to build up a counter electromotive force.

Fig. 21 shows a type of rheostat made of cast-iron grids in two rows, each row supported by two insulated rods. The rods pass through **U**-shaped openings in the grid lugs, so that any grid can be removed and replaced as indicated. The adjacent terminals of two grids are in contact with each other

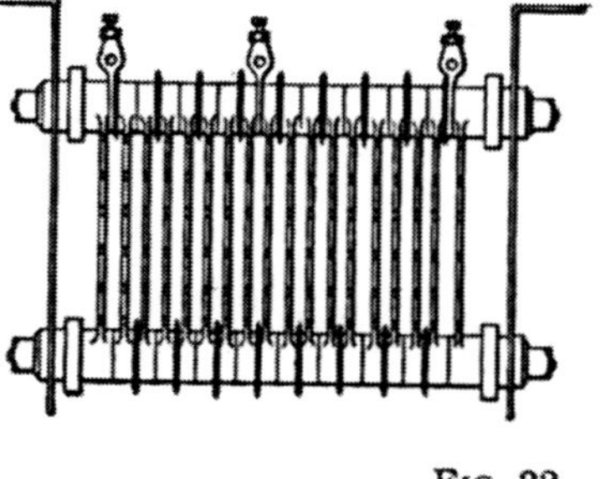

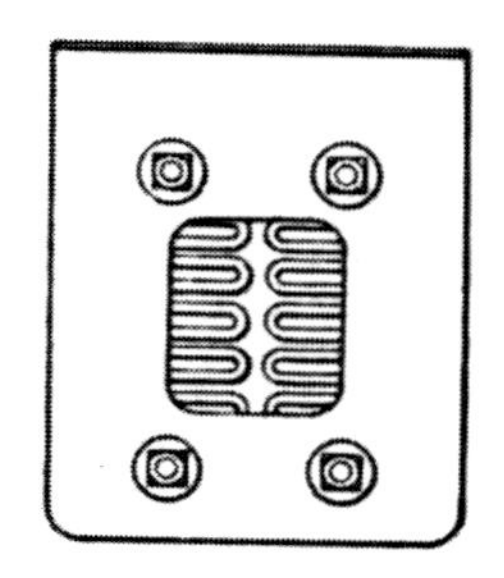

Fig. 22

on one rod and insulated from each other on the other, thus causing the current to take a zigzag path through the rheostat.

Fig. 22 indicates the method of mounting the grids on one side of the rheostat. The rheostats are usually mounted under the car in a position where water will not readily reach them.

LIGHTNING ARRESTERS

23. Carborundum-Block Arrester.—Fig. 23 shows a type of carborundum-block arrester. The spark gap is formed between a terminal plate shown at the right and the carborundum block. The other terminal of the arrester is a plate connected directly to the left side of the block. The discharge current due to lightning jumps the gap and spreads itself over the carborundum block along a number of minute discharge paths. The nominal voltage of the line is not sufficiently high to maintain arcs across these small air gaps after the lightning discharge has ceased. The arrester then assumes its normal condition.

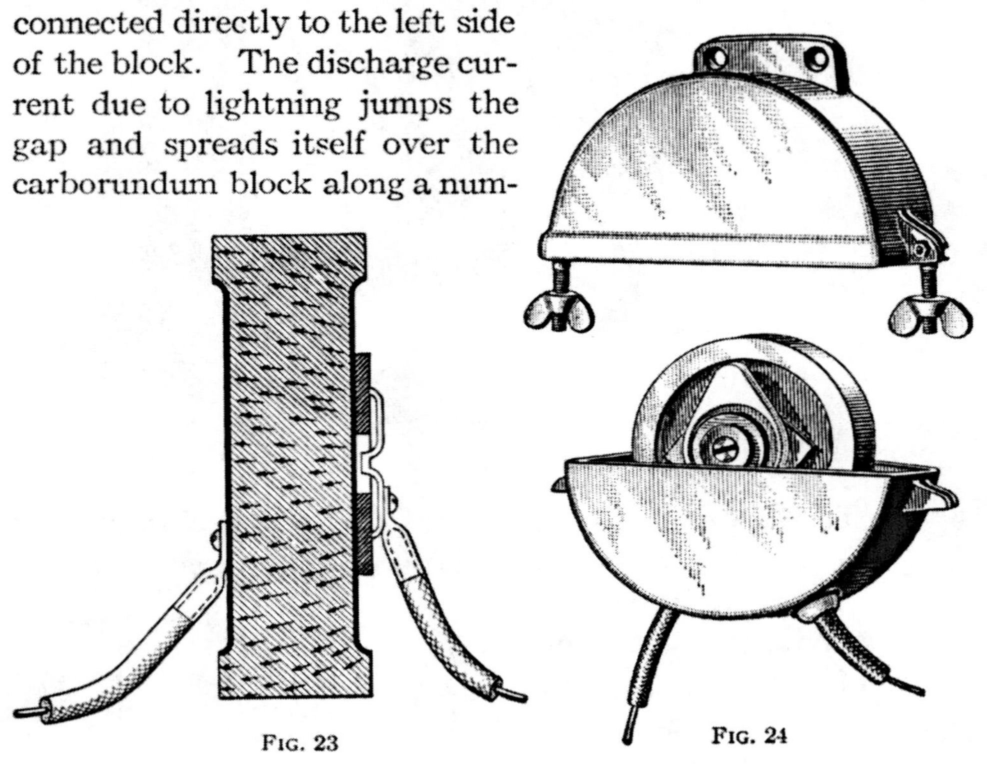

Fig. 23

Fig. 24

Fig. 24 illustrates the arrangement of the parts of the carborundum-block arrester and the case in which it is enclosed.

24. Magnetic Blowout Arrester.—Fig. 25 shows an arrester in which the arc due to the trolley current following the lightning discharge across the air gap is extinguished by means of a magnetic blowout coil. The terminal pieces for the spark gaps *a* are mounted on the cover and when in position lie between the pole pieces *b* of the magnet *c*. The magnet coil is in parallel with a portion of the resistor *d*.

Fig. 26 shows the connections of the magnetic blowout arrester. The action of the choke coil in the motor circuit

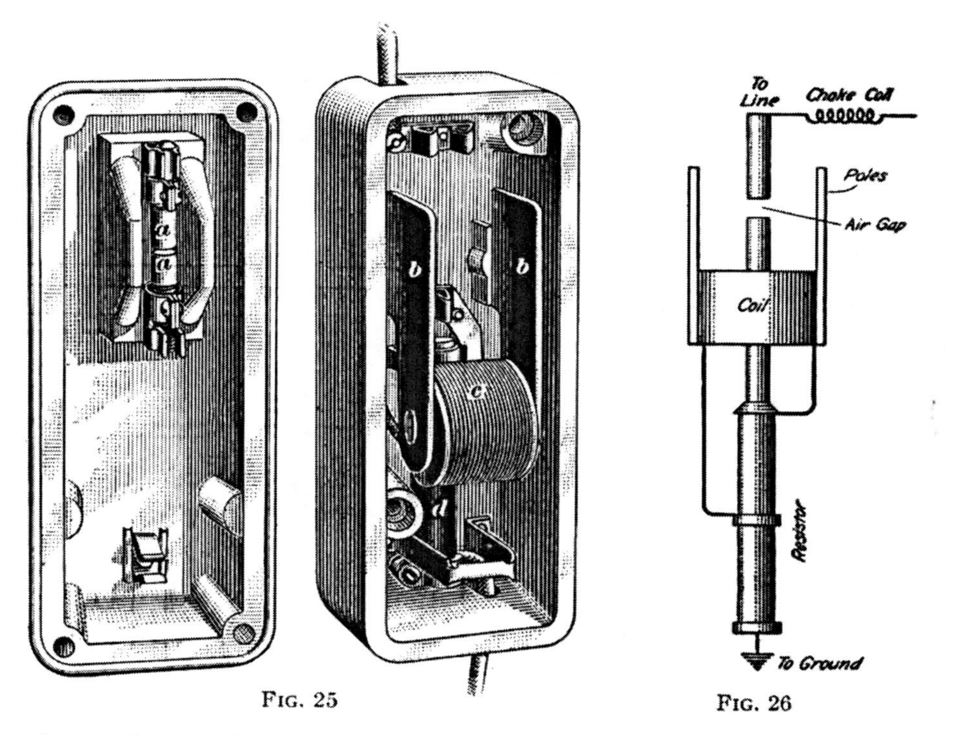

Fig. 25

Fig. 26

tends to divert the current of the lightning discharge across the air gap and directly through the whole length of the resistor to a ground connection on the motor frame. When the trolley current follows through this discharge path, some of it energizes the blowout coil and extinguishes the arc at the air gap. The arrester is then in readiness for another lightning discharge.

25. Arresters Opening the Ground Circuit.—Fig. 27 shows a type of arrester in which the arc is disrupted by opening the ground circuit. The line wire is shown at *a* and the ground wire at *b*. The coil *c* is connected in parallel with the lower portion of the resistor *d*. At the upper end of *d* is a fixed air gap. A plunger armature in coil *c* rests on a fixed grounded contact in an insulating tube shown below the coil *c*. The current of the lightning discharge passes across the fixed air gap and through the resistor and plunger to ground. When the trolley current follows, some of it energizes the coil *c*, thus lifting the plunger and opening the circuit at the lower fixed contact in the tube. The arc is extinguished, the

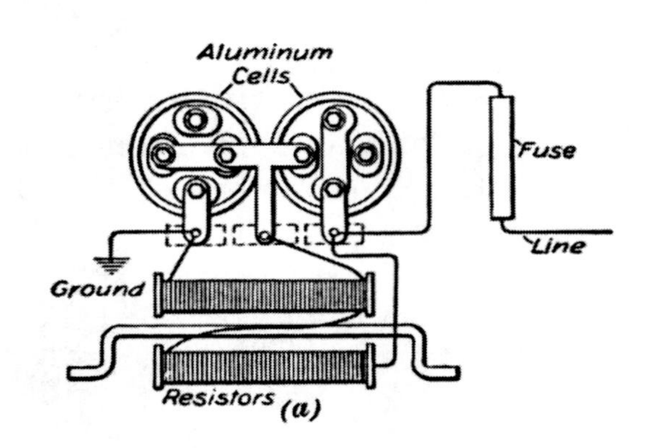

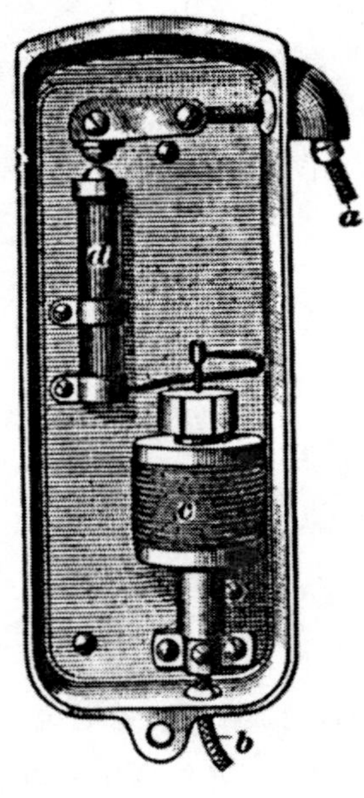

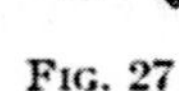

Fig. 27

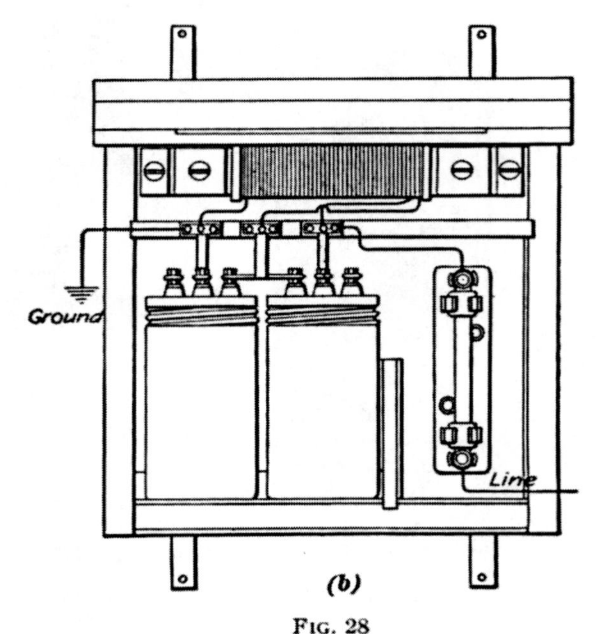

Fig. 28

plunger falls back in position, and the arrester is ready for further service.

26. Aluminum Electrolytic Arresters.—When aluminum is treated in certain electrolytes, a film forms on its surface. This film offers very high resistance to the passage of low-voltage electricity between the aluminum and the electrolyte and very low resistance to high-voltage electricity, especially if it is alternating at high frequency, as is the case with lightning discharges. By nesting together a series of cup-shaped aluminum trays with intervening spaces, and filling each tray with electrolyte, a very effective lightning arrester is formed. Each assembly of this kind forms a cell, or unit.

Fig. 28 gives end (*a*) and side views (*b*) of two cells connected in series with a fuse between a trolley line and ground. Each cell bridges a gap and each gap is also bridged by a resistor. A lightning discharge breaks down the resistance of the film and the electricity flows readily to ground; but as soon as it has passed, the film reforms and prevents the escape of electricity at the normal voltage, except a small leakage current, which serves the useful purpose of keeping the film in proper condition. If the cells are to be left idle for a considerable time, the electrolyte should be withdrawn. When the cells are again placed in service, the electrolyte should be put in and the film reformed by connecting the cells in series with five 32-candlepower, 120-volt incandescent lamps across a 600-volt circuit. The lamps will usually brighten for an instant and then darken, as the film is formed. Inspection should be made at regular intervals.

LIGHTING, HEATING, AND AUXILIARY APPARATUS

ELECTRIC-CAR LIGHTING

27. Lighting-Circuit Diagrams.—The lighting circuits on a car include the lamps for interior use, the platform lamps, the headlights, the signal lamps, and switching arrangements for cutting one or more of these devices into or out of service, as desired. The signal lamps should be provided with a source of energy independent of the trolley system to insure their operation when needed.

28. Fig. 29 shows a very simple lighting-circuit diagram, including interior lamps *a*, platform lamps *b*, headlights *c*, and a single-pole, double-throw, knife-blade switch *d*. With the switch in the upper position, the headlight at the left end and the platform light at the right end are connected in series with the interior lamps across the line. With the switch in the lower position, the other platform lamp and headlight are active.

29. Fig. 30 shows a car-lighting diagram for a large side-entrance car. The circuits are numbered from *1* to *5*, inclusive Each circuit is provided with a fuse and all circuits are controlled by a main switch *a*. The headlights which are in circuit *5* are controlled by a single-pole, double-throw switch *b*. Circuits *3* and *4* include only interior lamps. Each of circuits *1* and *2* includes four interior lamps and one sign lamp *1 c* or *2 c*. Circuit *5* includes two interior lamps, two sign lamps *5 c*, switch *b*, and one of the two headlights *5 d*, depending on which way switch *b* is closed.

Each of the sign lamps is in series with lamps of an interior circuit; therefore, any derangement of a sign circuit is indicated by the lamps within the car going out.

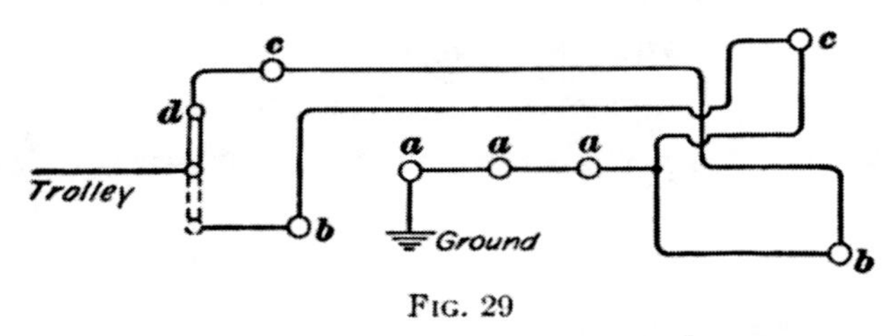

Fig. 29

30. Fig. 31 shows a wiring diagram for signal lanterns that may be operated from either the trolley circuit or from a storage battery. There are two trolley signal lamps and two battery signal lamps at each end of the car. The main switch is shown at *a*; one double-pole, double-throw switch at *b*, and one at *c*; a storage battery at *d*, and a relay at *e*. When the trolley lamps are in series, this relay forms a connection between terminals *1* and *2*, and when the battery lamps are in service, between *1* and *3*, the ground connection formed through *1* and *2* then being cut out.

Switches *b* and *c* serve to cut in or out of circuit the signal lamps on their respective ends of the car. While the trolley circuit is active and the relay connection remains between *1* and *2*, these switches control the trolley signal lamps only; if the trolley current fails, the relay connection is automatically transferred to points *1* and *3*, and switches *b* and *c* then serve to control the battery signal lamps. With the switches *b* and *c*

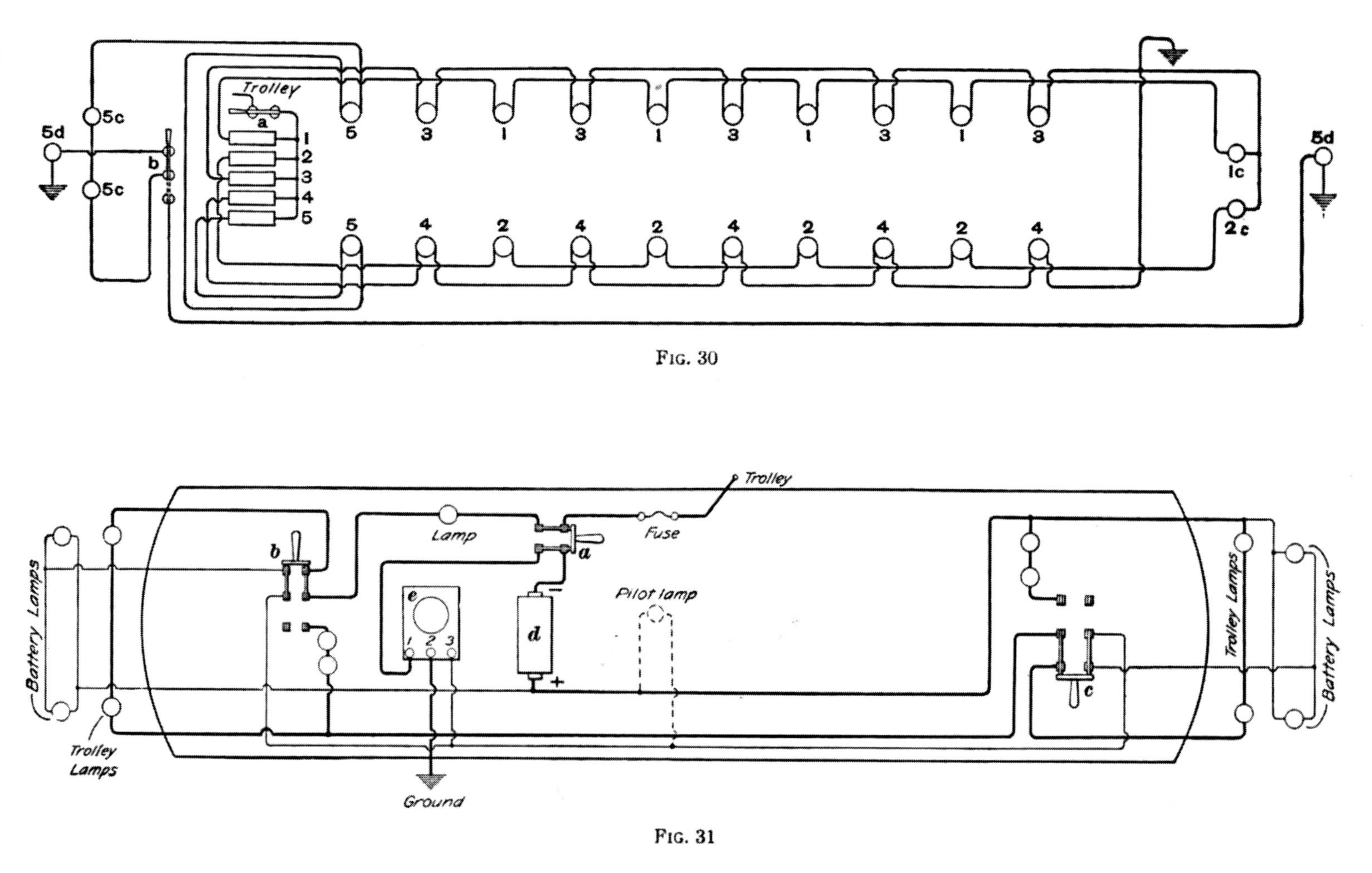

Trolley
a
b
5d
5c
5c
1
2
3
4
5
1c
2c
5d
Battery Lamps
Trolley Lamps
Lamp
Fuse
Trolley
Pilot lamp
Ground
e
d
c

Fig. 30

Fig. 31

in the positions shown and with the trolley current off, the battery signal lamps at both ends are active. The pilot lamp is also active, showing that the battery is discharging. When the trolley circuit again becomes active, the relay connects the trolley lamps in circuit and cuts out the battery lamps.

The battery is charging when the trolley lamps are active, as there are five lamps and the storage battery in series across the circuit.

31. Fig. 32 shows a signal lantern and the relative positions of the trolley and battery signal lamps. The lens may be of

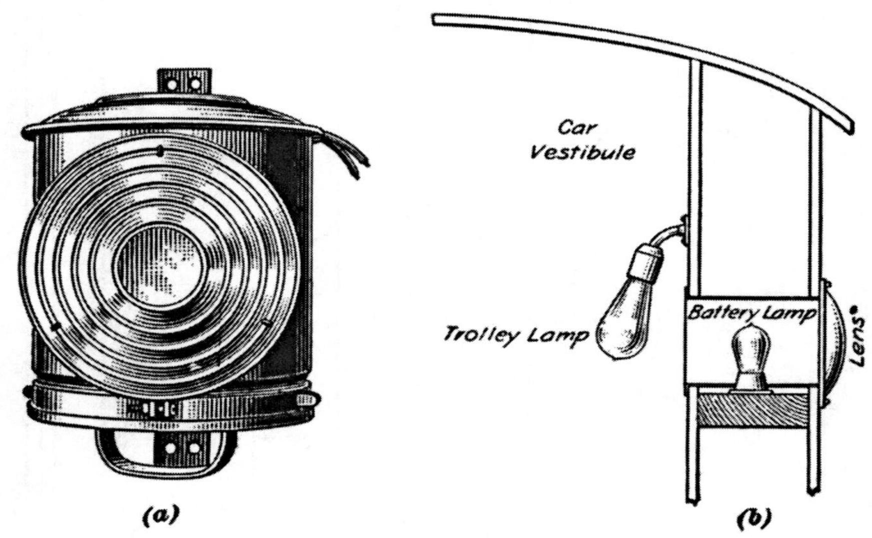

Fig. 32

the desired color to act as a danger signal or to indicate the classification of the car. Usually one lantern of a pair is used as a *tail-light* and the other as a *marker* to show the kind of service or the destination.

HEADLIGHTS

32. The illumination of headlights for electric cars or trains is usually provided by an incandescent lamp, an arc lamp, or both. A high-speed interurban train requires a considerable distance in which to stop after detecting an obstruction on the track; therefore, a headlight containing an arc lamp that will project a strong light for 1,000 feet or more, is required. Such a light is undesirable for city work because of its blinding effects on drivers of other vehicles and pedestrians. An incandescent lamp provided with a reflector is commonly employed in headlights for city service. Cars operating in both city and interurban service sometimes have a combination headlight in which are installed an incandescent and an arc lamp. Occasionally separate devices are used, the incandescent headlights are mounted permanently on both ends of the car and the arc headlight is of the portable type. The headlight is usually carried on the front of the dash of cars operating singly and on the top of the bonnet of cars used in train service.

33. Incandescent Headlights. Incandescent headlights may be of the portable or permanent type, and set within a hole in the dash, mounted on a bracket in front of the dash, or installed on the car bonnet.

Fig. 33 shows a type of portable incandescent dash light consisting of an incandescent lamp, a reflector, and an enclosing case. The case has an extension contact piece that may be inserted in a receptacle S attached to each end of the car. A cover C protects the interior of the socket when the dash light is not in service.

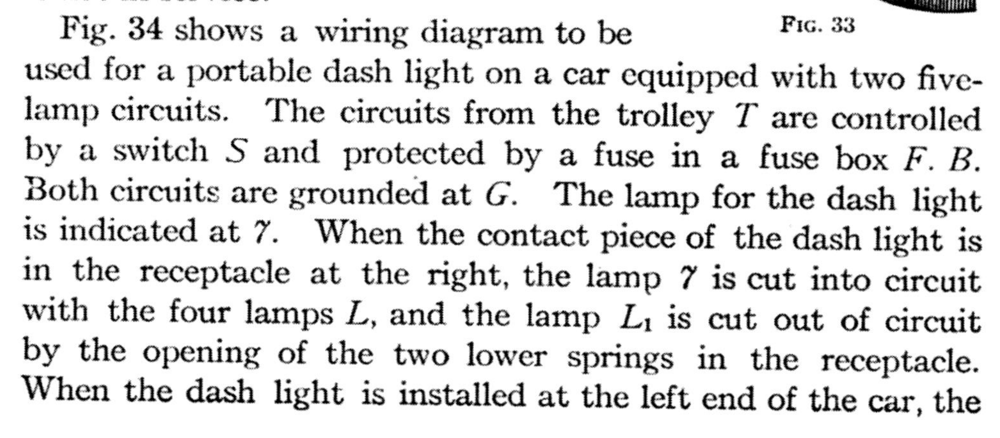

Fig. 33

Fig. 34 shows a wiring diagram to be used for a portable dash light on a car equipped with two five-lamp circuits. The circuits from the trolley T are controlled by a switch S and protected by a fuse in a fuse box $F. B.$ Both circuits are grounded at G. The lamp for the dash light is indicated at 7. When the contact piece of the dash light is in the receptacle at the right, the lamp 7 is cut into circuit with the four lamps L, and the lamp L_1 is cut out of circuit by the opening of the two lower springs in the receptacle. When the dash light is installed at the left end of the car, the lamp L_2, which, as shown is in circuit, is cut out and the dash light substituted.

34. Arc Headlights.—Fig. 35 is a general, and Fig. 36

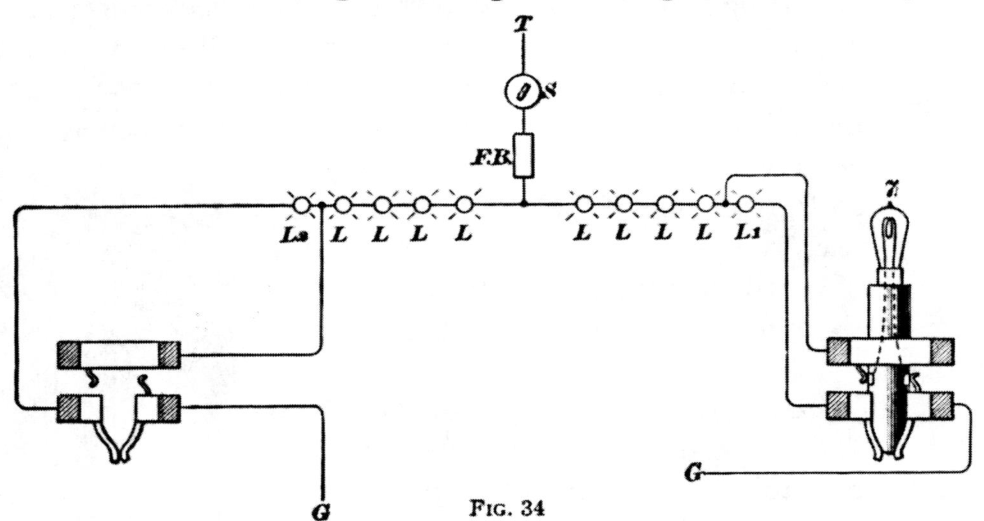

Fig. 34

a sectional, view of a combination incandescent and arc head-light. An incandescent lamp is attached inside the door and

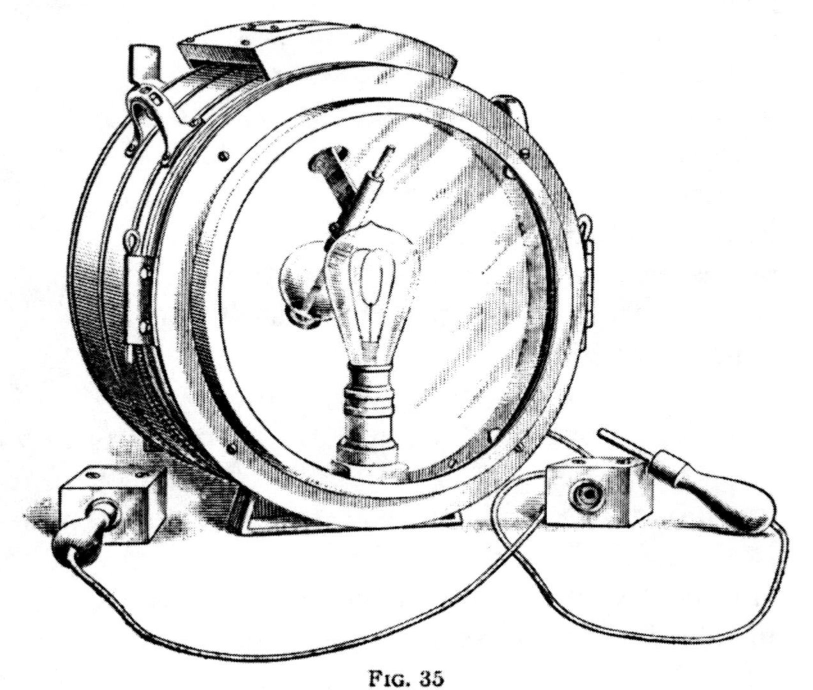

Fig. 35

may be used to provide light for city work. When the arc lamp is used, the incandescent lamp is cut out of circuit, but the lamp itself remains in the front of the headlight. The arc-light carbons are inclined at an angle of 45° with the perpen-

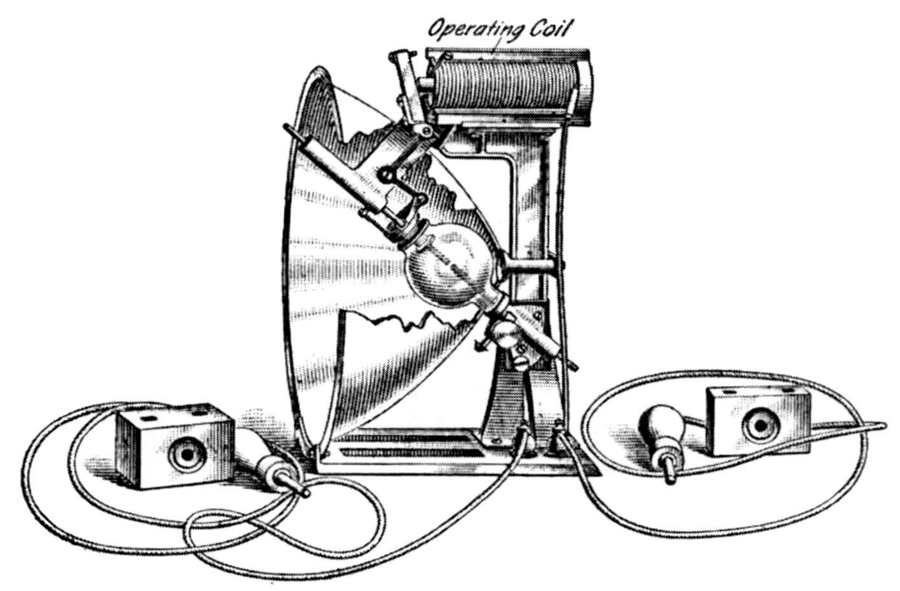

Fig. 36

dicular, to present a large area of crater to the center of the reflector. The initial drawing of the arc or its reestablishment and the feeding of the carbon are automatically controlled by a coil in series with the arc.

Fig. 37 shows a wiring diagram for a headlight of this type.

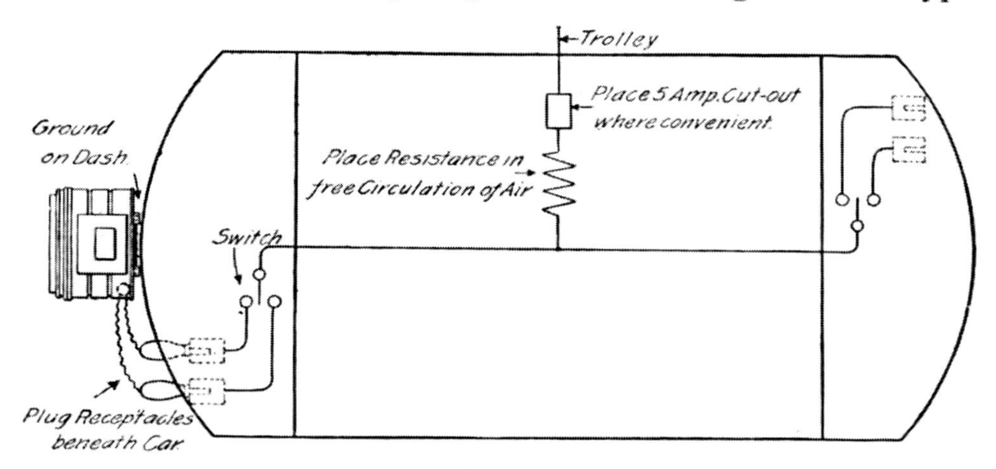

Fig. 37

The resistance used in series with the arc light is such as to admit a current of $3\frac{1}{2}$ amperes at a line pressure of 550 volts, the drop across the arc then being about 75 volts. Both the incandescent attachment, supported on the door, and the arc-lamp circuits have plugs and receptacles of their own, and in order to obviate the necessity of pulling the plug of the lamp circuit that is not wanted, both receptacles are connected to a two-point switch, the position of which determines which circuit shall be supplied with current. When the headlight is changed from one end of the car to the other, both plugs must be removed from the receptacles.

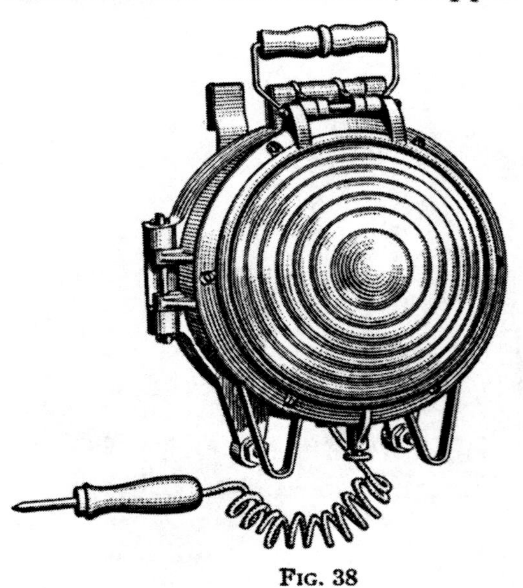

Fig. 38

35. Fig. 38 shows the exterior of a portable luminous arc headlight used in suburban and city service; Fig. 39 illustrates the interior and Fig. 40 the electrodes. The operating mechanism and the electrodes are placed in separate chambers in order to protect the coils from the arc gases. The upper or positive electrode is of copper and will last from 1,000 to 2,000 hours; the lower or negative electrode is an iron tube filled with a mixture of magnetite, chromium, and titanium and it will last from 150 to 175 hours.

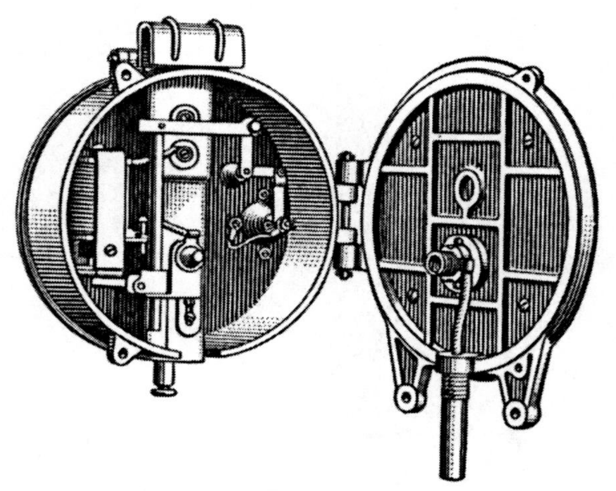

Fig. 39

The arc takes a current of 2 amperes at about 90 volts. The negative side of this headlight is grounded to the casing; therefore, only one plug and socket is required to connect the headlight in series with a resistor across the circuit.

36. For high-speed interurban service a luminous arc headlight taking a current of 4 amperes may be used. In order to dim the light when desired, an incandescent lamp placed in the case may be cut into circuit and the arc lamp cut out; or the current through the arc lamp may be reversed in direction and caused to pass through an additional resistor. With the copper electrode acting as a negative, the light will be lessened.

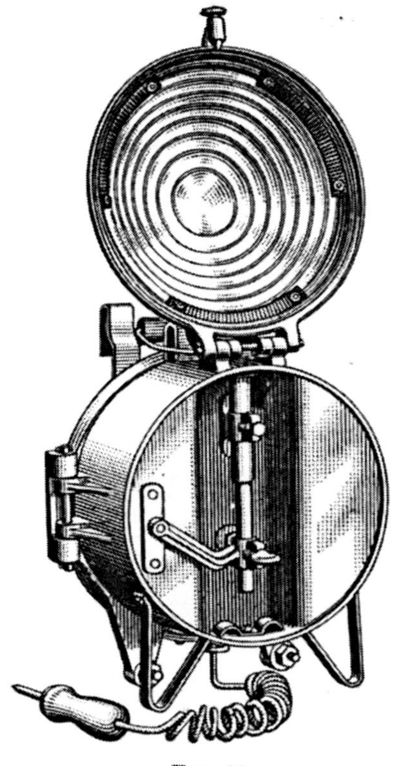

Fig. 40

Fig. 41 shows a wiring diagram for a 4-ampere, two-terminal headlight, arranged for reversing the current through the arc lamp by means of the double-pole, double-throw switches. The extra resistance in circuit when the copper is used as a negative electrode decreases the operating current to about $2\frac{1}{2}$ amperes.

ELECTRIC CAR HEATING

37. Electric heaters are extensively used for electric car heating because of their convenience and cleanliness. An

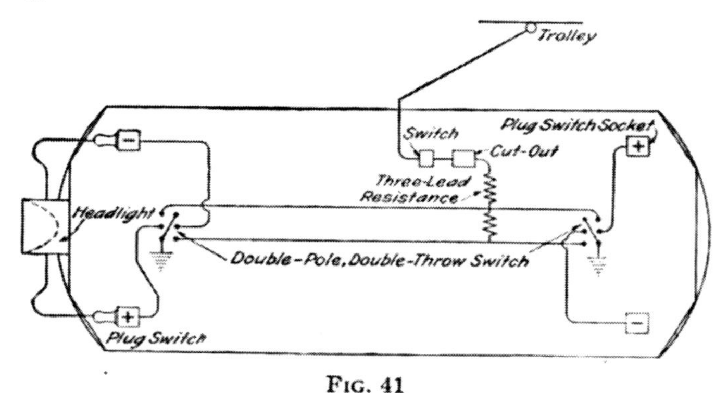

Fig. 41

electric car heater consists of resistance wire in an enclosing case. The wire is heated by the current, and the case keeps the clothing of passengers away from the hot wires while allowing a circulation of air. From 4 to 20 heaters may be required to a car, depending on the size of the car, the climate, the kind of traffic, and the make of heater. A number of small heaters symmetrically placed produce an even distribution of heat. To keep a 20-foot closed car comfortable during average winter weather in the vicinity of New York City requires about 10 amperes at 500 volts, or about 8.3 amperes at 600 volts; this is between 6 and 7 horsepower per car. A rough estimate for winter weather in a cold climate is that one-third of the total energy required for the operation of the car is utilized by the heaters. The reason for this high proportion is that the heating current is used continuously and the propulsion current intermittently.

Heaters are usually arranged to provide three degrees of heat by means of a switch and division of the heaters into groups. On some roads the heating of the car is left to the discretion of either the conductor or the despatcher, and in cther cases, the heating is controlled automatically by means of a circuit-closing thermometer.

CONSTRUCTION OF ELECTRIC HEATERS

38. Fig. 42 shows a type of heater in which the active wire formed into a spiral is wound in the grooves of a porcelain

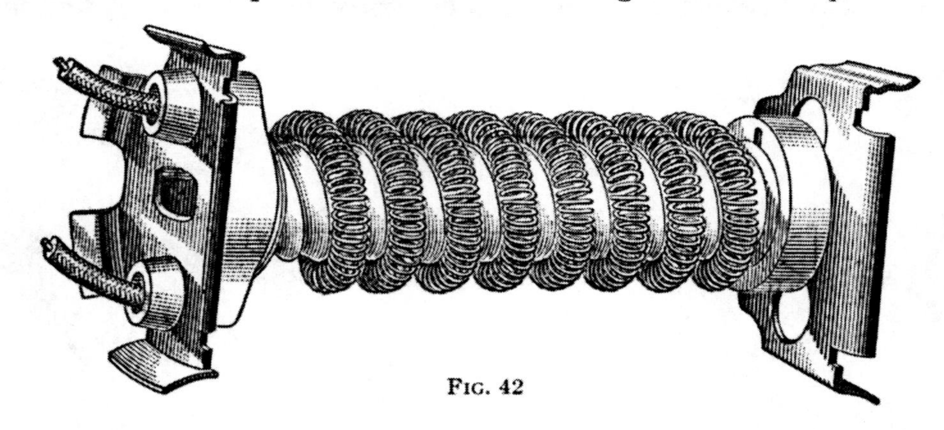

FIG. 42

tube. The tube is supported on a steel rod extending between metal heads. The terminals are brought out through porcelain bushings at one or both ends, as desired. One, or more, of these wire spirals is enclosed in a case, a portion of which, as shown in Fig. 43, is perforated to allow circulation of air.

39. In some heaters, the spiral of wire is supported on

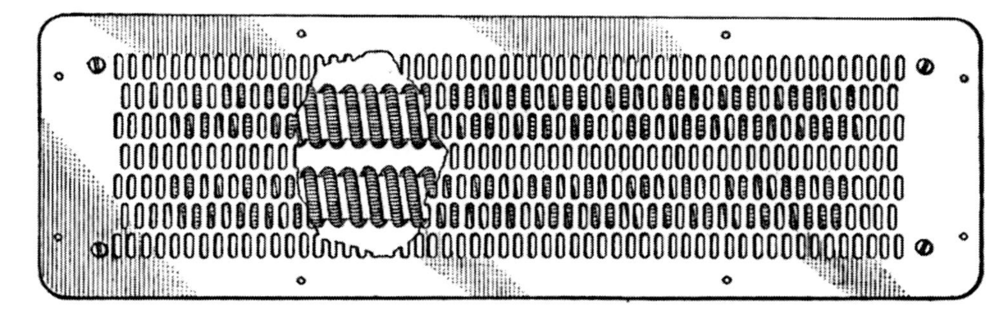

FIG. 43

knobs *a*, as indicated in Fig. 44. Knobs *b* serve as anchors for the terminals of the heating element.

Another method of mounting the spiral is shown in Fig. 45.

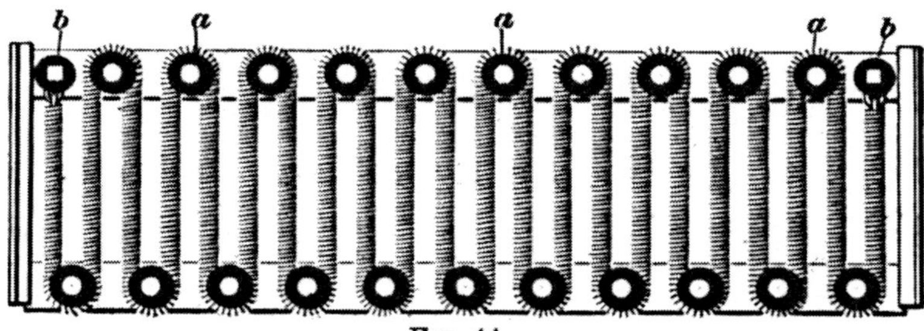

FIG. 44

The support consists of a $\frac{1}{4}$-inch curved, enamel-covered steel rod *R*, which touches the spiral *S* in only a few places.

The object of the spiral construction of bare wire is to allow the air free access to nearly the entire surface of hot wire. The

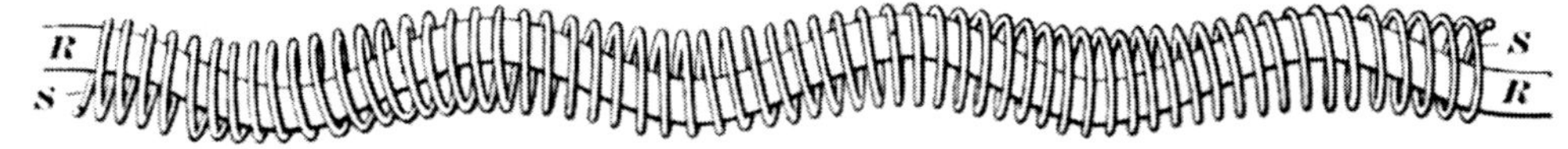

FIG. 45

tension on the turns, due to the method of mounting the spiral, tends to keep them from touching and thus cutting out of circuit some of the resistance.

HEATER CIRCUITS

40. All manually operated car electric-heater systems employ practically the same method of connecting the heaters and regulating the heat. Fig. 46 shows circuits for four heaters,

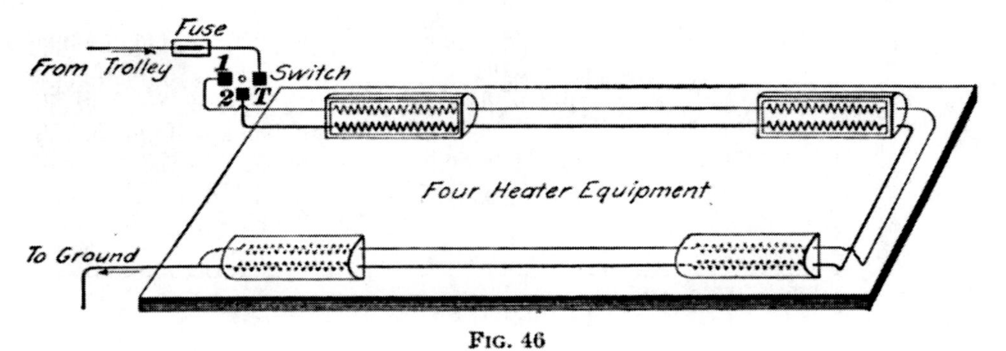

FIG. 46

each heater having two coils of unequal resistance. The top coils are connected in one series and the bottom coils in another. The switch can be closed so that either series will be in circuit alone or so that the two series will be joined in parallel across the circuit, thus providing for three degrees of heat.

41. Fig. 47 illustrates a switch used in heater circuits. The trolley terminal is shown at *T* and terminals *1* and *2* are

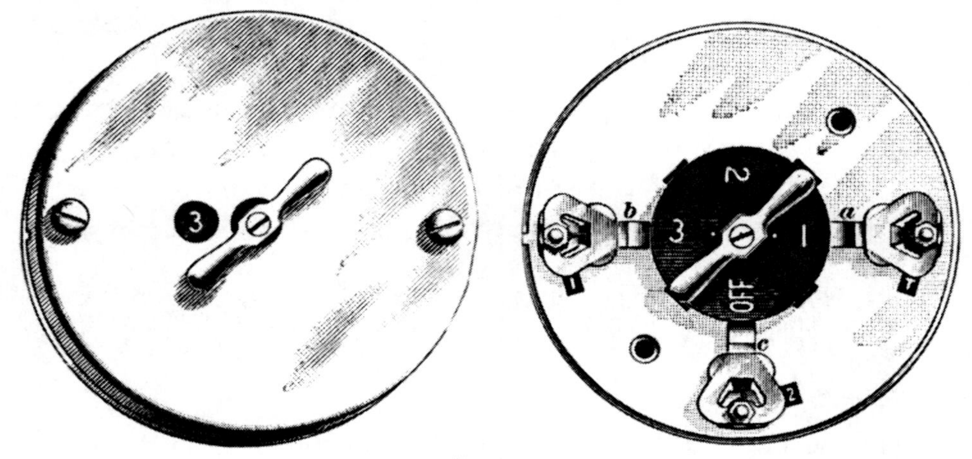

FIG. 47

connected to the two heater circuits, as indicated in Fig. 46. Arms *a*, *b*, and *c*, Fig. 47, rotate with the central part of the switch; the positions of the switch are indicated by the numbers and the word "off" marked on the black disk, as seen through the hole in the cover. In the position shown the top and

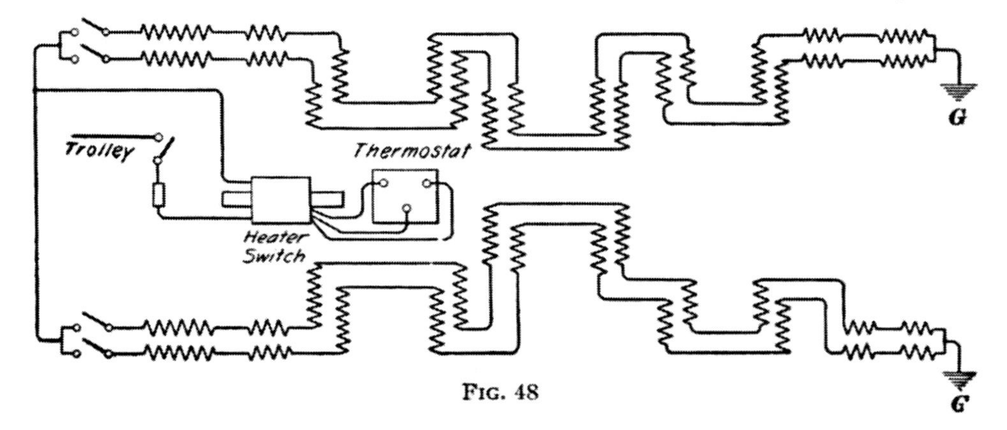

FIG. 48

bottom heater circuits are in parallel and the heaters are giving maximum service.

42. Fig. 48 shows, for a large center-entrance car, two independent circuits that contain ten heaters each, with two elements in each heater. The temperature of the car is controlled by means of a thermostat and an automatic switch. This thermostat is adjusted to cause the heater circuit to open if a certain temperature is exceeded and to close if the car temperature falls below this fixed point. One or more of the heater circuits, as the weather conditions require, are placed in service by means of small switches and the thermostat then regulates the temperature of the car.

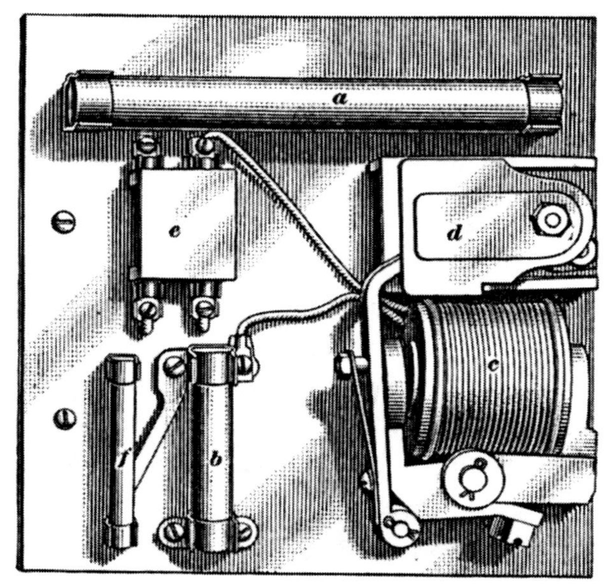

FIG. 49

HEAT REGULATION

43. Fig. 49 shows a type of automatic switch for the control of heating circuits, and Fig. 50 indicates the connections of the switch to a circuit-closing thermometer. In both figures resistors are shown at *a* and *b*, the operating coil of the heater switch at *c*, a blow-out coil at *d*, a relay coil at *e*, a fuse at *f*, and in Fig. 50, a circuit-making thermometer at *g*. The mercury of the thermometer as it rises makes a connection between a fixed and an adjustable contact. The operating coil *c* is energized by current through *a*, *c*, and *b*, to ground, Fig. 50. Coil *c* closes the switch, and the heaters receive current through the blow-out *d* and switch. When the temperature of the car rises to the point for which the upper thermometer terminal is adjusted, the relay is energized by current through *a*–*c*–*f*–*e*–*g*–ground. The relay contacts shown above coil *e* close, thus short-circuiting the operating coil *c*, which causes the switch and the heater circuit to open. When the temperature falls below the desired point, the short circuit is removed from coil *c*, which acts to close the heater circuit.

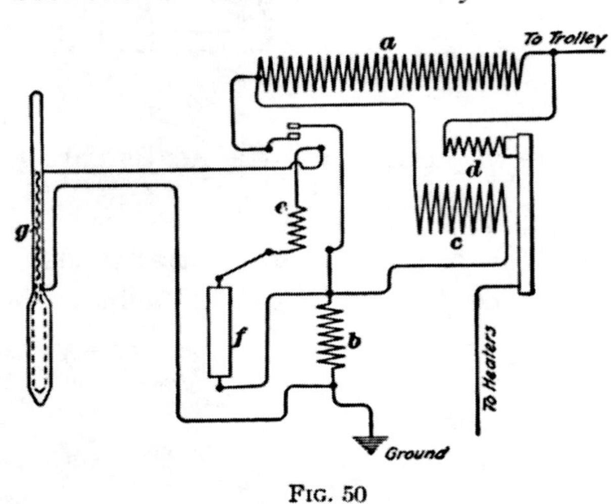

FIG. 50

The upper contact on the thermometer can be set to regulate the temperature at any desired value from which, through the action of the automatic switch, but a small variation is allowed.

COMBINED HOT-AIR AND VENTILATING SYSTEMS

44. In some systems for heating and ventilating cars, fresh air is drawn from the roof, side, or bottom of the car, heated by passage through a small coal furnace, and then discharged through a distributing duct usually placed on one side of the car. The movement of the air stream is set up by a blower mounted on the furnace. Fig. 51 indicates the general features of a heating and ventilating system of this type. The air enters at *a*, passes through the preliminary heating chamber *b* into the blower, and is then forced through the final heating chamber *c* into the distributing duct. The duct has slots of varying widths that provide proper distribution of the heated air throughout the car.

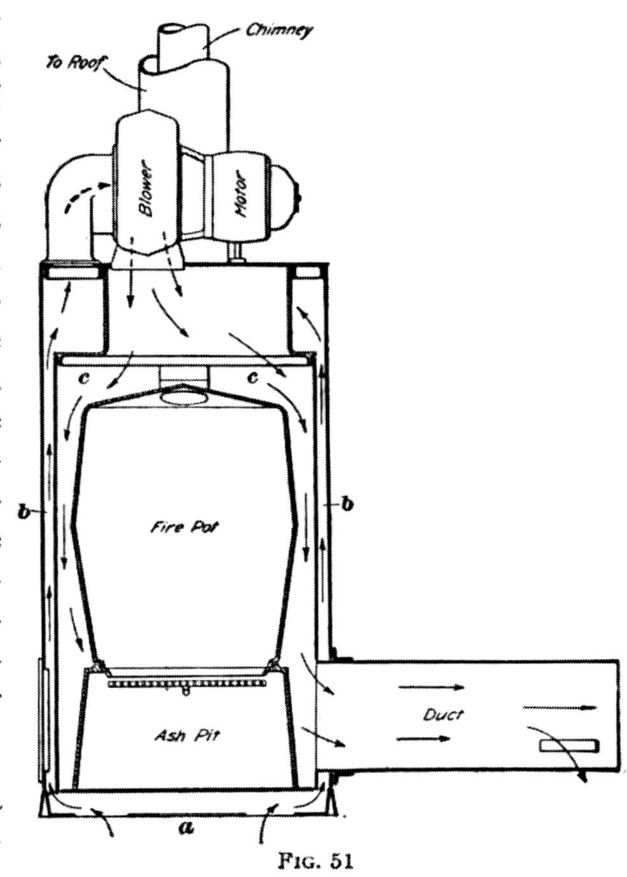

Fig. 51

45. In place of a coal furnace, an electric heater is sometimes used to warm the air and a blower provided to force the air through the heater and distributing duct. Fig. 52 shows

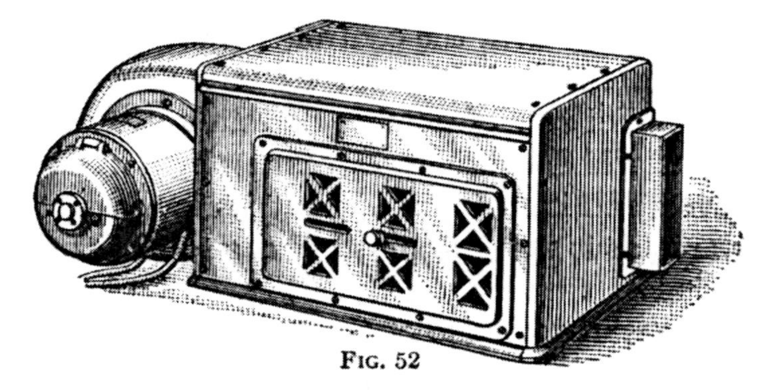
Fig. 52

the exterior of an electric heater and blower and Fig. 53 indicates the air passages. The heater elements, composed of wire wound in spiral form, are arranged in groups so that all or part may be connected in circuit. Automatic control by means of a thermostat serves to close or open the heater circuit in accord with the temperature conditions.

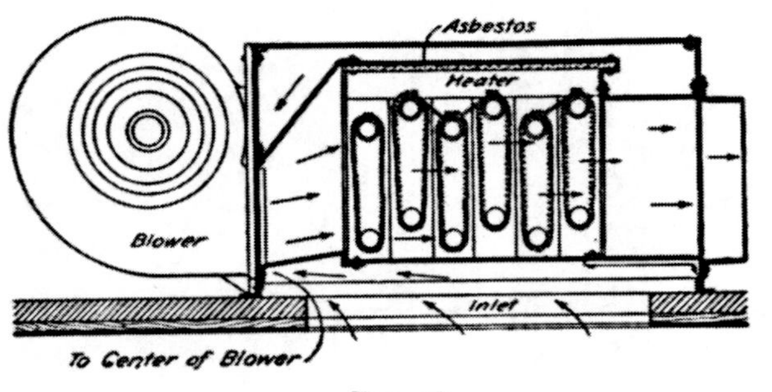

Fig. 53

46. In the hot-water system of car heating, the water flows through a closed circuit of piping including a pipe spiral placed within the fire-box of a small coal furnace. To prevent damage by freezing care must be taken to drain the pipe of water in case the car is to be out of service for a considerable time.

AUXILIARY CAR DEVICES

CONTROLLER REGULATORS

47. The **controller regulator** is a device attached to the top of a controller to retard the movement of the handle toward the running position, as too rapid movement would damage the electric apparatus. Fig. 54 shows the exterior of one type of a controller regulator called an *automotoneer*. The lower casting *a* is fixed to the top plate of the controller and the upper casting *b* turns with the controller shaft.

Fig. 55 shows the inside of the lower *stationary casting a*, containing an inner row of notches and an outer row of stops. The dog *b* is attached to the upper casting by a ball-and-socket joint or bearing, and is so mounted that its lower part is free to swing outwards as it slides along the inner row of notches on *a*. In the position of the dog shown in Fig. 55 its lower end has been forced outwards by one of the inner notches, so that the side of the straight stem of the dog engages with one of the stops in the outer row. This action blocks further forward movement of the dog, of the upper casting, and of the controller shaft until the pressure of the motorman's hand is released from the handle, when the dog swings inwardly by gravity through the path *c*, thus clearing the stop and allowing the handle to be turned again until the dog engages with the next outer stop.

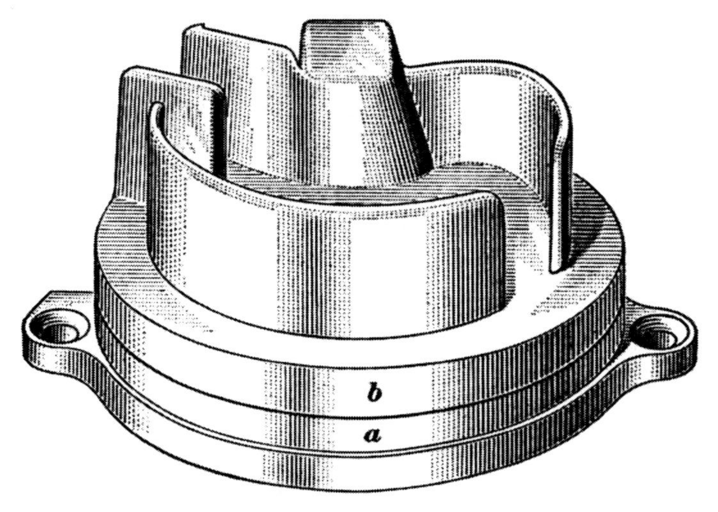

Fig. 54

These operations are repeated for each notch on the controller. The number of stops on the automotoneer is such that its notching movement corresponds to that of the controller on which it is mounted. The handle must therefore be advanced in a series of alternating steps and pauses, which is precisely

Fig. 55

the method necessary for proper acceleration of the motor speed. The controller can be thrown to *off* position as quickly as desired, for the dog will then assume an inclined position and slide over the stops and the notches on the stationary casting.

The upper and lower castings are locked together by means of a ring of balls placed between them. Half of the ball race is cut in a boss on the upper casting and half on the lower casting. Steel balls which are poured into this raceway when the parts are in position while allowing rotation of the upper casting, prevent its removal.

BUZZER AND BELL CIRCUITS

48. Fig. 56 shows the connections of buzzers *a* and bells *b* on a center-entrance car. The current for the two circuits is obtained from the trolley through a switch, a fuse, and a resistor. Twenty push buttons are connected to the buzzer circuit. The

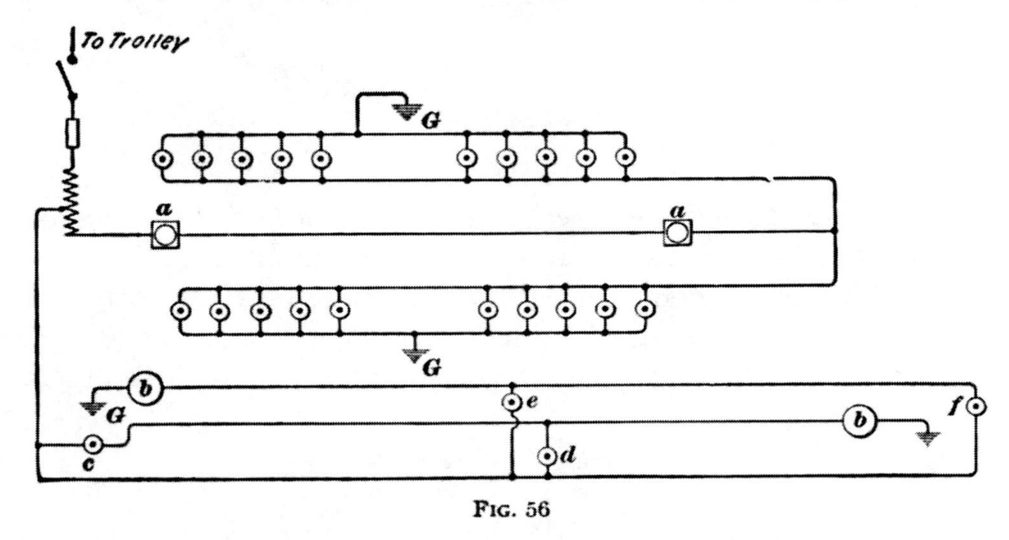

FIG. 56

passenger may signal for a stop by pressing any one of these buttons, and both buzzers, which are in series, will operate.

The bell circuit is for the use of the motorman and the conductor. If button *c* or *d* is pressed, bell *b* on the right operates; if button *e* or *f* is pressed, bell *b* on the left operates. Single-stroke bells are used and the ordinary one-, two-, or three-bell signals may be given.

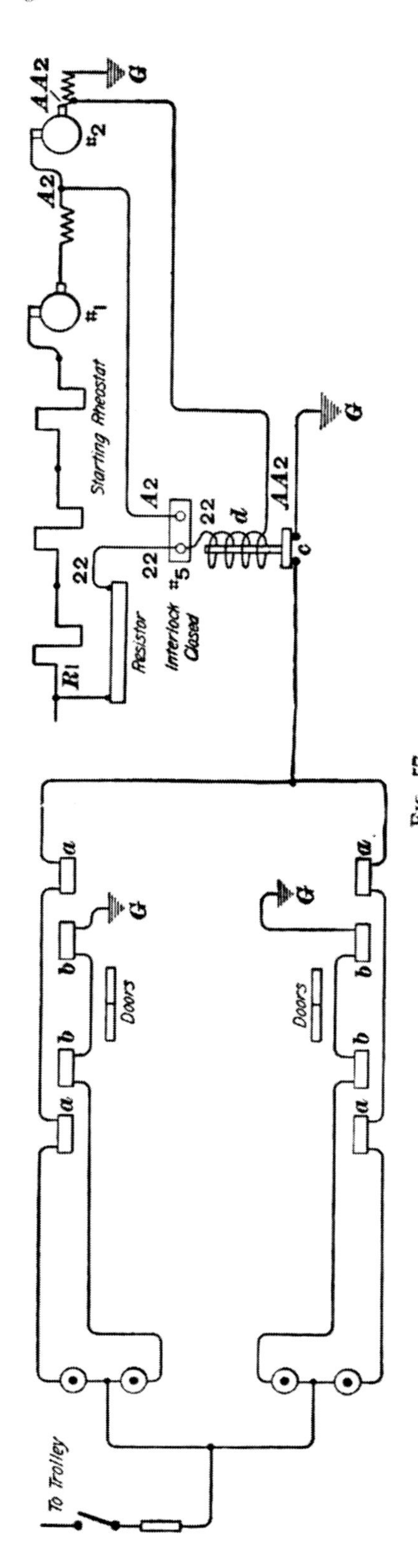

Fig. 57

DOOR-OPERATING CIRCUITS

49. Fig. 57 shows the electric circuits for the operating magnets that control the valves of the air cylinders used to open and close the two double doors on some center-entrance cars.

The magnets are energized by the trolley current. The circuit of the magnets *a* for opening the doors can be completed only when a push button is pressed and relay contacts *c* are closed, which occurs only when there is no car movement. The circuit of the magnets *b* for closing the doors can be completed at any time by pressing another push button.

While the car is running and the motor circuit is complete from the trolley to ground, main switch *No. 5* is closed and its interlock is open. Relay coil *d* then receives current through a resistor, and contacts *c* are kept open. When the motor circuit is opened, main switch *No. 5* opens and its interlock closes. As long as the car continues to coast, relay coil *d* receives current due to the electromotive force generated by the armature of *No. 2* motor rotating in a residual flux. Thus contacts *c* remain open until the car stops, and only then can the doors be opened. The *No. 5* interlock and the switch for the *c* contacts are

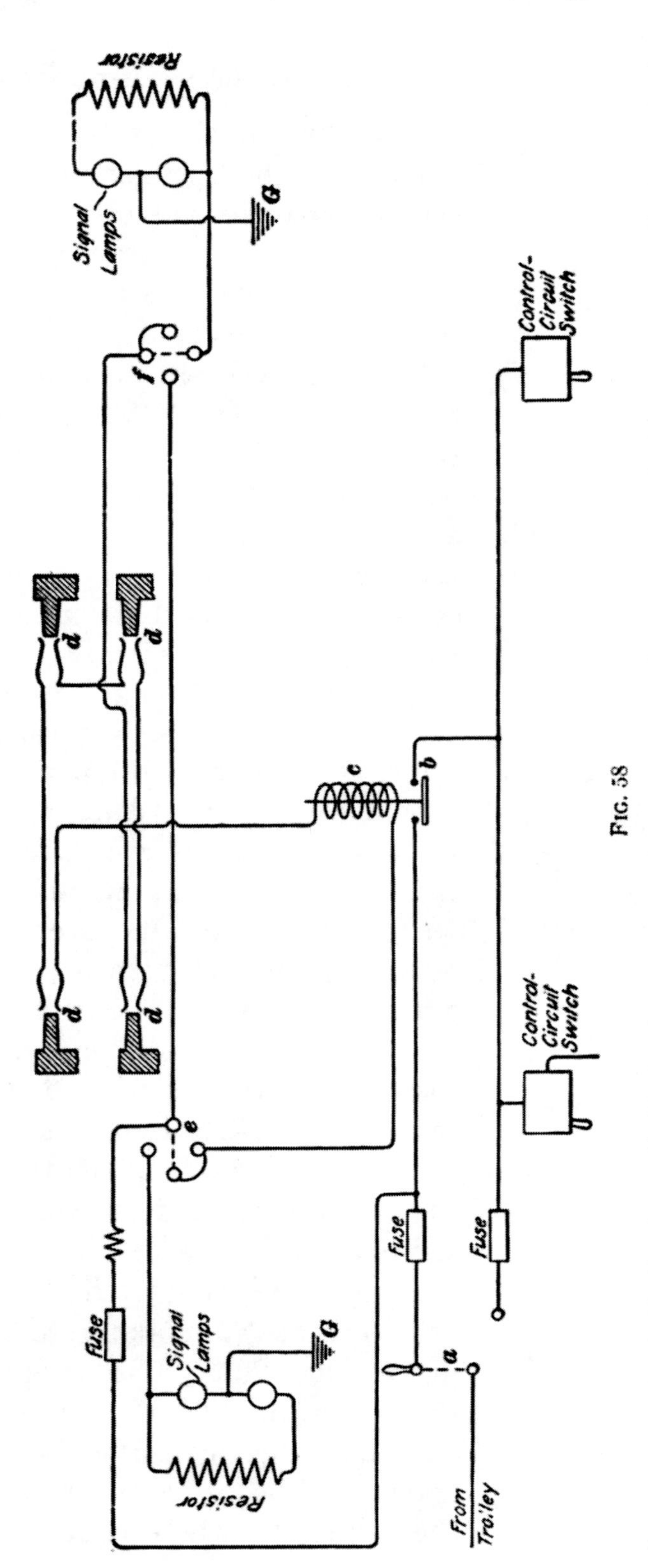

Fig. 58

shown in the positions assumed when the car has stopped. The reference letters and numbers shown near the motor circuit relate to the various parts of a car-wiring diagram, which subject is treated in another Section.

DOOR-SIGNAL LAMPS

50. Fig. 58 shows the wiring diagram for signal lamps that indicate whether or not the car doors are closed, and for the devices that make it impossible to start the car until the doors are closed. By means of a double-throw switch *a*, both these features can be made inoperative.

With switch *a* in its upper position, current for operating the master controllers must pass through the relay contacts *b*, and with the switch in its lower position, these contacts are out of circuit. The relay contacts are closed only when coil *c* is energized, and this occurs only when all of the door switches *d* are closed.

Two-way switches *e* and *f* serve to make the signal lamps at either end operative. If the doors are closed, with the switch *a* in the position shown, a circuit is completed through switch *a*–fuse–*e*–coil *c*–four door switches *d*–*f*–signal lamps at the right–*G*. By turning the signal-lamp switches *e* and *f*, the signal lamps at the other end of the car can be used.

A resistor in series with one signal lamp of each pair makes the brightness of the two lamps differ and helps to insure against failure by burning out of both lamps simultaneously.

REGISTER-RINGING DEVICE

51. Fig. 59 shows the wiring plan for an electrically operated cash register. When the push button *a*, placed on a pedestal near the conductor's station, is pressed, the operating coil *b* within the register rings the signal bell and rotates the counting mechanism. If the button is held down a device *c* in the register automatically opens and closes the circuit, repeating the signal and registering any desired number of fares.

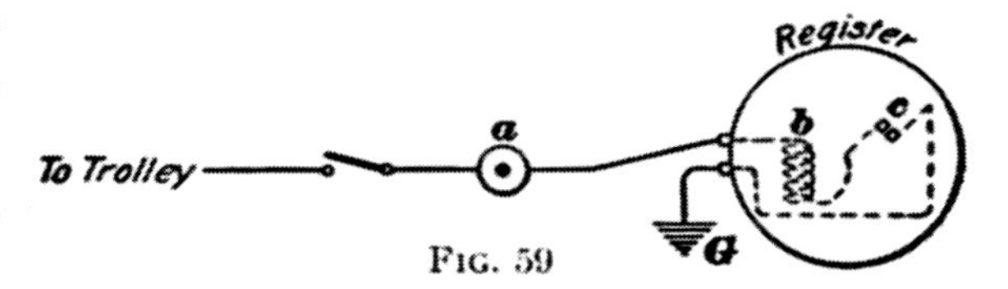

FIG. 59

CAR SIGNS

52. The main requirements of a good car sign are that it is always readable, indicates many destinations, is easily changed, and is durable. Fig. 60 (*a*) and (*b*) shows the general features of a car sign of the curtain type that with slight modifications may be mounted on the hood, the dash, or the side of the car. Light in the rear of the curtain makes the letters readable from the outside of the car. For side installation, the car lamps will usually provide sufficient illumination. For car-end installation where direct light from the car is not available, lamps within the sign case are usually employed. An indicator near the handle of the sign-rotating mechanism, or lettering on the

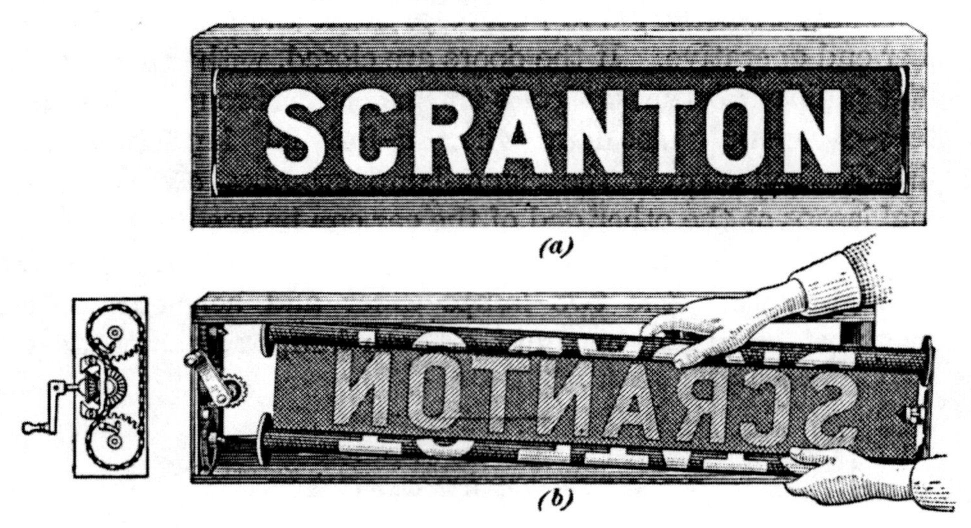

(a)

(b)

Fig. 60

back of the curtain, used in connection with a pointer, shows what sign is exposed and if it is centered.

INSTALLATION OF CAR WIRES

53. The wires installed on a car may be divided into two groups: those connecting the devices of the motor circuit, such as trolleys, car motors, controllers, rheostats, circuit-breakers, lightning arresters, etc., and those connecting the devices of the lighting, heating, and auxiliary circuits and referred to independently as light wiring, heater wiring, signal wiring, etc.

Car wires have been installed in canvas hose, wooden molding, or pressed boards made of insulating material. On modern equipments of the best type, however, iron conduits are extensively used. With platform controllers that handle main-motor current, the motor and rheostat wires are carried from one controller to the other either in one pipe, which runs down the center of the under side of the car between the two center sills, or in two pipes, one on each side of the car. In either case the taps from the wires are taken out through *condulets*. All trolley and ground wires and their taps are best run in separate pipes to lessen damage that may occur from lightning discharges.

In cars equipped for the multiple-unit system of control, the small control wires from the master controller to the box containing the motor switches are drawn into a single conduit. The group of main wires from each motor to the switch box are included in single conduits. Other conduits are provided for the heater, lighting, and signal wires.

54. In order to limit the effect of self-induction when alternating current is employed, some form of non-metallic conduit should be used when only a single wire is installed; but if a metal conduit is used both sides of a circuit must be installed within it. If a metal conduit should be used for a single wire carrying an alternating current, and the conduit touches any metal portions of the car in such manner as to form a closed circuit, the conduit may act as the secondary of a transformer, the primary of which is the single wire. The conduit may thus be heated and injure the insulation of the enclosed wire. Care should be taken, if alternating current is used, to prevent a closed circuit of piping, even if the wires of both sides of the circuit are installed in the conduit. The metal conduit is grounded so that if a wire within the pipe comes in contact with it, the fuse protecting that circuit will blow.

55. The "National Electrical Code," issued by the National Board of Fire Underwriters, contains rules relating to the installation of car wiring. These rules are subject to revision and the latest edition of the "Code" should be consulted by any one having charge of car wiring.

56. The current used in determining the size of motor, trolley, and resistance leads is taken as a percentage of the full-load current required by the motors, based on 1 hour's run. Table I shows percentages. The full-load current is equal to

$$I = \frac{746 \times n \times H.P.}{E \times a}$$

in which n = number of motors;
$H.P.$ = horsepower of each motor;
E = line voltage;
a = efficiency, expressed decimally, of geared motors.

For estimating current in order to select conductors, the approximate efficiency of the motors with gears may be assumed to be 85 per cent.

Table II indicates the safe carrying capacity of rubber-covered wires.

TABLE I

PERCENTAGE OF FULL-LOAD CURRENT IN LEADS

Size of Each Motor Horsepower	Motor Leads Per Cent.	Trolley Leads Per Cent.	Resistance Leads Per Cent.
75 or less	50	40	15
Over 75	45	35	15

TABLE II

ALLOWABLE CARRYING CAPACITIES OF WIRES

B. & S. Gauge	Current Amperes	Size of Wires Circular Mils
8	35	16,510
6	50	26,250
5	55	33,100
4	70	41,740
3	80	52,630
2	90	66,370
1	100	83,690
0	125	105,500
00	150	133,100
000	175	167,800
0000	225	211,600

EXAMPLE.—Determine the size of: (*a*) trolley leads, and (*b*) motor leads to be used with an equipment of four 50-horsepower motors, the line voltage being 500 and the efficiency of the motors with gears 85 per cent.

SOLUTION.—(*a*) In the formula, $n=4$, H. P.$=50$, $E=500$, and $a=.85$: then

$$I=\frac{746\times4\times50}{500\times.85}=351 \text{ amp.}$$

From Table I, the size of the trolley leads is based on 40 per cent. of the full-load current 351 amp., or $351 \times .4 = 140$ amp. From Table II, the safe size of wire for 140 amp. is No. 00 B. & S., having a cross-section of 133,100 cir. mils. Ans.

(*b*) The approximate full-load current for each motor is one-fourth of 351 amp., or 88 amp. From Table I, Motor Leads, 50 per cent. of 88 amp. is 44 amp.; from Table II, the wire to be used is No. 6 B. & S. Ans

---

COMBINED FARE BOX AND REGISTER

57. Many devices have been developed for the purpose of receiving and counting the money, tickets, and transfers delivered by the passengers. Fig. 61 shows one type of combined fare box and register. The hopper is mounted above the coin-examination box and has separate openings for coins and for tickets. The coins are deposited through a series of perforations and fall through a channel into the glass enclosed examination box. They then pass through the counting mechanism into the cash drawer, from which they may be removed by the conductor for making change. The insertion of a ticket through a separate slot moves a trigger that makes an electric contact and causes inked rollers to cancel the ticket, which is then deposited in the ticket box. The cranks shown at the right serve to operate the releasing and registering mechanism.

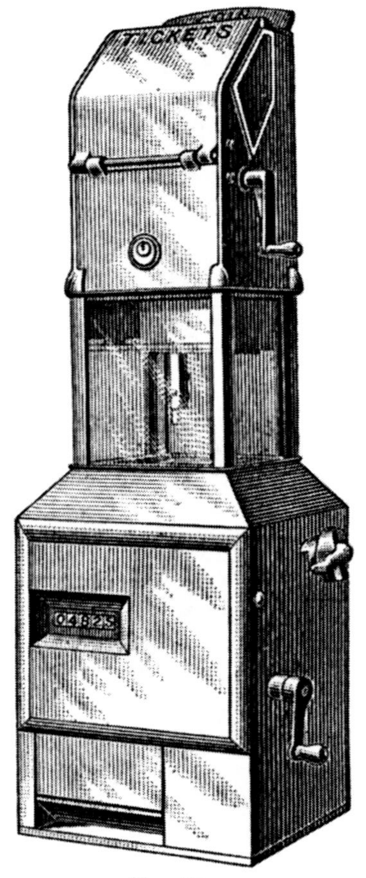

FIG. 61

58. In some systems of fare collection, an ordinary register is used in connection with a fare box. The fare box records the cash fares and cancels the tickets and the register records the total number of fares of all kinds. The difference between the total number of fares on the ordinary register and the sum of the canceled tickets in the fare box and the transfers taken by the conductor must be the number of cash fares.

SAND BOXES

59. Lever-Operated.—Application of sand to the rail

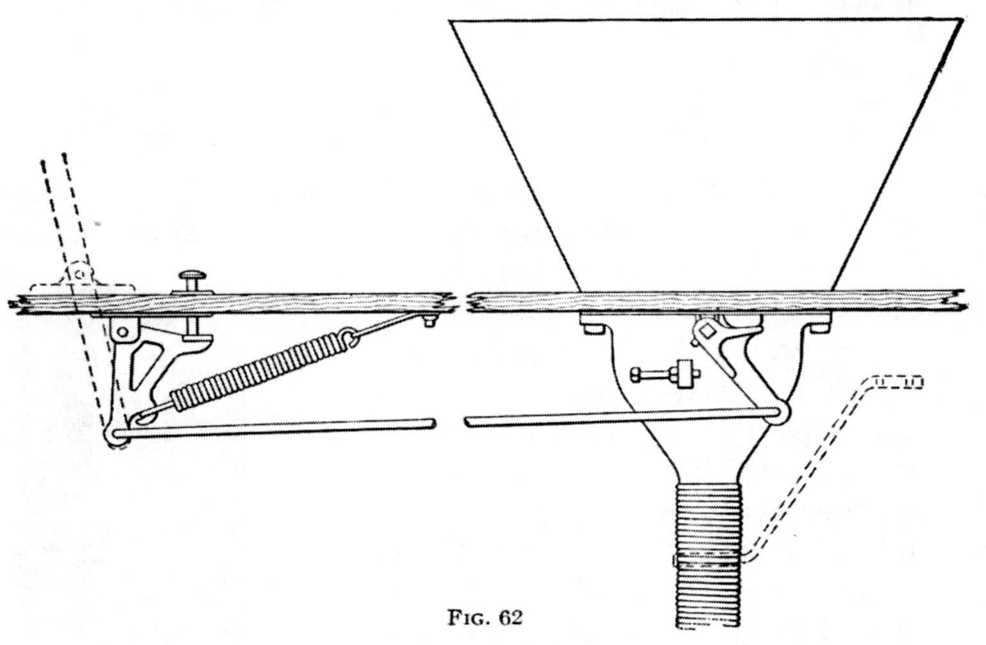

Fig. 62

will, under some conditions, greatly increase the friction between the drivers and rails and thus aid the car to climb a grade, control the speed when descending a hill, and lessen the time taken to bring the car to a full stop.

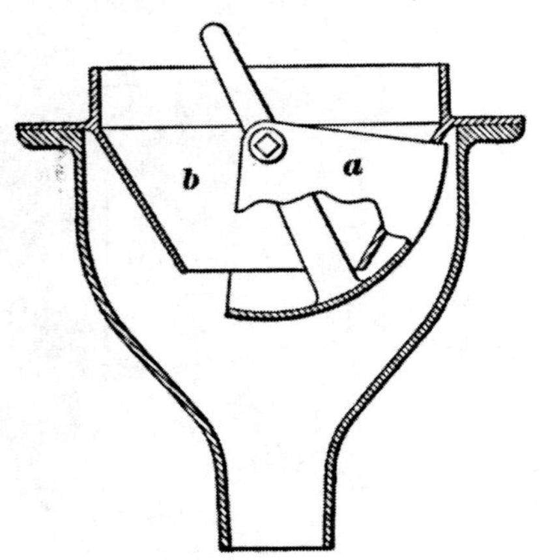

Fig. 63

Fig. 62 shows one type of sand box that may be operated through levers moved by hand or foot.

Fig. 63 illustrates the method of releasing the sand. A rocker casting *a* when in normal position is below the mouth of the sand hopper *b*. On pouring sand into the box, the rocker casting is first filled and this sand blocks the mouth of the hopper which may then be filled. To operate the box, the rocker casting is swung to one side and the sand pours through the pipe leading to the rail in front of the drivers.

60. Air-Operated.—Sand boxes on cars equipped with air brakes may be operated by an air blast directed on the sand in a trap attached to the lower part of the hopper.

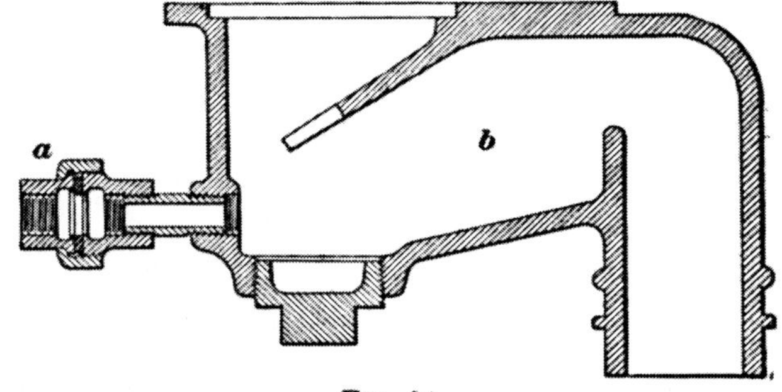

Fig. 64

Fig. 64 shows the trap portion of an air-operated sand box. Pipe *a* is connected to the air-brake system of the car. The lower part of trap *b* is nearly filled with sand which normally blocks the flow of sand from the hopper. When the box is to be operated, a valve in pipe *a* is opened and a jet of air blows the sand from the lower portion of the trap through the sand pipe and on to the rail. While the air blast continues, the sand will flow from the hopper to the trap and then to the rail.

AIR GONGS

61. Fig. 65 indicates a method of operating a platform signal gong by air as well as by the usual foot-lever. The mechanism for foot operation is shown at *a*. When the button, which projects slightly above the floor of the platform, is depressed, the hammer strikes the gong.

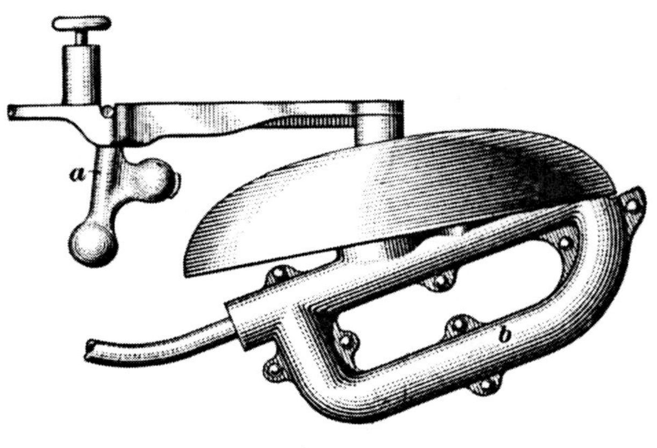

Fig. 65

The pneumatic part *b* is a **D**-shaped casting which forms a raceway for a steel ball. When air is admitted, the ball is driven around the raceway, the upper end of which is cut away, so that the ball hits the gong.

Fig. 66 indicates the piping connections for a foot-operated air valve to a gong. A double air valve may be used in a

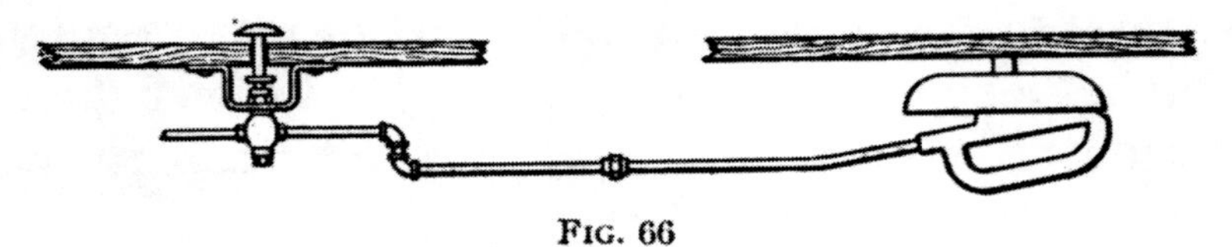

Fig. 66

similar manner for the operation of the sand box in combination with either a whistle or a signal gong.

TRACK SCRAPERS

62. The purpose of track scrapers is to keep the rails clear of snow until the sweeper cars get into action. Fig. 67 shows one type of scraper the plates of which are spring-supported and may be forced into contact with the rails by a hand-operated shaft or by compressed air. Scrapers may be installed at one or both ends of the cars, to conform to service requirements.

Fig. 67

CAR BODIES

63. Car bodies are made in a very large variety of styles to conform to operating conditions. Body lengths range from 18 to 80 feet, and the construction may be open for summer service, closed for winter service, or semiconvertible for use throughout the year.

The modern tendency in body design is to reduce the weight of the cars per seated passenger, install ball or roller bearings, shape the ends of the bodies to lessen air resistance, and otherwise to reduce to a minimum the energy required for operation.

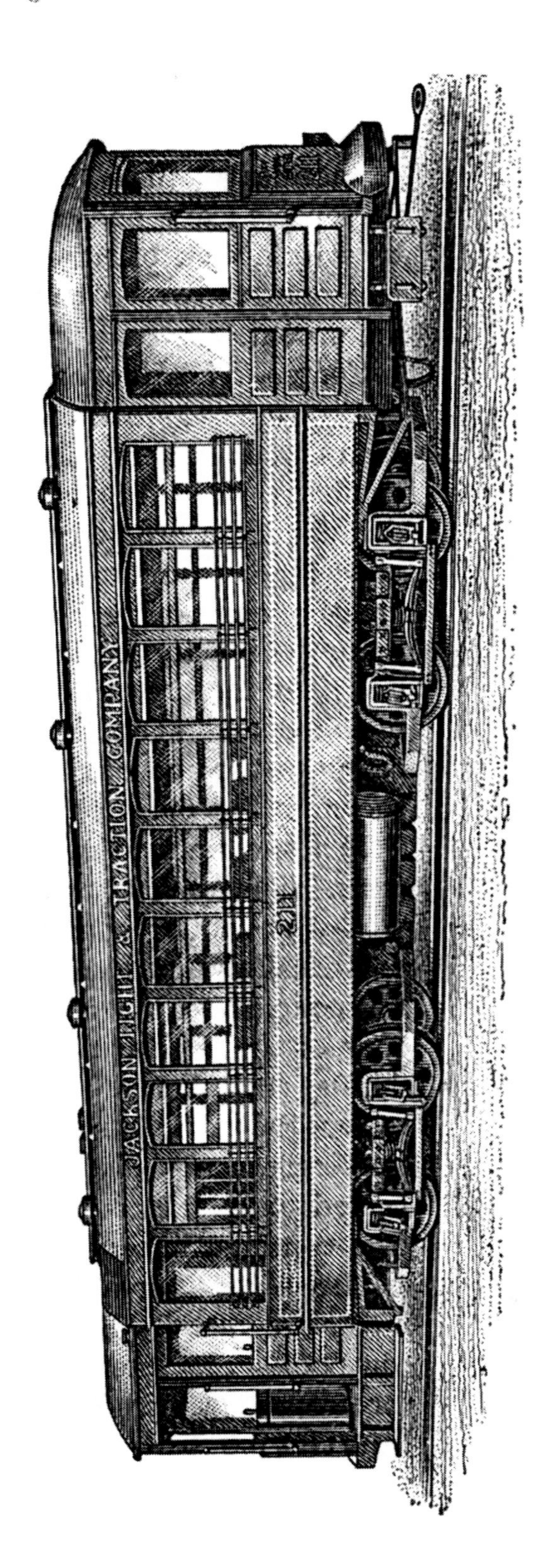

FIG. 68

Car bodies over 20 or 22 feet in length are usually equipped with double trucks because a long rigid single truck will not readily pass around curves at street corners.

If a semiconvertible car body is used, the trucks and motor equipment remain with the body during winter and summer, thus obviating the necessity of keeping both closed and open car bodies and the labor of changing the motor equipment from one to the other twice a year. In some semiconvertible bodies, the window sashes during the summer can be pushed into pockets in the car roof. In other cases, the window panes are removed and stored; heavy curtains then protect the passengers from rain. In either case plenty of ventilation is secured.

64. Car bodies of the type shown in Fig. 68, have platforms arranged for the entrance and exit of passengers by different routes and are used where the prepayment system of fare collection is employed.

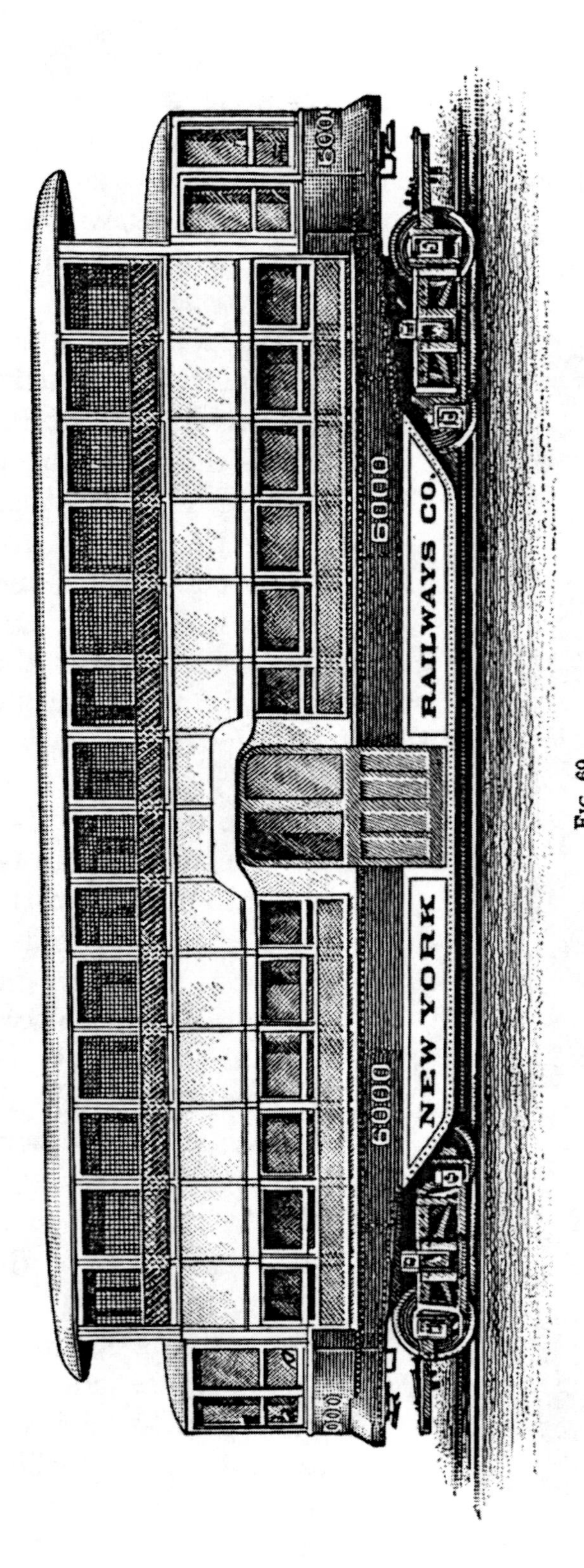

Fig. 69

This system of construction and collection tends to diminish the number of accidents, since the passengers leave and enter within view of either the conductor or the motorman.

Center-entrance car bodies are sometimes used and where the traffic is very heavy, double decks are sometimes provided, thus greatly increasing the seating capacity of the cars. One type of double-deck, stepless, center-entrance car is shown in Fig. 69.

In some railroad systems, light-weight cars without motor equipment, called *trailer cars*, are coupled to the rear ends of the regular cars during rush hours. As soon as the traffic lessens, the trailers are disconnected and the motor cars continue in service.

The seating arrangements are varied. Some car bodies are intended for cross-seats, some for lengthwise seats, and others for a combination of the two seating systems.

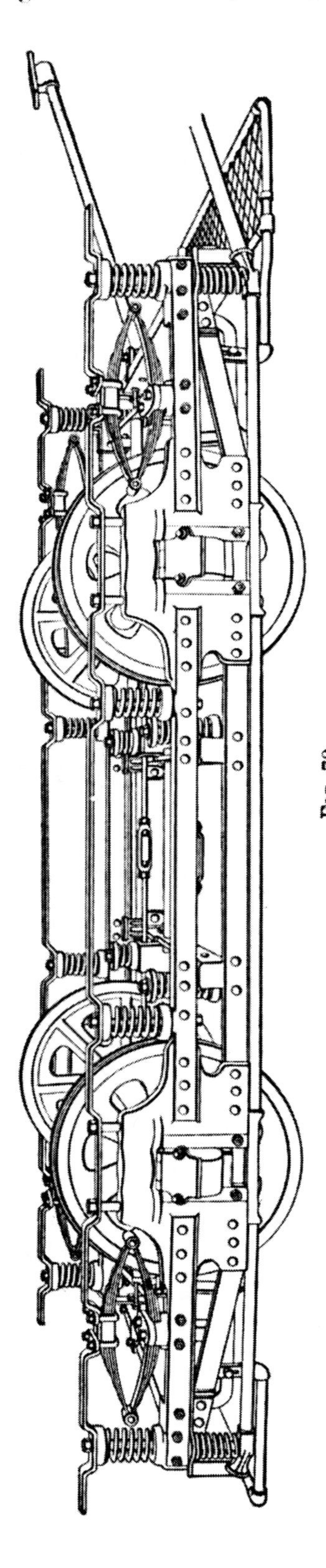

Fig. 70

TRUCKS

65. Classification of Trucks. A **truck** may be defined as a set of wheels in a framework designed to support the whole or part of the weight of a car body and equipment. The main requirements of a good truck are that it is easy riding, durable, of simple construction, the wearing parts are easily renewable, and the main members cannot be distorted under service conditions. The trucks should be self-contained, that is, one framework must include the wheels and axles, the brakes, the motors, and the driving gear.

Trucks may be divided into two general types of construction: The *single*, or *rigid truck*, in which the car body is bolted to sides of the truck framework; and the *swiveling truck*, two of which are required per car; the ends of the car are then supported through bearings placed at or near the center of each truck. The latter type may be subdivided into the ordinary truck and the maximum-traction truck.

The rigid truck and the ordinary truck have four wheels of equal size and may carry one or two motors each. The maximum-traction truck has two large and two small wheels and carries only one motor, which is geared to the axle with the large wheels. The bearing of this truck is not in the center, but is so placed that about 70 per cent. of the weight supported by the truck rests on the large driving wheels. The weight on the small wheels is sufficient to keep them on the rails as the car is rounding a curve.

66. Experiments have shown that for a given car weight, maximum-traction trucks require less energy expenditure than rigid trucks with a 7-foot wheel base. The wheel base is the distance between the point of contact of the forward wheel of a rigid truck with the rail and the corresponding contact point for the rear wheel of the truck on the same rail. The rigid truck is more apt to bind on curves than the maximum-traction truck with its shorter wheel base.

The ordinary truck with one motor has the disadvantage that the driving power is all exerted from one axle, while the

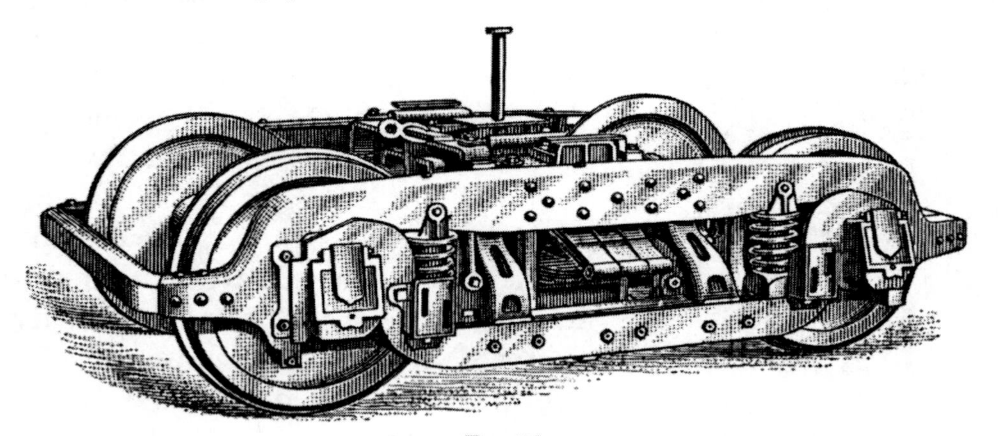

FIG. 71

weight that limits tractive effort rests on two axles. The result is that the drivers tend to spin when heavy duty is impressed on the motors because the friction between the drivers and the rails is not sufficient to prevent wheel slipping. When a smaller motor is put on each axle of the truck, spinning is less likely to occur.

Large cars for interurban service are usually of the double-truck type. In some cases one truck is designed to support two motors and the other truck, which is of much lighter construction, is not equipped with the motors. For heavy service a motor is usually installed on each axle of both trucks.

67. Types of Trucks.—Fig. 70 shows one type of single, or rigid, truck, Fig. 71 an ordinary truck, and Fig. 72 a maximum-traction truck. On a single-truck car, the car body is bolted to the upper framework of the truck. Trucks for

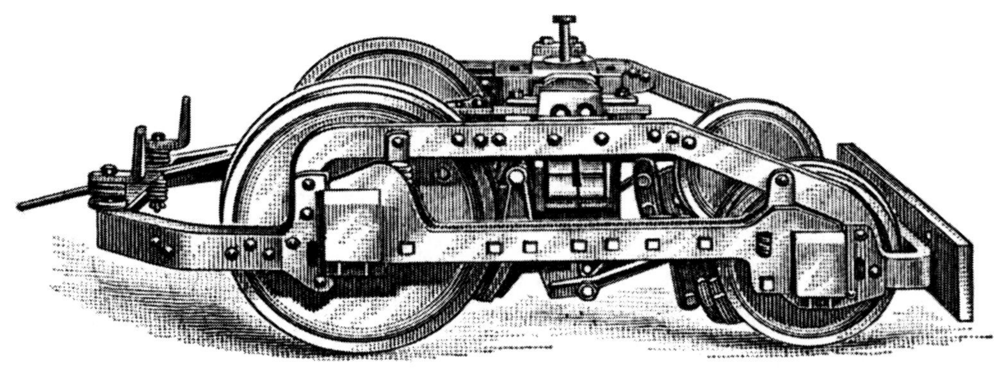

Fig. 72

double-truck cars are attached to the car body by means of bearings placed at or near the center of the trucks, and king pins around which the trucks may turn as the car passes around a curve.

68. Fig. 73 shows a type of center bearing employing balls to lessen the friction. The upper member is mounted on the car body, the lower member on the truck. Part of the car weight is sustained and the car body kept balanced by rub plates, or side bearings, which are circular bronze plates, one set mounted on the truck and a similar set on the car body. These plates are kept well lubricated. The upper member of the center ball bearing and the upper rub plates are shown in position underneath a car body in Fig. 74.

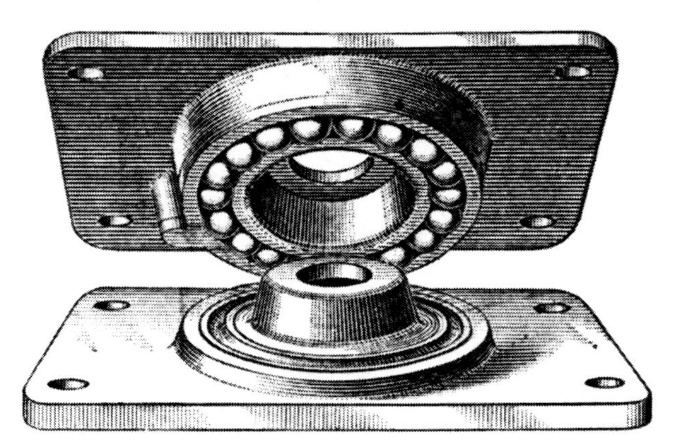

Fig. 73

A car body mounted on two trucks sits higher above the rail than a closed car body using only one truck because the wheels on a double-truck car must clear the body when the trucks turn around their king pins as occurs when the car is on a curve.

69. Axle Bearings.—Fig. 75 shows one form of bearing construction used on rigid trucks. The bearing surface at the end of the axle on which the wheels are mounted is shown at *a*, the bearing brass at *b*, and the box casting at *c*. The weight

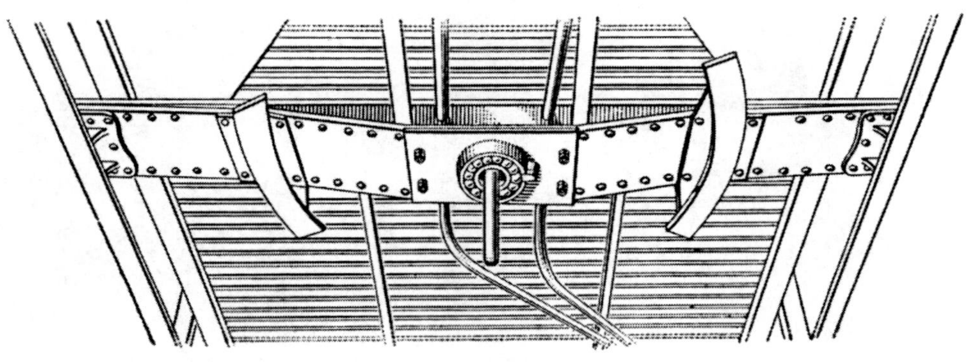

Fig. 74

resting on the truck transmits pressure through the spring *s*. held in a socket of frame *f*, to the box casting and the bearing brass. Since the pressure is always downwards, the bearing brass extends only part way around the shaft.

If the piece *d* is removed the frame may be lifted clear of the

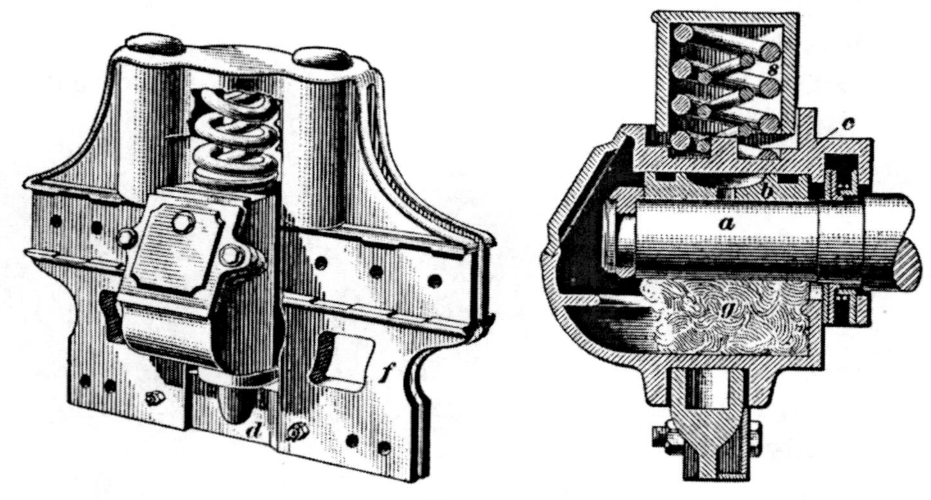

Fig. 75

axle and bearing casting. The axle is lubricated by oil-soaked waste *g* placed in the lower part of the bearing casting.

70. Wheel Bearings.—Fig. 76 shows a roller bearing for a car wheel that rotates on a fixed axle. The vertical stress is

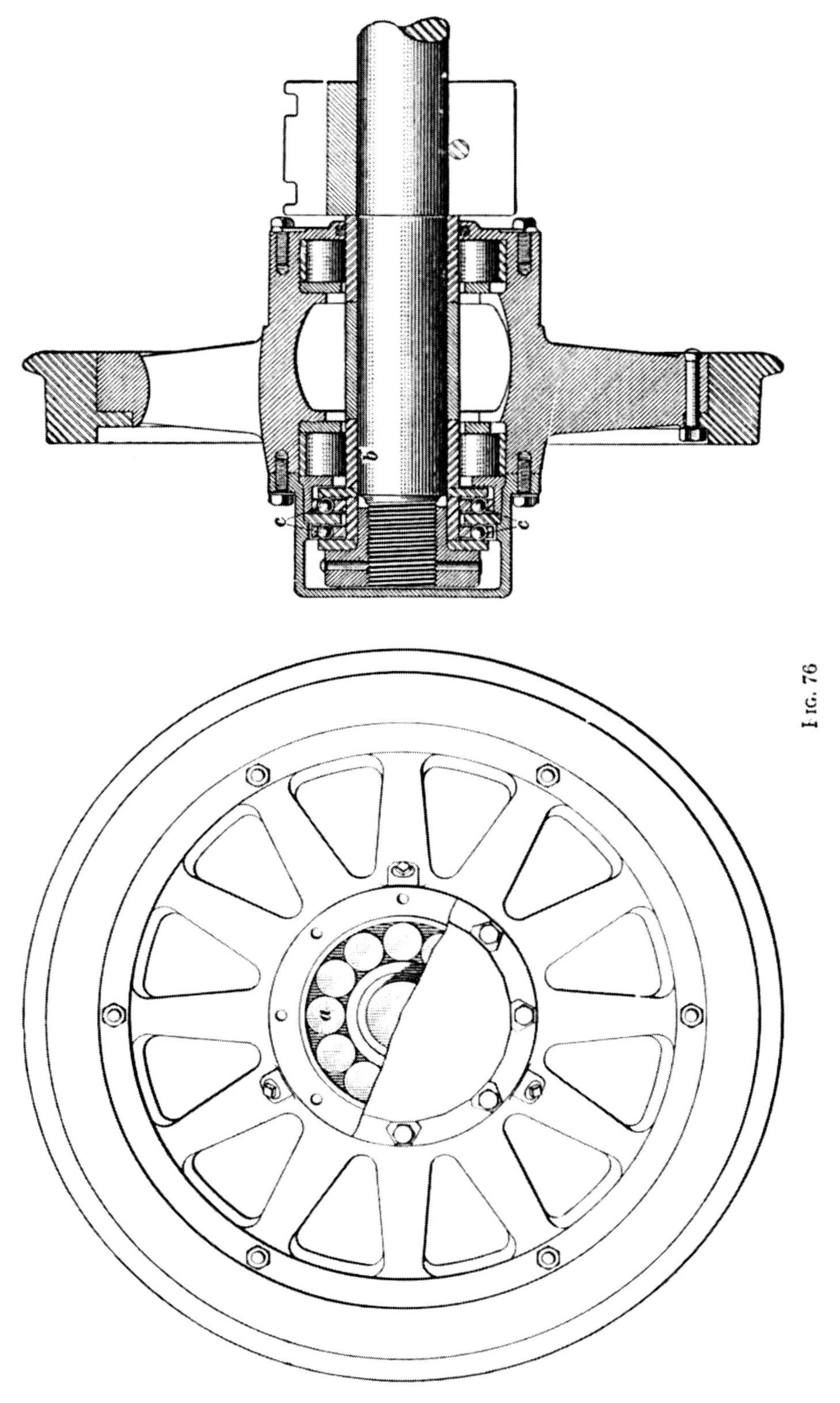

Fig. 76

carried by rollers *a* turning on a sleeve *b* on the axle; the end thrust is taken up by ball bearing *c*. The rollers and balls are carried in retainers.

LENGTH OF WHEEL BASE

71. The wheel base should be long enough to support the car body so as to prevent excessive oscillation of the car, but its length should not cause the wheels to bind on the shorter track curves. A car body that demands a wheel base exceeding 8 feet should be provided with double trucks. A car having an excessive wheel base requires considerable energy to pass around curves of short radius, and causes excessive wear of rails and wheels. On curves of longer radius the energy requirements and the wear are greatly lessened.

To enable cars to round curves with the least effort and to save the wheels and rails, guard-rail flanges at curves should be kept clean and well greased. In laying out a road, all the curves should be made of as long a radius as possible and when purchasing trucks for an existing road, the radii of the curves should be considered.

72. Wheels.—The wheels used on electric cars vary from 30 to 36 inches in diameter; on ordinary street cars, the diameters are usually from 30 to 33 inches. Somewhat larger wheels must be used for heavy work, so as to give more clearance for the motors; therefore, diameters of 33 to 36 inches are quite common. For light cars operating at low speed, cast-iron wheels with chilled treads are sometimes employed. For high-speed interurban work and for heavy city traffic at low speed, a wheel provided with a tire made of rolled open-hearth steel is often used.

The tire can be fastened to the cast-iron center by bolts, or retaining rings, but the usual method in wheels for electric cars is to fuse or cast-weld the tire to the center. The tire is heated, placed in the mold, and the iron center poured; the melted iron fuses the tire and a perfect joint between the two results. The main advantages of steel-tired wheels that compensate for their high cost as compared with chilled cast-iron wheels are: (1) Greater strength and security; these wheels are not likely to fly in pieces, however high the speed may be or however severe the stresses due to rough track or very cold weather. (2) They are much less liable to develop flat spots. (3) They are less liable to slip, since the wrought steel tire has, with the steel rail, a much higher coefficient of friction than a chilled cast-iron wheel and the action of the brakes is more effective.

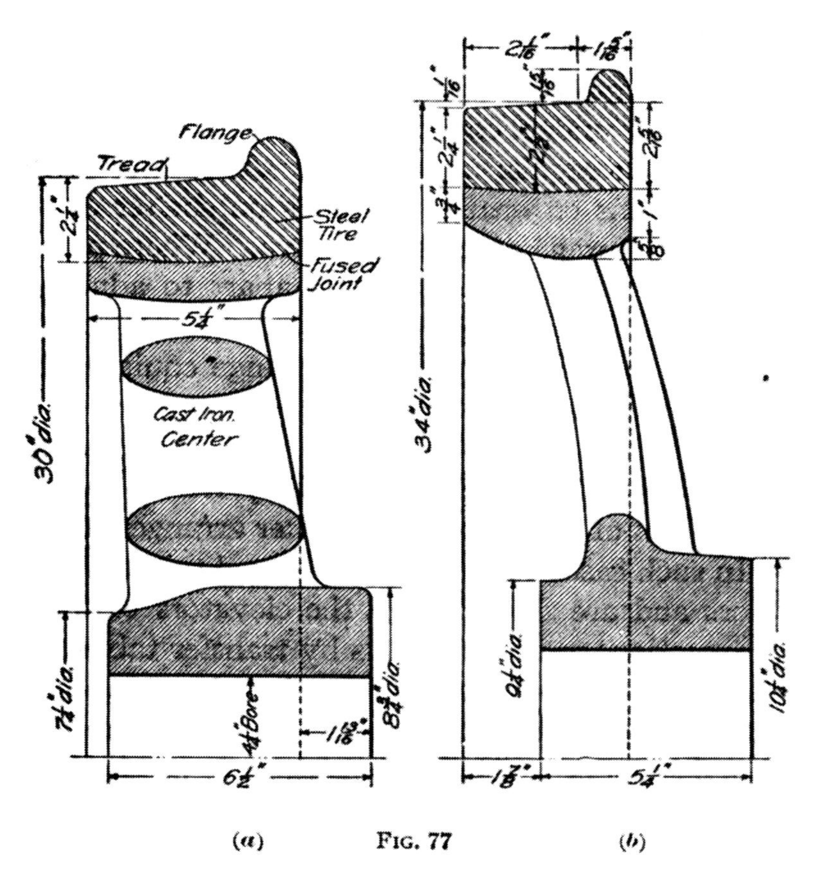

(a) Fig. 77 (b)

(4) They avoid trouble due to chipped or to broken flanges. (5) The rim can be made thick so that the wheel will wear a long time before becoming unsafe; with chilled wheels the depth of chilled metal is limited.

Fig. 77 shows sections of two fused steel-tired wheels; view (*a*) is a 30-inch wheel weighing 650 pounds, used on trail cars for an elevated road; view (*b*) is a 34-inch wheel, weighing 688 pounds, for an interurban road.

On the treads and flanges of the wheels depends greatly the ease with which a car will take a curve. The treads should not be so wide that they run on the paving outside of the track, and the shape, depth, and width of the wheel flange should be governed by the corresponding dimensions of the rail groove.

CAR HOUSE

73. The **car house,** or **car barn,** is a building used for storing cars that are not in use, for inspection, and for minor repairs. Storage under cover is usually provided only for cars that are not in regular commission. A car fitted to operate all day in all kinds of weather requires no shelter over night, especially, when storage room is expensive. Where practicable, car-barn tracks should be far enough apart to admit of easy passage between rows of cars, and if the light in the aisles is good the labor and time required to change equipment at the end of the seasons may be much reduced.

In some car houses, the storage room may be all on one floor; in others, as, for example, in New York City, it may be on several floors, as the cost of land makes any other arrangement impracticable. In such cases the cars are conveyed to the upper floors by elevators and are shifted from the elevators to the several storage, inspection, or repair tracks by transfer tables. Where cars must be handled by elevators, stripped, out-of-season car bodies are usually stored on the upper floors, where they may be supported on horses or barrels, as there is no chance for the cars to be run out in case of fire. Where the storage room is on a level with the street, the car bodies should be supported on temporary trucks. Where possible, every storage track should lead to the street. In some car houses, the tracks are graded toward the street, so that the cars may be easily run out in case of fire. Sprinkler systems, the efficiency of which has often been demonstrated, are installed in all modern car houses.

PIT ROOM

74. For inspection of trucks and motors, pits about 4 feet 8 inches deep should be located directly under the tracks, no pit to be shorter than any car to be placed over it. The amount of pit room to be allowed per car depends on how much trouble the equipments give, on the class of work done at the car house, and on the ability of the car-house organization to so lay out the work that minimum pitting time per car is realized. A value based on long experience under varying working conditions with different equipments and different track layouts is 1 linear foot of pit room for each car that runs into a depot; that is, to accommodate two hundred cars, a car house should have pit room under at least 200 feet of track. This allowance would be scant for a car house full of crippled cars; but it is liberal where trucks are run from under and all overhauling is done above the floor line. This practice is popular because a more thorough job can be done when all parts are in full view in good light.

The pits should have cement bottoms with natural drainage; they should be well equipped with both stationary and extension lights. The floor level between tracks should be covered with removable boards. Underneath the floor line, the pit room should be open from wall to wall. A couple of shelves and a row of small bins in each pit, to hold bolts, nuts, and washers of the commoner sizes, will save much time. If the pits are arranged for heating in the winter and for cooling in the summer, the efficiency of the work done is increased.

The well-equipped pit room is supplied with motor- or air-driven dismantlers, that raise the car body from the trucks, which may then be run to their destination by hand or by connecting the motors to the circuit through a rheostat. A pole with wire attached may be used to make connection with the trolley wire. The same air supply may be employed for blowing out motors and controllers, and for operating small air lifts.

REPAIR SHOP

75. All heavy repairs, alterations, and manufacturing operations pertaining to cars should be made in the **repair shop.** A well-appointed repair shop includes a machine shop, carpenter shop, mill, blacksmith shop, paint shop, winding room, commutator room, controller room, store room, external oil room, and a wheel-grinding annex. In the machine shop, all general machine work is done, such as fitting bearings, turning commutators and shafts, cutting keyways, recutting bolts, etc. In the winding room, field coils, armature coils, armatures, heater coils, governor coils, rheostats, etc. are repaired or constructed. In the commutator room, the parts of commutators are assembled and the finished commutators are tested. In the controller room, controllers, switches, circuit-breakers, brush holders, etc. are repaired and assembled. There is usually work enough on small roads to keep one blacksmith and helper busy on brake rods, levers, hangers, fenders, trucks, and on line, track, station, and building construction work.

The shop should be a fireproof structure designed to afford maximum cleanliness and light. The best shops are so laid out that a car can progress from the pit room to and through the paint shop without retracing its route. The ideal arrangement is to separate the body and truck in the pit room, send the body on temporary trucks to the carpenter shop, and to lay out all repair work on body and truck so that they can be brought together and the car as a whole completed and tested before running it into the paint shop, from which it emerges ready to operate.

MACHINE SHOP

76. The machine-shop tools should be placed in such positions that there will be good light on the work and also to occupy minimum floor space. The number and kind of machines depend on the class of work. Enough machines should be provided to keep up with the work. The following machines would probably be required: One lathe for axles with wheels mounted, one smaller lathe for armatures and bearings, one speed lathe, two sizes of drill press, one boring mill, one metal saw, one power saw, one planer, one shaper, one bolt cutter with right- and left-hand dies, one milling machine, one wheel press, one axle straightener, one grindstone, emery wheels, one punch press, one ratchet drill. On a small road the preceding list can be considerably modified by a good mechanic who knows how to do almost any kind of work on almost any kind of machine available.

WINDING ROOM

77. Where floor space is limited, the winding may be done in a gallery built around the wall above the machine shop; with this disadvantage, however, that heavy parts and material must be elevated. Winders should have direct window light where it is practicable. For a road operating one hundred cars or over, from 6 to 8 square feet of floor space per car should be sufficient winding space. For a small road, more space would be required per car. The outfit of every winding room in which coils are made and installed should include the following: One armature banding machine, one field-coil winding machine, armature-coil winding machines, taping machines, coil forms for each type of armature used on the road; brick-enclosed gas stove for heating soldering irons; a gasoline stove can be used if the tank is removed to a safe place; device for taking off and pressing on commutators, pinion puller, stands for armatures in course of winding, racks for completed armatures, racks for insulation stock, insulation-cutting machine, coil-pressing machine, glue pots to melt glue for holding coil papers in place, ample facilities for dipping coils in varnish or compound, and a good oven, unless air-drying varnishes are used.

The winding room should have duplicate substantial patterns of every standard piece of insulation used; one set should be hung in a convenient place and duplicate sets should be stored in a fireproof vault; every piece should be clearly and indelibly marked. Time and labor are saved if all armatures are handled with buggies and cranes. The journals should be protected with sleeves of fiber or of paper.

COMMUTATOR ROOM

78. The commutator room should be in the care of a good mechanic who understands commutators. It should contain a lathe, drill press, milling machine, and oven for baking the commutators. If it is the practice to groove commutator mica, either a mica-grooving machine or a milling machine adapted to that work should be provided. The commutator room should carry a full line of mica gauges and ring and plug gauges used to fit commutators to their seats on the armature shafts. A suitable press, for subjecting a commutator to external pressure while tightening it, should be supplied so that the job can be done without twisting the commutator bars out of alinement. An adequate supply of assembly rings and a full line of wrenches for adjusting them, are also essential. No emery wheel should be allowed in the room, because particles of emery trapped between bars cause trouble. The commutator room is best located next to the armature room; it should be enclosed and should have good light and ventilation. A reliable test for locating connections between bars or between a bar and the shell should be installed.

CONTROLLER ROOM

79. The controller room should be located where the dust incident to blowing out controllers, will not reach the commutator room, insulating bench, or coil winders; adjacent to the machine shop is a good place. On the brush controller holder repair bench there should be suitable gauges to insure the correctness of brush holders made or repaired.

MILL AND CARPENTER SHOP

80. The mill is the room in which the wood-working machines are placed and the carpenter shop is where the cars are run in for general body repairs. Both rooms may be within the same enclosure—the mill at one end and the carpenter shop at the other. The best place for them is near the machine shop, pit room, and paint shop, a line of single or double track running through, so that a car can enter at one end of the building and go out at the other. In the mill there should be a planer, boring machine, lathe, band saw, circular saw, and grindstone.

PAINT SHOP

81. The paint shop should be at the extreme rear of the main shop and should have free access to the street; it should be provided with as many doors on the street side as there are tracks, so that in case of fire the cars can be run out without any shifting or transferring. The paint shop should receive only cars that have been repaired and are ready to run on the road except for the painting. Each track in the shop should have a trolley wire over it, the whole system of trolley wires being kept cut out by means of a switch except when they are to be used. Under no circumstances should the car bodies be set on horses or barrels in the paint shop; the risk of fire is too great. They should always be on temporary trucks, and where possible, at the head of each line of cars should be a car fully equipped, so that in case of fire they can be coupled together and towed out of danger. Another good plan is to have the tracks down grade out of the house, so that when the brakes are released or the chocks removed from the wheels, the cars will run out by gravity. The great fire-risk incidental to the storage of so many inflammable materials, oils, varnishes, etc., demands an absolutely fireproof wall between the paint room and the rest of the shop, communication between the two shops being only through self-closing fireproof doors. As a prime precaution against fire, the building should be of brick, with a fireproof roof and a cement floor. The floor should be graded to gratings that lead to the sewer or to a cesspool and the roof should be designed to give the best possible light and ventilation. All inflammable materials should be kept in a small, absolutely fireproof room that will admit barrels, etc., without trucking them the entire length of the paint shop. The question of fire-risk in a paint shop is a serious one, as the shop is generally full of cars that will burn quickly if once started.

BLACKSMITH SHOP

82. The blacksmith shop must be located where the coal dust and gases from the forges cannot reach the paint shop. It should contain at least two forges, anvils, and a blower. One forge should be provided with an ordinary bellows all ready to be connected on, in case of accident to the blower or its motor. Besides the usual complement of forge tools, there should be a machine hammer, shears, and a drill press.

GRINDING ROOM

83. If the brakes on a trolley are applied too hard on it for any reason the car skids along the track, flat spots, or **flats,** as they are called, are found on the tread of the wheel. These flats cause the wheels to pound on the rails, and unless they are removed by grinding or a new wheel put on, the trouble is likely to increase. In many car wheels of cast iron, the tread is chilled in the molding so that the iron is very hard to a depth of $\frac{3}{8}$ or $\frac{1}{2}$ inch. If the wheel is worn down so that the chilled portion is ground through, it is useless, as the soft iron under the chilled part will last only a short time. Small flats can often be removed by substituting for the regular brake shoe a special wheel-truing shoe provided with emery, carborundum, or similar abrasive.

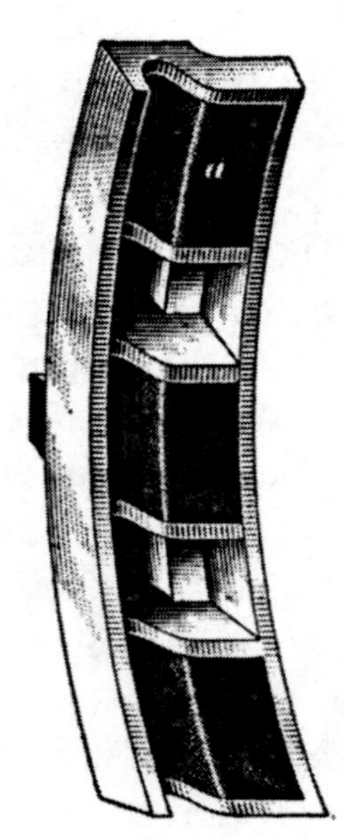

FIG. 78

Fig. 78 shows one of these shoes; it simply replaces the regular brake shoe and in the course of a few hours' run the abrasive blocks *a* grind the wheel true. A bad flat is removed by a regular grinder, which is a device for holding a revolving emery wheel against the tread of the wheel to be ground. The wheels may be ground either in place on the car or separate from the car. The better practice is to use the car-wheel grinder at one of the depots, if the wheels are to be ground on the car. Where the wheels are taken out to be ground, extra means must be provided for driving the axle; whereas, if ground on the car, one of the car motors can do the work. In either case, the car wheels should make from 20 to 40 revolutions per minute, and the speed of the rim of the emery wheels should be about 5,000 feet per minute. Steel-tired wheels are trued up by turning in a lathe.

QUESTIONS AND ANSWERS

84. The following questions and answers are introduced in order to call attention to some of the more important points brought out in the preceding instruction. Because of their great instructive value, the answers should be studied carefully and comparison made with the articles in the text to which reference is made. Thought expended on a statement fixes the fact contained therein more firmly in the mind than does a hasty reading of the answer or of the text to which it refers.

QUESTIONS AND ANSWERS RELATIVE TO MOTOR-CIRCUIT AND AUXILIARY APPARATUS

QUESTION.—Name the main parts of a trolley.

ANSWER.—The trolley, which is the device mounted on the roof of a car for collecting current from the trolley wire, consists of the wheel, the harp, the pole, and the stand. Art. **6.**

QUESTION.—What is the usual range of pressure of the wheel against the trolley wire?

ANSWER.—The pressure usually ranges from 20 to 40 pounds, depending on the tautness and alinement of the trolley wire, the speed of the car, and the condition of the roadbed. Art. **11.**

QUESTION.—Why is it sometimes desirable to install a trolley guard at a crossing of a steam and an electric road?

ANSWER.—So that in case the trolley wheel leaves the wire, the guard will lead the wheel back into its proper position on the wire and the car will not lose current in what may be a dangerous situation. Art. **14.**

QUESTION.—How may the circuit-breaker shown in Fig. 16 be operated by hand so as to open the circuit that it controls?

ANSWER.—By pressing the large button shown near the top of the case. A latch is unlocked and the switch contacts immediately open. Art. **18.**

QUESTION.—What is the purpose of punching a hole in the middle of a fuse of the type shown in Fig. 19?

ANSWER.—The cross-section of the fuse strip is thus decreased at the center and the metal will melt at that point in case an excessive flow of electricity occurs. There is less liability of a terminal clamp being injured if the arc forms at the center than if near the terminal. It is, therefore, desirable to locate the probable position of the arc midway between the terminals. Art. **20.**

QUESTION.—What prevents a considerable flow of electricity through an aluminum electrolytic arrester when the arrester is connected beween the trolley and the ground and the normal line voltage is impressed on its terminals?

ANSWER.—The film that is formed on the surface of the aluminum trays prevents all but a very small flow of electricity under normal conditions. If a very high electromotive force is impressed on the arrester, the resistance of the film breaks down and current passes to the ground. Art. **26.**

QUESTIONS AND ANSWERS RELATIVE TO LIGHTING, HEATING, AND AUXILIARY APPARATUS

QUESTION.—Why is it advisable that the signal lamps on a car be provided with a source of energy independent of the trolley system?

ANSWER.—In case of loss of trolley current at night and consequent stopping of the car, the car signal lamps should be illuminated so that warning may be given to another car that may be approaching through the use of current from an adjacent live section of trolley wire or by coasting. Sometime the signal lamp is supplied with current from a small storage battery and in other cases oil-burning signal lamps are employed. Arts. **27, 30,** and **31.**

QUESTION.—In the portable luminous arc headlight shown in Fig. 38, of what material is made: (*a*) the positive electrode? (*b*) the negative electrode?

ANSWER.—(*a*) Copper. Art. **35.** (*b*) An iron tube filled with a mixture of magnetite, chromium, and titanium. Art. **35.**

QUESTION.—With reference to Fig. 41, in what position should the double-throw switch be in where: (*a*) the luminous arc headlight is to be used with full brilliancy? (*b*) with dimmed brilliancy?

ANSWER.—(*a*) Closed in its upper position. Art. **36.** (*b*) Closed in its lower position. Art. **36.**

QUESTION.—State how three degrees of heat may be obtained with two heater circuits of unequal resistances.

ANSWER.—The high-resistance circuit alone is active for the lowest heating effect; the low-resistance circuit alone is active for the intermediate heating effect; and the two circuits connected in parallel provide the greatest heating effect. Art. **40.**

QUESTION.—Why are car bodies that are over 22 feet long usually equipped with double trucks instead of single trucks?

ANSWER.—A long rigid single truck will not readily pass around curves at street corners. Art. **63.**

QUESTIONS AND ANSWERS RELATIVE TO THE CAR HOUSE

QUESTION.—What estimate is sometime used to indicate the relation between the number of cars stationed at a depot and the linear feet of pit room to be allowed for repair and inspection work on them?

ANSWER.—One linear foot of pit room for each car that runs into a depot; that is, if 200 cars are stationed at the depot, there should be pit room under at least 200 feet of track. Art. **74.**

QUESTION.—What special shops are included in a well-equipped railway repair shop?

ANSWER.—A machine shop, carpenter shop, mill, blacksmith shop, paint shop, winding room, commutator room, controller room, store room, external oil room, and a wheel-grinding annex. Art. **75.**

QUESTION.—Name the equipment that is commonly used in a commutator room.

ANSWER.—A lathe, drill press, milling machine, commutator press, a mica-grooving machine, a supply of assembly rings, and wrenches, gauges, and a test for detecting faulty connections between bars and between a bar and the shell. Art. **78.**

QUESTION.—Because of fire risk in a paint shop, what precaution should be taken in regard to the mounting of car bodies?

ANSWER.—The car bodies should be mounted on trucks and so located that the cars can be quickly run out of the paint shop in case of fire. Art. **81.**

QUESTION.—(*a*) What are flats on car wheels? (*b*) What causes flats?

ANSWER.—(*a*) Flat spots on the tread of the wheel. Art. **83.** (*b*) When by means of the brakes or other causes the wheel is held from rotating, and the momentum of the car forces the wheel to slide along the track, flats are likely to be formed on the wheel. Art. **83.**

SPEED CONTROL

METHODS OF SPEED CONTROL

GENERAL REMARKS

1. The subject of speed control for electric-railway cars deals with the methods of starting cars from rest and controlling their speed while in motion. Series motors are usually employed to drive the cars, and change of motor speed is obtained by varying the voltage impressed across the motor terminals. A low voltage is at first impressed on the terminals of each motor, and its armature need revolve only at low speed in order to generate sufficient counter electromotive force to limit the motor current to the value required for the load condition. A higher voltage is then impressed and the motor speed increases until the counter electromotive force is sufficient to limit the motor current to that required by the new load condition. Finally, full line voltage is impressed and the motor runs at maximum speed for the voltage and load conditions, provided the field-coil connections remain unchanged.

The three methods commonly used for varying the voltage impressed on the motors are: (1) the introduction of an adjustable rheostat, or resistor, in series with the motors; (2) changing the motors from series to parallel connection across the circuit combined with the use of an adjustable resistor; and (3) the connection successively of the motor terminals to the taps of an autotransformer. The first two methods are for direct-current and the last for alternating-current operation.

RHEOSTATIC CONTROL

2. Simple Form of Controller.—The switch that is generally used to regulate the speed of an electric car is called a **controller,** and in most cases one of these devices is installed at each end of the car. A very simple form of rheostatic controller is shown in Fig. 1.

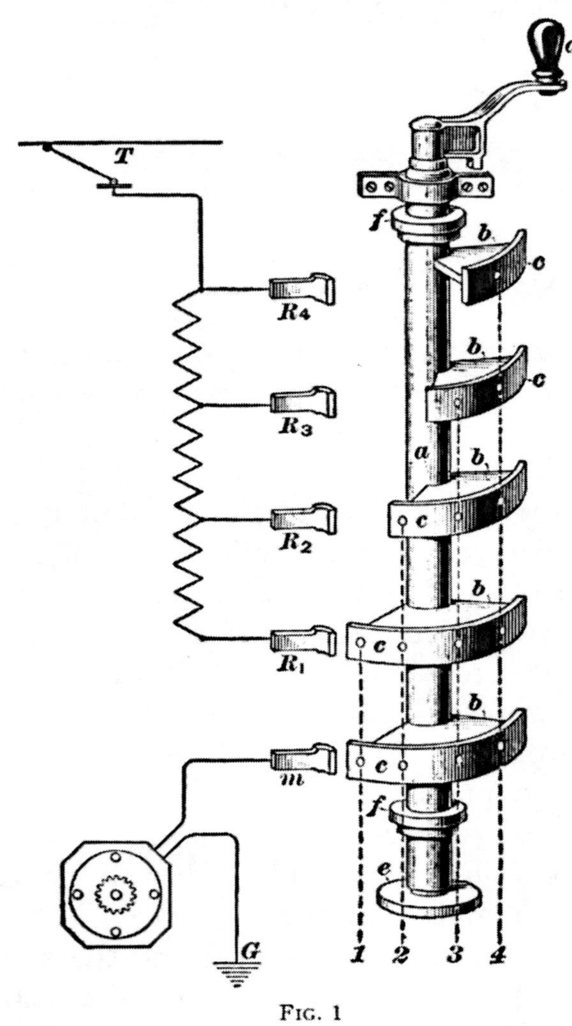

Fig. 1

An iron shaft *a* has projecting from it lugs *b*, on which are mounted copper segments *c*. The two lower segments are of the same length, but the others are of different lengths. The shaft is insulated from the handle *d* and the base *e* by insulating couplings *f*. The *brushes*, or *fingers*, which, when the shaft is turned, make contacts with the segments, are indicated at R_1, R_2, R_3, R_4, and *m*. The fingers marked *R* are connected to the three resistor sections. The trolley connection *T* is made at finger R_4, and the motor connection is made at finger *m*. The motor is grounded at *G*. The dotted lines *1*, *2*, *3*, and *4* indicate the relative positions of the fingers on the segments when the segments are turned into contact with the fingers on the first, second, third, and fourth *notches* of the controller.

The off-position is indicated in Fig. 1. The circuit is open between fingers R_1 and *c* and *m* and *c*. At notch *1*, fingers R_1

and m are in contact with their segments c and as all of the segments are connected together through their supporting lugs and the shaft, the motor circuit is closed and all of the resistor is active. At each notch a separate section of the resistor is short-circuited by the segments and shaft so that on notch *4* all of the resistor is short-circuited and this is a *running* notch. The resistor is not usually intended to be left in circuit; therefore, notches *1*, *2*, and *3* are not running notches.

3. Theory of Rheostatic Control.—When the whole resistor is active, a considerable part of the line voltage is expended in forcing current through the resistor and, therefore, the voltage impressed on the motor is low, resulting in a low speed. As the sections of the resistor are short-circuited, a larger part of the line voltage is impressed on the terminals of the motor and its speed increases accordingly. At the running notch, practically the full line voltage acts on the motor and its speed is maximum under given voltage and load conditions.

Speed control by rheostat alone is used for mine locomotives, hoists, cranes, etc., but not usually for railway cars.

SERIES-PARALLEL CYLINDER CONTROL

DISTRIBUTION OF VOLTAGE

4. Series-parallel control is common for direct-current railway work; it requires less resistance and is more economical than rheostatic control. Two or four motors are generally used with series connections for low speed and parallel connections for high speed, thus giving the name series-parallel control.

For example, with two motors, Fig. 2, in series across a 500-volt circuit, the pressure across each motor is only 250 volts and each will run at about half of its full speed without resistance in circuit. For higher speed, the connections are changed from series to parallel by means of the controller, thus impressing on each motor 500 volts, as indicated in Fig. 3, and resulting in full speed. Some resistance is used in starting and in passing

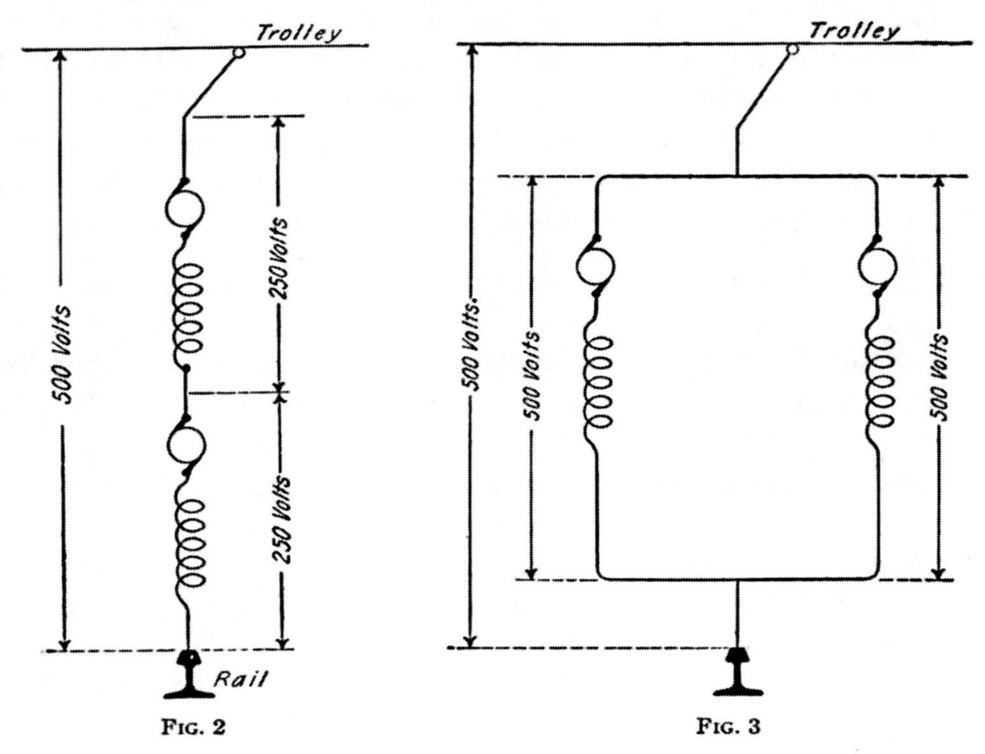

FIG. 2

FIG. 3

from series to parallel connections, but no resistance is in circuit on the running notch of the series or parallel position.

SERIES-PARALLEL CONTROLLER

5. In Fig. 4 is shown a type of controller intended to control the speed of two motors by the series-parallel method. The main controller shaft *a* supports six castings. All of the castings are insulated from the shaft and from one another and all of the segments on each casting are electrically connected. The cylinder of the reverse switch *b* carries the segments used for reversing the direction of rotation of the motors. The row of fingers for the main controller segments are shown below *c* and the row for the reverse segments below *d*. The small coils shown below *e* are blow-out magnet coils. Their action is to assist in breaking the arcs that tend to form when the segments leave the fingers. The arc forms a movable current-carrying conductor. The current in the blow-out coil establishes through iron extension plates and in the air near the fingers a magnetic flux. The arc conductor moves across the flux and the arc is blown out between partitions *f*, called *arc deflectors*, which in their running position partly enclose the fingers. The parti-

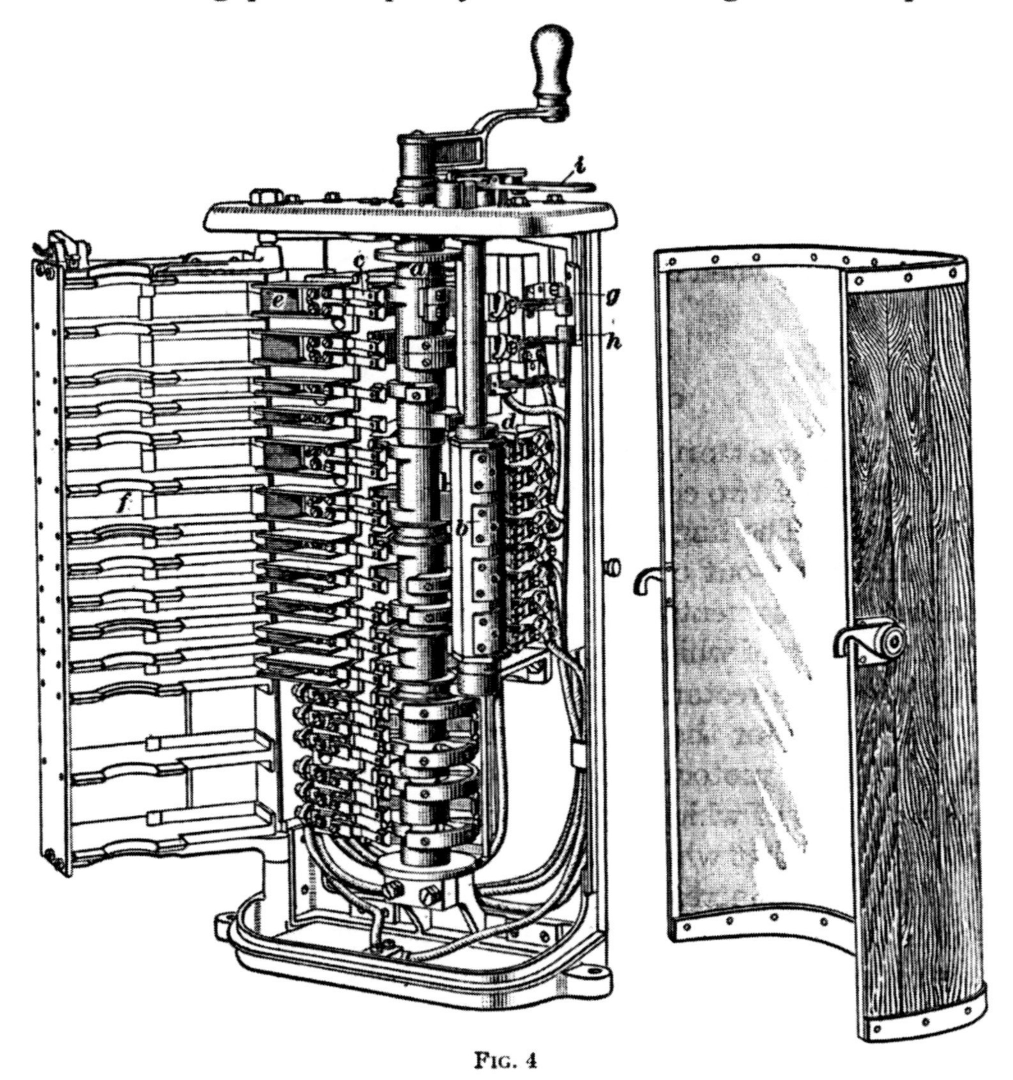

Fig. 4

tions of insulating material prevent the hot gases of the arc from forming short circuits across adjacent connections. Cut-out switches *g* and *h* serve to cut out of circuit either motor in case of a fault in one motor.

The main controller shaft should take up a definite position corresponding to each notch; therefore, a device is provided, usually in the form of a notched wheel, called the *star wheel*, mounted near the top of the shaft. The star wheel turns with the shaft, and a small roller is drawn by a spring into the notches, making the movements positive and definite.

An interlocking device prevents the reverse-switch shaft from being turned unless the main shaft is at off-position. This prevents reversing the motion of the car when current is in the motors. The reverse-switch handle *i* cannot be removed except at the off-position of the reverse switch, at which position the motor circuit is open.

When the reverse-switch handle points ahead, the car runs forwards, and when it points back, the car runs backwards.

CAR-WIRING DIAGRAM

6. Explanation of Marks on Diagram.—Fig. 5 shows the connections of two controllers and two motors for series-parallel operation. The fingers are indicated by the small black circles and their blow-out coils by the zigzag lines to the left of these circles; the segments, by the black rectangles; the controller castings, each of which supports a group of connected segments, by the outline rectangles *1*, *2*, *3*, *4*, *5*, and *6*; and the positions of the fingers for the controller notches, by the vertical dotted lines. These motors are provided with commutating poles, the windings of which are not indicated in Fig. 5, and have taps *M1* and *M2* that when active cut out portions of the main field windings, thus increasing the speed of the motors. The cut-out switches are shown at the right of the upper portion of the controller and the reverse switch just below.

There are many hundreds of wiring diagrams showing methods for the speed control of railway cars. It is the purpose here, however, to treat only of the general principles of the different methods of speed control and to familiarize the student with the usual manner of representing motor and controller connections so that any ordinary wiring diagram may be read.

7. Controller Steps.—Fig. 6 shows in detail the method of tracing the circuits of a car-wiring diagram. The blow-out

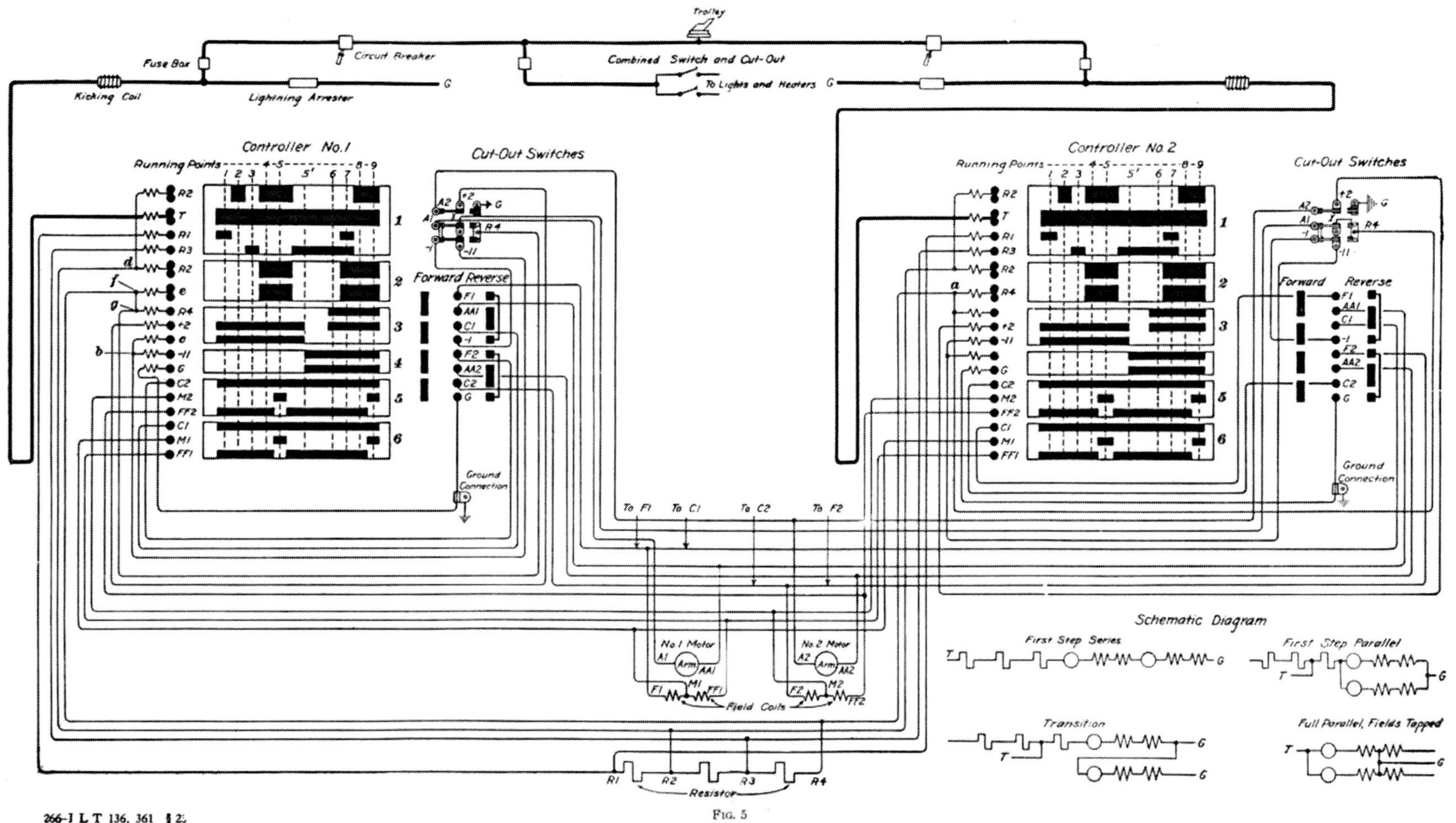
Trolley
Fuse Box
Circuit Breaker
Combined Switch and Cut-Out
Kicking Coil
Lightning Arrester
To Lights and Heaters
Controller No. 1
Controller No. 2
Running Points
Cut-Out Switches
Forward Reverse
Ground Connection
To F1
To C1
To C2
To F2
No. 1 Motor
No. 2 Motor
Arm
Field Coils
Resistor
Schematic Diagram
First Step Series
First Step Parallel
Transition
Full Parallel, Fields Tapped

FIG. 5

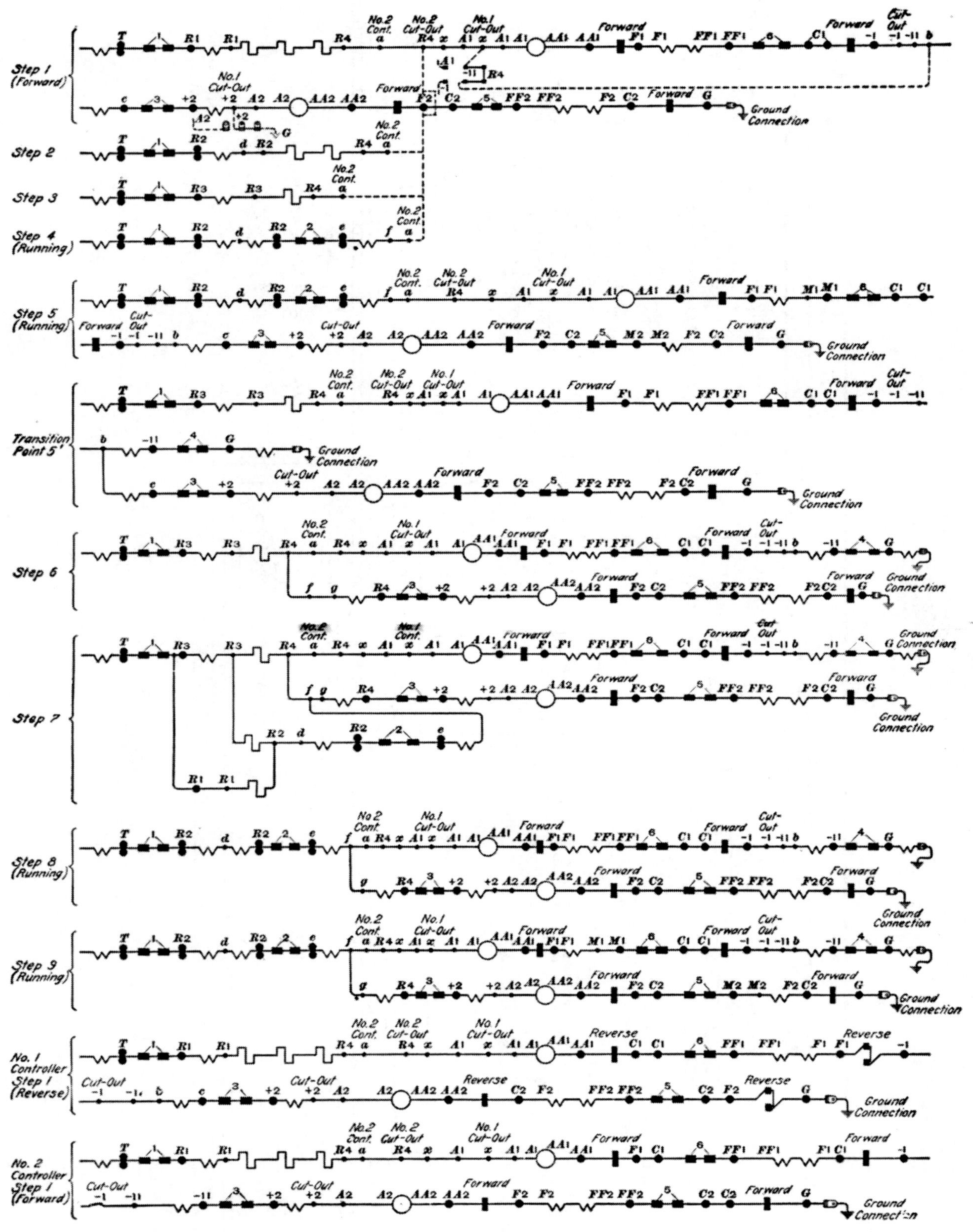
Step 1 (Forward)
Step 2
Step 3
Step 4 (Running)
Step 5 (Running)
Transition Point 5'
Step 6
Step 7
Step 8 (Running)
Step 9 (Running)
No. 1 Controller Step 1 (Reverse)
No. 2 Controller Step 1 (Forward)
No. 2 Cont.
No. 2 Cut-Out
No. 1 Cut-Out
Forward
Reverse
Cut-Out
Ground Connection

FIG. 6

coils, fingers, segments, resistor, field coils and armatures are represented the same as in Fig. 5, but Fig. 6 shows the connections of these parts on the successive steps of the controller. *The student should trace each path on Fig. 5 as it is shown on Fig. 6.* This is important in order to gain facility in tracing wiring diagrams.

Any diagram can be traced in this manner and a simple schematic diagram made to show connections on each step.

For example, with the circuit-breaker of No. *1* controller, Fig. 5, closed, the reverse switch in forward position, the cut-out switches closed to the left and the drum of controller No. *1* turned until the fingers rest on the *first series step 1*, the current path is as shown in step *1*, Fig. 6. Fingers *T* and *R1* rest on segments of casting *1*; fingers *+2* and *c* on segments of casting *3*; fingers *C2* and *FF2* on segments of casting *5*; and fingers *C1* and *FF1* on segments of casting *6*. The armatures and field coils of both motors and all the sections of the resistor are in series, as shown in the schematic diagram, Fig. 5, and the *first step*, Fig. 6.

On the *second series step*, an inspection of the development of the controller segments shows that the only change in the segments caused by the movement from step *1* to step *2* is such that finger *R1* is now inactive and upper finger *R2* is active. One resistor section is cut out, since the current can pass from finger *T* through finger *R2* to *R2* on the resistor.

On the *third series step*, the change in segments is such as to make finger *R2* inactive and finger *R3* active. Finger *T* is now connected through finger *R3* to *R3* on the resistor and there is only one section of the resistor in circuit.

On the *fourth series step*, which is the first series running point, upper and lower fingers *R2*, and finger *e* of the first part of the circuit are active. Finger *T* is now connected through fingers *R2*, *R2*, and *e* to *a* on the No. *2* controller and all of the resistor is cut out.

On the *fifth series step*, which is the second series running point, the change in segments is such as to cut out portions *M1–FF1* and *FF2–M2* of the motor field windings. Fingers *M1* and *M2* and taps *M1* and *M2* are now active. The speed on the fifth step is higher than on the fourth step because the weakened magnetic flux of the motors necessitates a higher speed, in order to generate the proper counter electromotive force for the load conditions.

On the *transition point 5′*, Figs. 5 and 6, a section of the resistor is cut into circuit and for an instant both ends of the No. *2* motor circuit are grounded, thus temporarily short-circuiting this motor.

On approaching the *sixth step*, which is the *first parallel resistance step*, fingers *+2* and *c* drop their contacts with casting *3* and the No. *2* motor circuit is opened. When the sixth step is completed, fingers *R4* and *+2* make contacts with casting *3*, and the armature end of No. *2* motor circuit is connected to point *R4* at the end of the last section of the resistor. The two motors are in parallel and one section of the resistor is connected ahead of the junction of the motors.

On the *seventh step*, which is the *second parallel resistance step*, the three sections of the resistor are in parallel ahead of the motors which are also in parallel.

On the *eighth step*, which is the *first parallel running point*, fingers *T*, *R2*, *R2*, and *e* are active and all of the resistor is cut out.

On the *ninth step*, which is the second parallel running point, fingers *M1* and *M2* and taps *M1* and *M2* are active and the motors are in parallel without the resistor and with portions of the field windings cut out. This is the high-speed point.

8. Motor Connections.—When two motors are placed on a single truck, it is customary to install them back to back, with the commutator ends of the two motors toward opposite sides of the truck. When the motors are so placed, the rotation of the two armatures must be in opposite directions, when viewed from the commutator end of each motor, in order for the motors to act in unison in driving the car. The controller connections must be so made that the current in the field coil of one motor must be in the opposite direction to that of the current in the field coil of the other motor, provided the armature currents have the same relative directions. In Fig. 6, first step, the current enters the *A1* end of the No. *1* armature and the corresponding *A2* end of the No. *2* armature, but the current enters the *F1* end of the No. *1* field coil and the *FF2* end of the No. *2* field coil. The rotation of the No. *2* motor is, therefore, opposite to that of the No. *1* motor.

9. Action of Reverse Switch.—If the reverse switch, Fig. 5, is thrown so that the reverse segments are in contact with the fingers, the direction of the current through the field coils of both motors is opposite to that when the forward segments are active and, therefore, the motion of the car is reversed. In the next to the last tracing of Fig. 6, the path of the current is indicated for the first reverse series steps. The relation between the directions of the armature and field currents is maintained throughout the reverse steps on the No. *1* controller. The directions of the field currents should be noted in Fig. 6 for the tracing for the first forward series step and for the first reverse series step of the No. *1* controller.

10. Braking by Reversing.—The car may be stopped suddenly by throwing the main controller handle to off-position, throwing the reverse handle to its reverse position, and then advancing the main controller handle. The motors will then tend to drive the car in the direction opposite to the direction that the momentum of the car is driving it; the car will come to a sudden halt and move backwards if the current is left on. This action causes very severe stresses on the apparatus, and should be resorted to only in case of an emergency.

11. No. 2 Controller Connections.—The No. *2* controller, Fig. 5, is similar to the No. *1* controller except that the short connecting wire, or jumper, near *x* is used on the No. *2*, but not on the No. *1*, and the field-coil connections of each motor are reversed because forward movement when referring to one end of the car is the opposite to the forward movement when referring to the other end of the car. It is necessary, therefore, that for reverse on No. *1*, the motors should rotate in the same direction as for forward on the No. *2* controller. The last tracing in Fig. 6 indicates the path and direction of the current for the first forward series step of the No. *2* controller. It should be noted that the direction of the current in the field coils of both motors is the same for the first reverse series step of the No. *1* controller (next to the last tracing) as for the first forward series step of the No. *2* controller (last tracing), thus the car motion that is backwards for the No. *1* is forwards for the No. *2* controller.

12. Action of Cut-Out Switches.—Cut-out switches serve to disconnect a faulty motor in a two-motor equipment or a pair of motors in a four-motor equipment in case one motor of a pair is faulty. The two motors of a pair are connected together in parallel and may be considered as one motor in regard to general connections. Suppose that No. *2* motor, Fig. 5, develops a fault and the blade of the upper cut-out switch on the No. *1* controller is thrown to the right, point +*2* on the cut-out switch, Fig. 5, and the first tracing, Fig. 6, is grounded at *G*, and there is an open circuit between points +*2* and *A2*. The No. *1* motor and three sections of the resistor are connected between the trolley and the ground connection of the upper cut-out switch and the circuit of the No. *2* motor is opened at that point, as indicated by the dotted connecting lines under point +*2*, first tracing Fig. 6.

If the No. *1* motor is faulty, the two blades of the lower cut-out switch are thrown to the right and the blades connected together by the double clip at terminal *R4*, Fig. 5. Point *x* is now connected to point −*11* and the No. *1* motor circuit is opened between points *x* and *A1* and −*1* and −*11*. The No. *2* motor circuit is completed through *x*–*11*–*b*–*c*, etc., Fig. 5. The path is indicated by the dotted connection near *x* on the first tracing of Fig. 6.

MOTORS USED AS BRAKES

13. Forward Movement of Car.—Under certain conditions, the motors of a car may be used to brake the car independently of the mechanical brakes and of the connection of the car to the trolley line. The ability of the motors to act as generators enables them to perform this duty.

The case first considered is when a car equipped with two motors is running forwards down a hill, with the brakes out of order and the trolley off the line. If the reverse switch is thrown to reverse position and the controller to any one of the parallel steps, preferably the first resistance parallel step, one of the motors will almost immediately pick up as a generator and the current from this generator will pass through the other motor, tending to drive that motor in opposition to the motion of the car. The machine acting as a generator requires energy to drive it and this is furnished by the moving car. The combined action of both machines is, therefore, to retard the speed of the car. As the speed slackens, the retarding effect decreases so that the car cannot be brought to a stop on a down grade by this method. The speed is greatly reduced, however, and it may be possible to stop the car by placing an obstruction on the rail.

14. In order for either one of the series motors to pick up as a generator, its circuit must be complete and the relative direction of rotation and of flux must be suitable. Throwing the reverse switch to reverse position gives the proper relation between rotation and flux, and moving the controller handle to a parallel position completes the circuit by grounding one terminal of each motor and interconnecting the other two terminals, as indicated by step *6*, Fig. 6. The relation between armature and field-coil connection would be, however, that indicated in step *1* (reverse), Fig. 6.

Both motors tend to start generating and to set up electromotive forces that directly oppose each other; but one machine starts to pick up a little before the other, owing to the fact that the residual magnetism of one machine is usually a little stronger than that of the other, as explained in connection with series generators in parallel in a previous Section. The stronger electromotive force overcomes the weaker electromotive force, and the first machine picks up as a generator and forces current through the other machine, which acts as a motor.

15. Backward Movement of Car.—In case the car starts to run backwards down a hill and the motors are to be used as brakes, the overhead switch should be knocked off, the reverse switch left on its forward position, and the controller drum turned to a parallel position. As the car is running backwards, the rotation of the armatures is reversed, so that the armatures and field coils are properly connected to make the motors generate and set up braking action.

16. Braking Action With Four-Motor Equipment. When there are four motors on the cars arranged in two pairs, the braking action will occur if, when the car is moving forwards, the reverse switch is thrown to reverse position and the controller left at the off-position. For backward movement. the braking action occurs when the reverse switch is thrown to forward position and the controller left at the off-position.

17. Precautions.—It is well to know how to use the motors as brakes in case a brake chain should snap and the line lose its power, thereby rendering both the brakes and reversing gear ineffective. It is not well, however, to make a practice of stopping a car in this way, for as the fuse or breaker is outside of the local circuit, consisting of the two motors and the reversing mechanism, the motors are not protected from overload. Again, the sudden reversing of the armature of the machine, acting as a motor, strains the pinion just as does regularly reversing, or "plugging," the car under headway.

SERIES-PARALLEL MULTIPLE-UNIT HAND CONTROL

MASTER CONTROLLER AND MOTOR CONTROLLER

18. When a car is intended to be used as a part of a train, some form of multiple-unit system is employed to control the speed of the train. In the multiple-unit system, a *master controller* distributes current through control circuits that include the operating coils of switches. The switches are in the main circuits of the car motors and resistor and act in response to movements of the master controller. The currents required for the control circuits of the operating coils are small and the master controller, therefore, is very compact. Its operation and general appearance is, however, similar to that of the larger type previously described.

By extending the control circuits from car to car by means

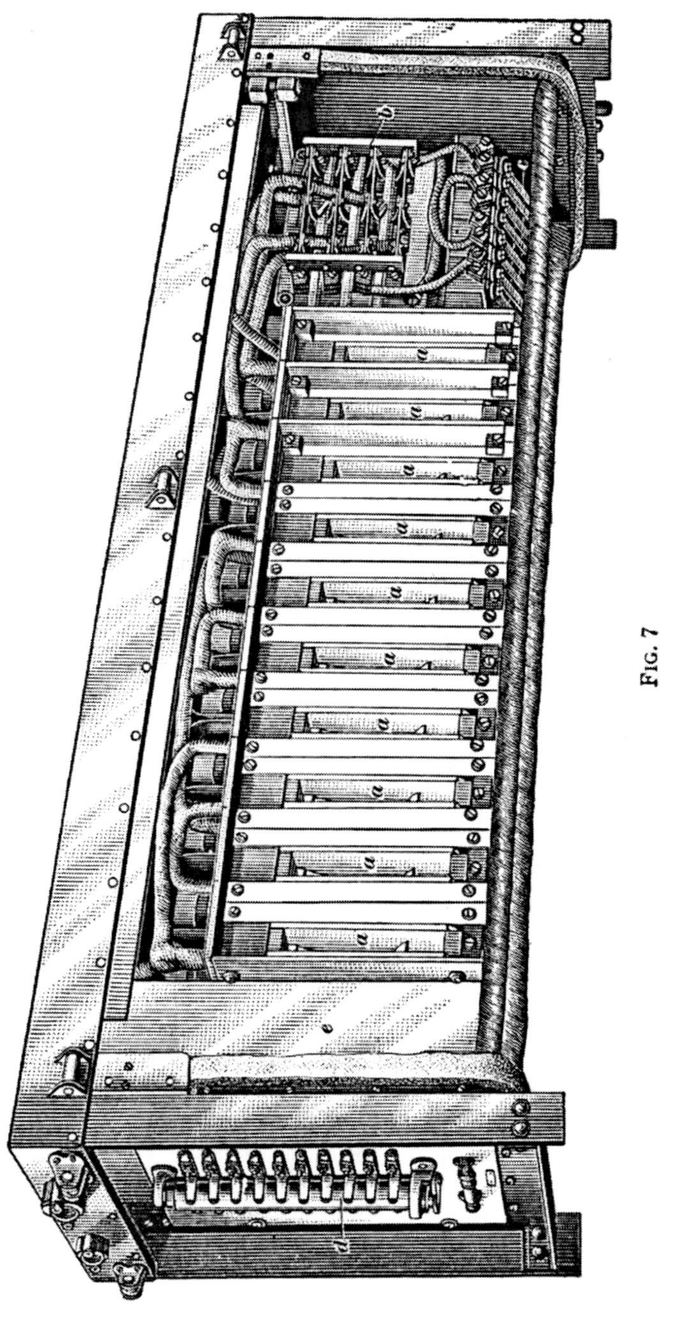

Fig. 7

of cables and jumpers, each car is operated as a unit in parallel with the other cars from any master controller, usually the controller on the front end of the first car of the train. The system is sometimes used on a car intended for heavy service and for single operation. The heavy wiring required for the motor circuits need not be brought to the master controller and much space on the platforms is gained by the use of the small master controller. The motor controller contains a group of main-circuit switches and is placed under the car. The operating coils of the motor controller are connected by the control cables to the master controller.

Fig. 7 shows a *motor controller*, or a *contactor box*. The switches, or contactors, are at positions *a*; the motor cut-out switches, at *b*; the motor reverse switches, at *c*; the cut-out switches for the control circuits, at *d*; and an overload relay, at *e*.

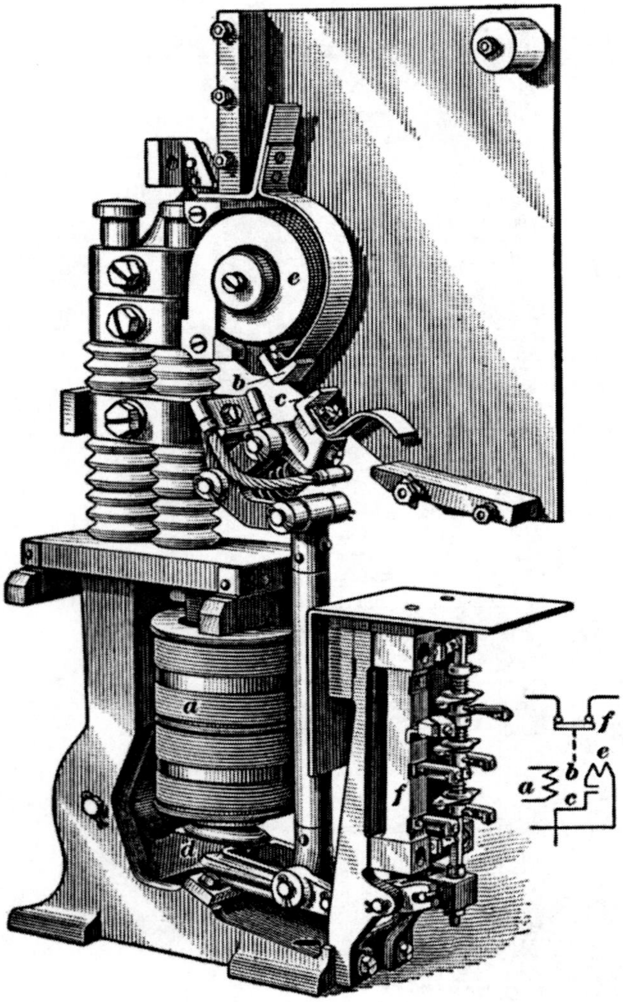

Fig. 8

CONTACTORS

19. The *contactors* vary in details of construction to suit different operating conditions, but their action is usually based on the same general principles. Fig. 8 shows the working parts of a contactor intended for 2,400-volt service. In contactors for lower voltage, the parts are arranged more compactly. When current passes through the operating coil *a*, the switch contacts *b* and *c* are closed by means of a connecting-rod extending to the armature *d* of the coil. When the current in the coil ceases, the switch contacts open and any arc that tends to form between them or their projecting horns is blown out by the blow-out coil *e*.

The movement of the armature *d* also causes the shaft of the interlock switch *f* to move. Small disks mounted on the shaft move into or out of contact with stationary terminal posts. The contacts *b* and *c* form a part of the main motor circuit and the contacts of the interlocks form part of the control circuits. The interlocks serve to arrange circuits for the operating coils of the contactors.

In car-wiring diagrams, the operating coil *a*, the contact points *b* and *c*, the blow-out coil *e*, and the interlock *f* are usually represented as indicated in the detail sketch.

CUT-OUT SWITCHES AND REVERSER

20. Fig. 9 shows the motor cut-out switches *a* and *b* and the reverser *c*. In a four-motor equipment, switch *a* cuts out

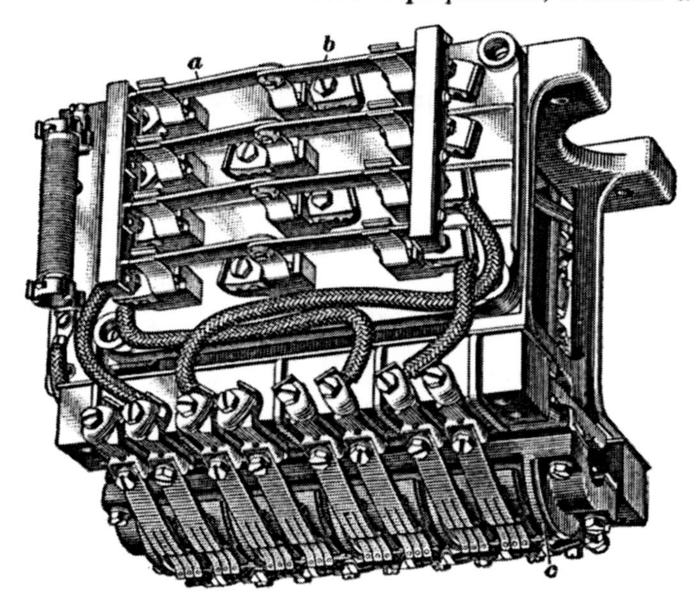

FIG. 9

one pair of motors and switch *b*, the other pair. The reverse switch is provided with two operating coils, both of which are included in the control circuits. One coil turns the reverse-switch cylinder to forward position and the other coil, to reverse position. The reverse switch reverses the connections of the field coils of the motors.

OVERLOAD RELAY

21. Fig. 10 shows one form of overload relay. The coil *a* is in the motor circuit. In case of excessive current, armature *b* is drawn up, moving latch lever *c* and releasing the armature *d* of the coil *e*. The action of armature *d* opens through levers *f* the contacts of three small switches of the control circuits. The blow-out coils of these switches are shown at *g*. The circuits of the switches include operating coils of active contactors and when the switches open, the motor circuits are opened by the contactors. The current in coil *a* then ceases and the armature *b* drops. The control switches may be closed again by passing current from the control circuit through the reset coil *e*. The armature *d* is then moved toward coil *e* and its action closes the switches.

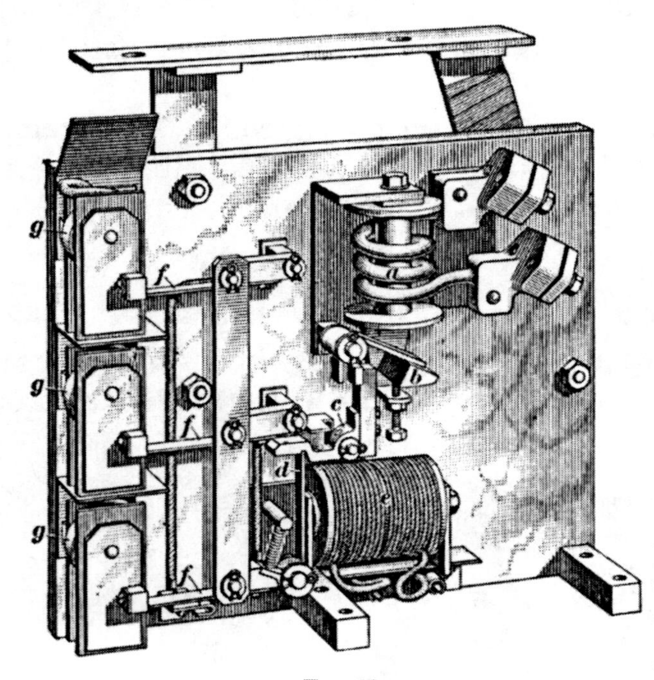

FIG. 10

CONTROL AND RESET SWITCH

22. Near the master controller is a small double-throw switch, one type of which is shown in Fig. 11. The switch in one position closes the supply circuit for the master controller, and in the other position closes the circuit to the reset coil of the overload relay in the motor controller. In the reset position, only temporary contact is desired and the handle is held against a spring pressure that will return it to off-position

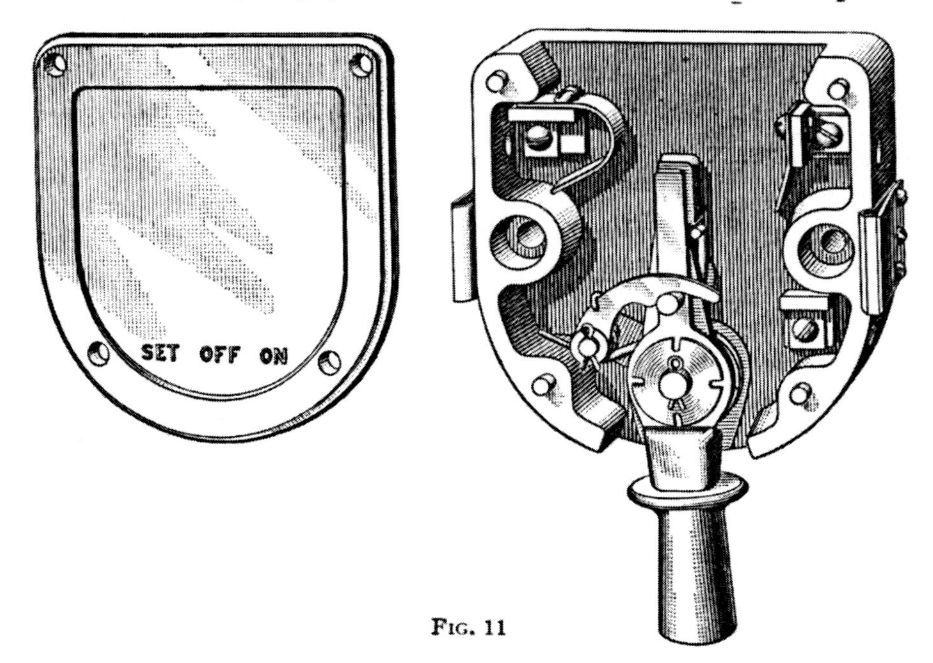

FIG. 11

when the hand is removed. The notches on the segment near the center of the switch allow the switch to remain at off-position or at closed position on the left contact.

CAR-WIRING DIAGRAM

23. In Fig. 12 is shown a car-wiring diagram for a multiple-unit system with hand-operated control (the Sprague-General Electric type M K). The diagram shows connections for a large car that is usually operated singly, but which, if desired, may form part of a train. At starting, the acceleration of the car depends on the time taken by the operator to move the controller handle through the different steps.

The wires for the control circuit are indicated by light lines and for the motor circuit by heavy lines. The small open circles located in the control circuits indicate resistors and the numbers near the circles indicate the resistances in ohms. One end of the control circuit is connected to the upper contact of the main switch and the other end to the ground. The full voltage of the line is impressed on the control circuit, but the resistors and the operating coils limit the current to a small amount (about 2 amperes per car). One hood switch controls the current for both master controllers and a controller and reset switch at each end of the car control the current for the master controller at that end and for the reset coil of the overload relay.

To start the car, the main control switch is closed and the control and reset switch for the master controller that is to be active is first thrown so as to make contact at the reset side, which energizes wire *10* and the reset coil of the relay, and is then finally closed to the controller side. The master controller reverse switch is thrown to forward position and the master controller is then moved through the steps.

24. Control Circuits.—In Fig. 13 are indicated the paths of the control currents through the circuits formed when the controller is moved through the steps. The devices, such as controller fingers, operating coils, interlocks, resistors, etc., in Figs. 12 and 13, are indicated in a similar manner. Each path, indicated in Fig. 13, should be traced on Fig. 12.

On the first step, Fig. 13, the control current passes through No. *1* casting of the controller, the forward segment of the controller reverse switch, the forward operating coil *0* of the reverser, the two lower interlocks attached to coil *0* and to ground. The operating coil immediately causes the two lower interlocks to open, the upper interlock to close, and the path is then indicated by the lower connections from the forward operating coil of the reverser. Contactors *7* and *9* of this path are closed. Another path is through wire *1*, contactor *1*, and wire *6*. Contactors *1*, *7*, and *9* are active.

When tracing the steps, the positions of the interlocks of the active contactors should be noted. In Fig. 12, the positions of the interlocks when the contactors are inactive are indicated. In Fig. 13, the interlocks are numbered the same as the contactors to which they are attached, as interlock *10* of the first step.

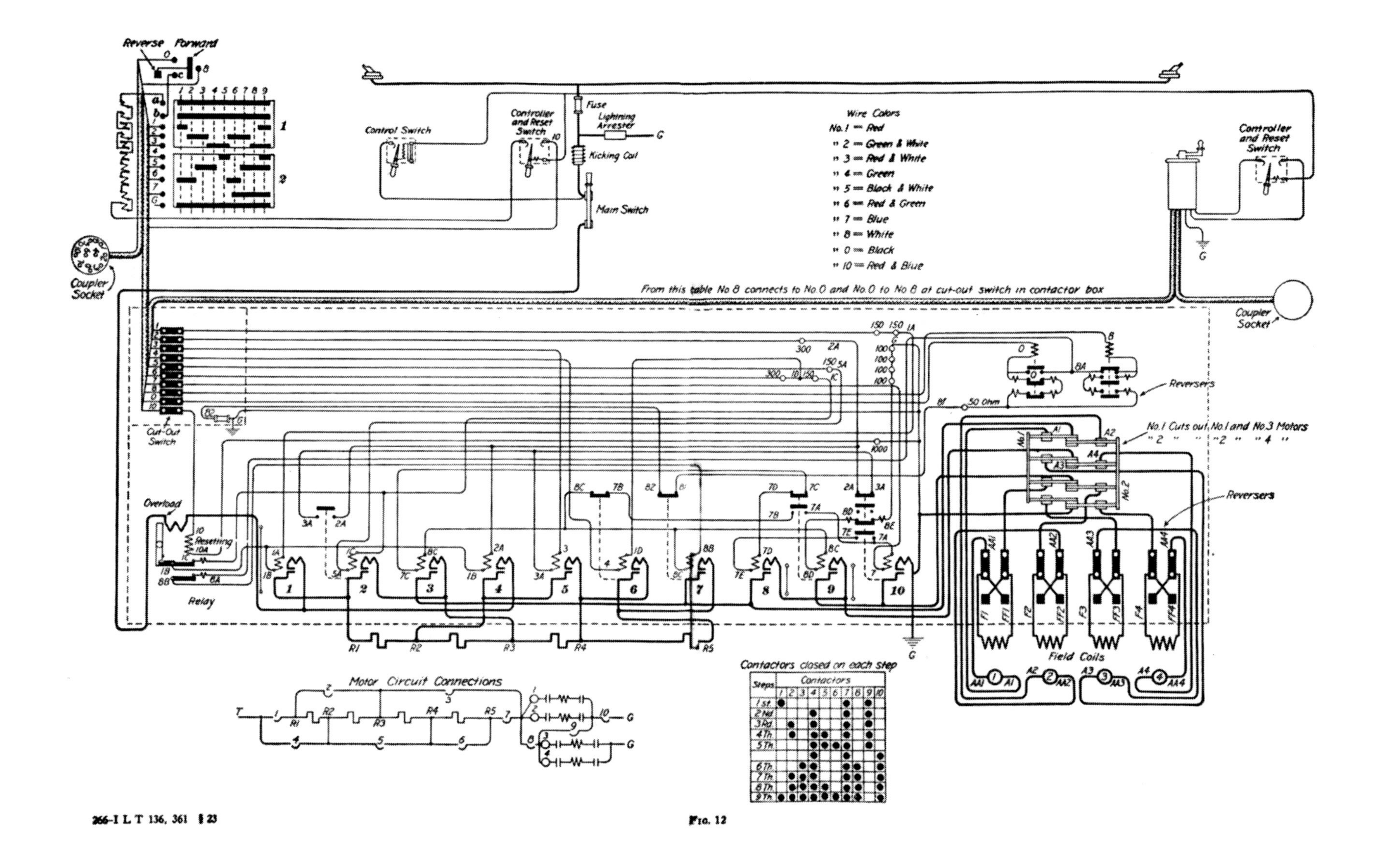

Steps	Contactors 1	2	3	4	5	6	7	8	9	10
1st.	●						●		●	
2Nd.				●			●		●	
3Rd.		●		●			●		●	
4Th.		●		●	●		●		●	
5Th.				●	●	●	●		●	
				●			●			●
6Th.			●	●			●	●		●
7Th.		●	●	●			●	●		●
8Th.		●	●	●	●		●	●		●
9Th.	●	●	●	●	●	●	●	●		●

FIG. 12

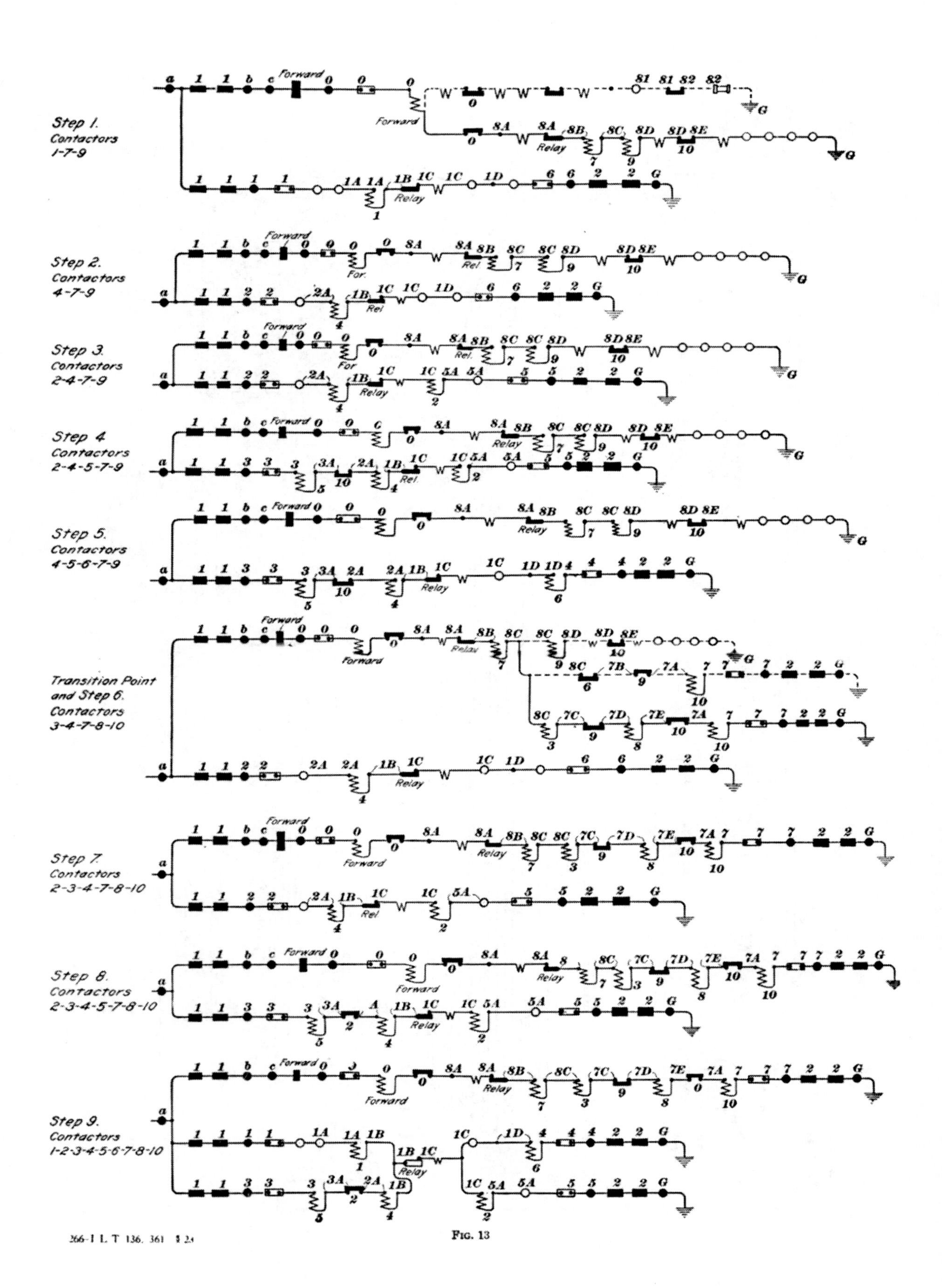

Fig. 13

266-I L T 136, 361 $ 2s

On the transition point and step *6*, the changes in the active controller segments and active interlocks cause the changes in the current paths noted in the figure. The portion of the paths connected by dotted lines indicates the preliminary paths. The full-line paths are for step *6*.

In all of the control circuits, except the preliminary circuit for the reverser, either of the two control switches used in the overload relay of this system is included in the circuit; therefore, in case of excessive current, these switches open, which causes the contactors to open the motor circuits.

25. Motor Circuits.—The sketch, Fig. 12, showing the motor circuit connections and the table showing the contactors closed on each step serve to indicate the motor connections at each step. On the first step, with contactors *1*, *7*, and *9* active, the main current passes through four sections of the resistor and two pair of motors in series. Each pair of motors consists of two motors in parallel. On the sixth step, with contactors *3*, *4*, *7*, *8*, and *10* active, the current passes through one section of the resistor *R2–R3*, and then through the four motors in parallel. The resistor sections are used either in series or in parallel or in combination on the different steps.

26. Reverser.—When all of the fingers of the reverser, Fig. 12, bear on the large segments, as shown, current in the field coils has one direction; when the reverser drum is turned so that only the upper row of fingers is on the large segments and the lower row on the small segments, current in the field coils is in the other direction. The direction of car movement is thus changed by turning the reverser drum.

When the forward segment of the controller reverse switch at either master controller is active, the operating coil of the reverser that is then energized moves the reverser to the position for forward motion of the car. When the reverse segment is active, the reverser is moved to the position for backward motion.

27. Couplers.—Cars to be used in multiple-unit trains are provided with sockets mounted near each platform and which are connected to the wires of the control cable on the car.

Two coupler sockets are shown in Fig. 12. A jumper cable with plugs at either end serves to connect the sockets of adjacent cars. The control system is thus extended throughout the train and any master controller will control all of the motors on the train. If the motors on one car are not to be used the control cut-out switch, shown below the left-hand master controller, Fig. 12, is thrown so as to open all of the control circuits.

SERIES-PARALLEL MULTIPLE-UNIT AUTOMATIC CONTROL

AUTOMATIC CLOSING OF SWITCHES

28. In automatic control, the controller handle may be moved slowly from step to step or the handle may be thrown at once to either full series or parallel running position and the switches in the motor circuit will close automatically and in the same order as for the gradual movement. A limit switch, the operating coil of which is included in the motor circuit, opens by means of its interlock certain of the control circuits in case the motor current exceeds the value for which the switch is set and delays the closing of further switches until the current decreases to a safe value.

UNIT SWITCHES

29. The motor switches in the Westinghouse system of multiple-unit control here considered are closed by compressed air taken from the air-brake system of the train. The magnet valves operating the switches are controlled by the currents in the control circuits and the proper distribution of these currents is made by the master controller.

Fig. 14 shows a unit switch. The magnet valve is shown at *a*; the air cylinder at *b* with part of the casing broken away to show the coiled spring inside; and the main switch contacts and their projecting horns at *c* and *d*. A blow-out coil is so placed in front of the arc shield *e* that its magnetic flux passes through the space occupied by the contacts *c* and *d*. If the unit switch is provided with an interlock, the segments of the interlock are mounted on the projecting arm *f* of the piston rod. The interlock fingers are fixed in position and the segments move up and down against the fingers, thus making or breaking the control circuits necessary to cause the progressive action of the unit switches. The general arrangement of the interlock fingers and segments is shown in Fig. 15.

When the magnet valve *a*, Fig. 14, is energized, the small air valve *g* is depressed, allowing air from the inlet *h* to pass to the bottom of the piston in the cylinder *b*. The piston and its rod are forced upwards, closing the switch and moving the segments of the interlock. The switch remains closed as long as the air pressure exists on the piston. If the circuit of the magnet valve is opened, the inlet valve *g* closes and an exhaust valve under the magnet opens, allowing the air under the piston to escape. The spring in the air cylinder then opens the switch, which also happens when the air pressure in the system fails from any cause.

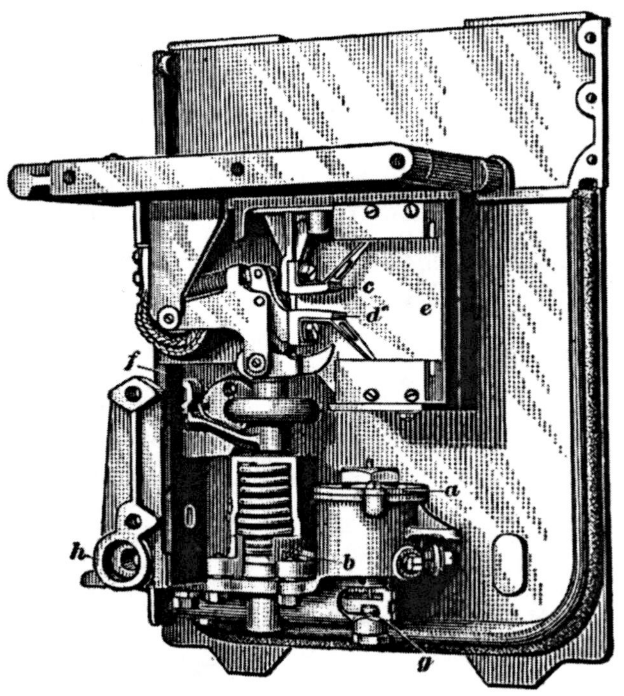

Fig. 14

The unit switches are arranged in a switch group, placed under the car, and connected to the control circuit, the motor circuit, and the air-brake system. Fig. 15 shows the interlock side of a switch group with the cover removed from the case. Control cables are shown passing through one end of the case, and bushed openings for the entrance of the motor leads are shown near the top.

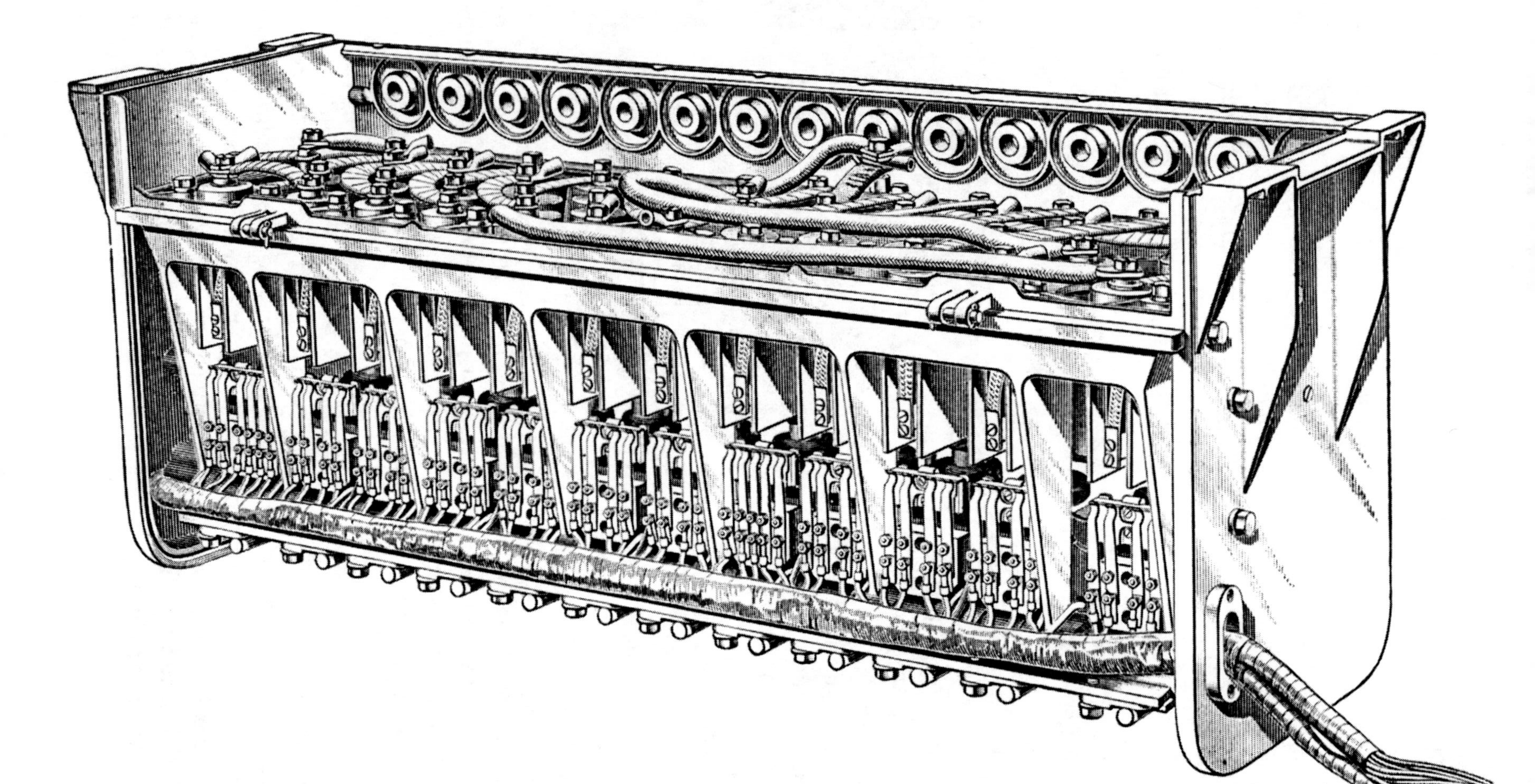

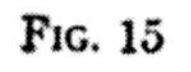

FIG. 15

OVERLOAD TRIP

30. Fig. 16 (*a*) shows the construction of the overload trip at one end of the switch group. The trip coil *a* is the blow-out coil of the adjacent unit switch. In case of overload, the plunger *b* is drawn into the trip coil, thus opening two control circuits by withdrawing contact disks *c* and *d* from stationary contacts *e f* and *g h*. At the same time a spring *i* is compressed and the end of the plunger of a reset magnet falls into a notch *k*, thus holding the control circuits open, as in (*b*). By closing a reset switch near the operator, the reset magnet *j* is energized, the plunger withdrawn from notch *k*, and the spring *i* closes the control circuits again, if the overload has been removed. The tripping point is adjustable by screwing in or out the small rod attached to plunger *b*. This rod turns within the shell that holds the disks.

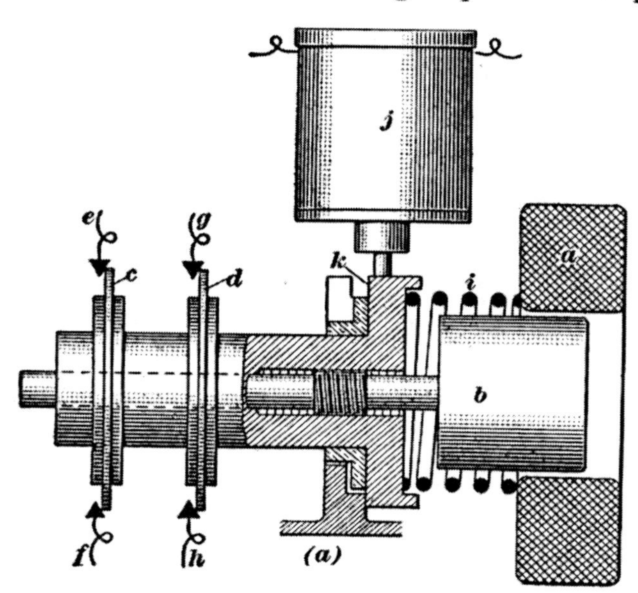

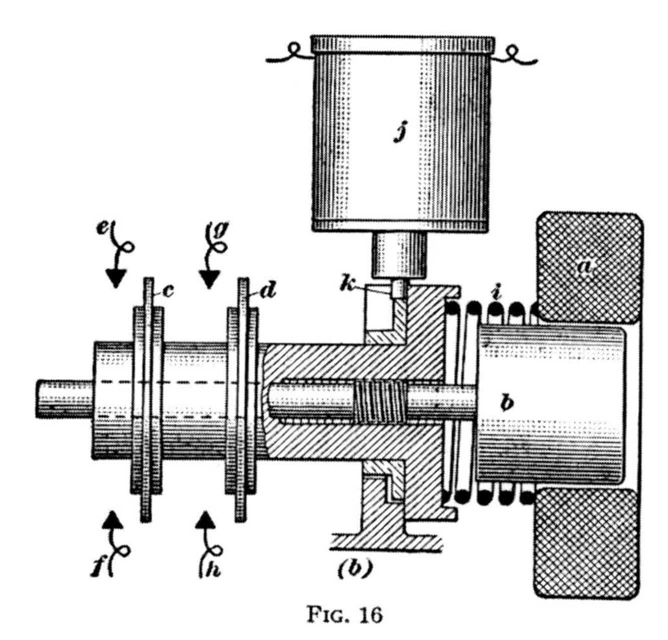

Fig. 16

CAR-WIRING DIAGRAM

31. Fig. 17 is a simplified, schematic, car-wiring diagram of the Westinghouse unit-switch automatic control. In this diagram, the parts are arranged where most convenient to show connections, regardless of their true arrangement. The master controller, the main contacts of the unit switches, and the motor circuits are shown above the train-line cable and the control circuits and the magnet valves of the unit switches below this cable. The operating coil of the limit switch is shown to the right of motors Nos. *2* and *4*, and its interlock is indicated near the disk marked *Limit*, in the control circuit.

The combined blow-out and trip coil of line switch *LS1* is shown below the main fuse. The control disks of the trip interlocks of the two line switches *LS1* and *LS2* are marked *Overload Trip* and are in circuit with the magnet valves of these switches as shown in Fig. 17. The reset coil of the overload trip is in circuit with control wire *8*, as shown at the right side of the diagram.

The segments and the fingers of the interlocks are indicated by the blocks and by the small circles. When the switches are *out*, the segments make contact with the fingers on a line even with the word *out*; and when *in*, on a line even with the word *in*. The main contacts of the unit switches are indicated in the motor circuit by short vertical parallel lines.

The voltage impressed on the magnet valves of the control circuits is that between points *6* and *7* or *7* and *G* on the control resistor connected between the trolley and the ground. In some equipments, the current for the control circuits is obtained from a storage battery.

32. With connections complete and the car ready to start, the controller handle can be thrown at once to notch *4*, Fig. 17, for forward movement of the car or to notch *2*, for backward movement. In the following, it is assumed that the handle is moved to the forward parallel-running notch *4*.

When tracing the circuits, the table of sequence of switches, shown near the right-hand upper corner of Fig. 17, should be

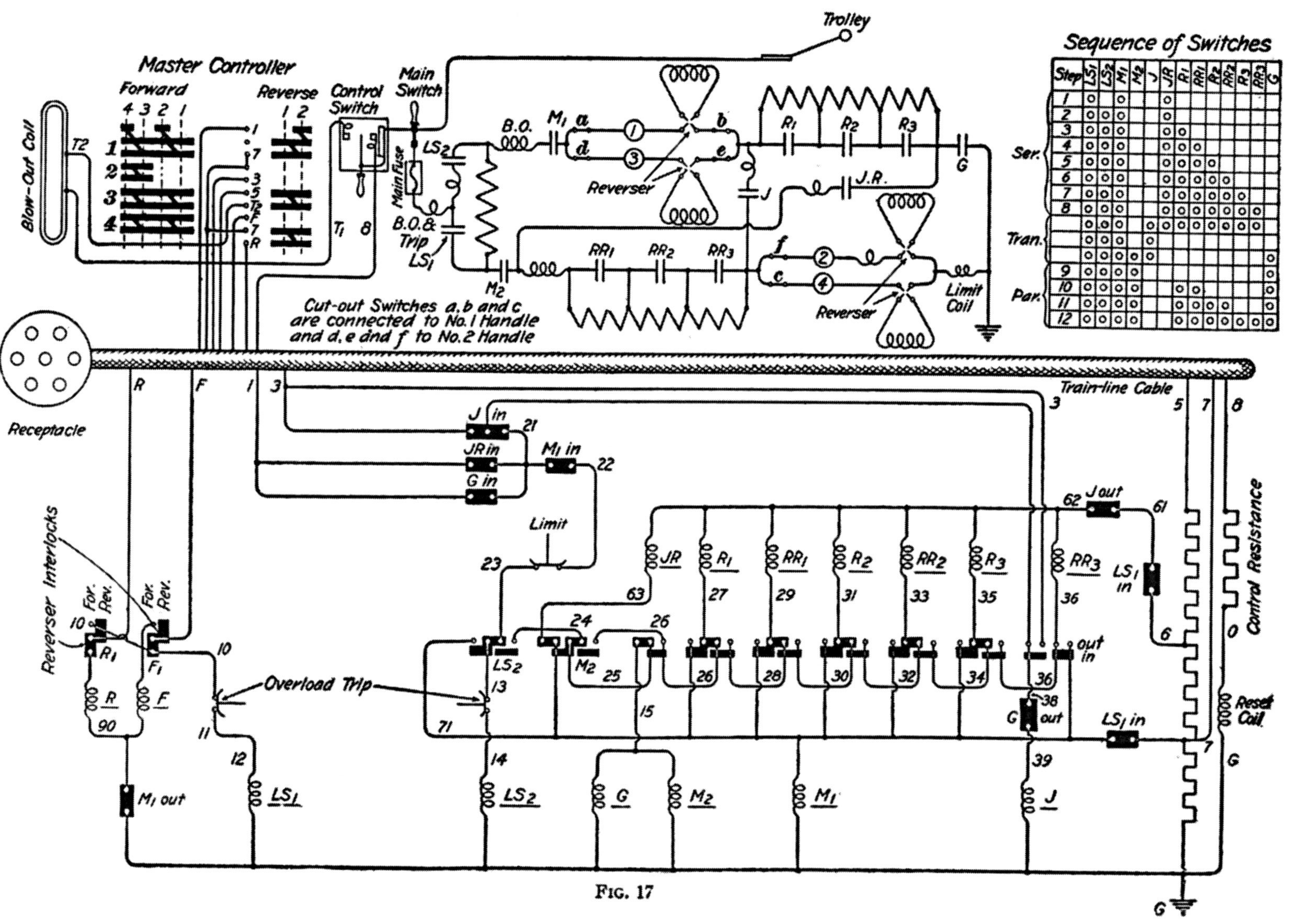
Master Controller
Forward
Reverse
Control Switch
Main Switch
Trolley
Sequence of Switches
Blow-Out Coil
Main Fuse
B.O. & Trip
Reverser
Limit Coil
Cut-out Switches a, b and c are connected to No. 1 Handle and d, e and f to No. 2 Handle
Receptacle
Train-line Cable
Reverser Interlocks
Overload Trip
Limit
Control Resistance
Reset Coil
Ser.
Tran.
Par.

FIG. 17

frequently consulted and the position of the interlocks of the switches noted for each new path. The movements of the interlocks prepare the control circuit for the closing or opening of the switches next to be affected, and for the transfer of some of the magnet valves from a pick-up circuit including the interlock of the limit switch to a retaining circuit that does not include it. The limit switch thus drops its control of the switches that have been closed and only applies its guarding features to switches about to close.

33. Fig. 18 indicates the control circuits that carry currents for energizing the different magnet valves, the positions of the interlocks (in or out) that are active in each circuit, the sequence of opening and closing of the switches, the methods of transferring some of the magnet valves from the pick-up to the retaining circuit, and which of the magnet valves are affected by the overload trip and by the interlock of the limit switch. Figs. 17 and 18 should be studied together when tracing circuits.

The first current tracing shows the circuit of the reset coil when the control switch is closed temporarily to the right. The next tracing shows the circuit when the control switch is closed to the left and the master controller is active. Current passes through the resistor and the wires *6* and *7* are energized.

The magnet valve of switch *LS1* is energized through the wire *7*, fingers *7* and *F* on the controller, and wire *F*. This switch closes, followed immediately by the closing of switches *M1* and *JR*.

The interlock of the forward magnet valve of the reverser is in forward position; therefore, the magnet *F* is not in circuit since the reverser is in proper position for forward movement of the car.

If the reverser is in the wrong position when the master controller handle is moved either way, one of its magnets *R* or *F* is automatically energized and the reverser is moved to the position corresponding to the movement of the master controller.

The operating coil of switch *LS2* is in series with an overload trip contact and the interlock of the limit switch and

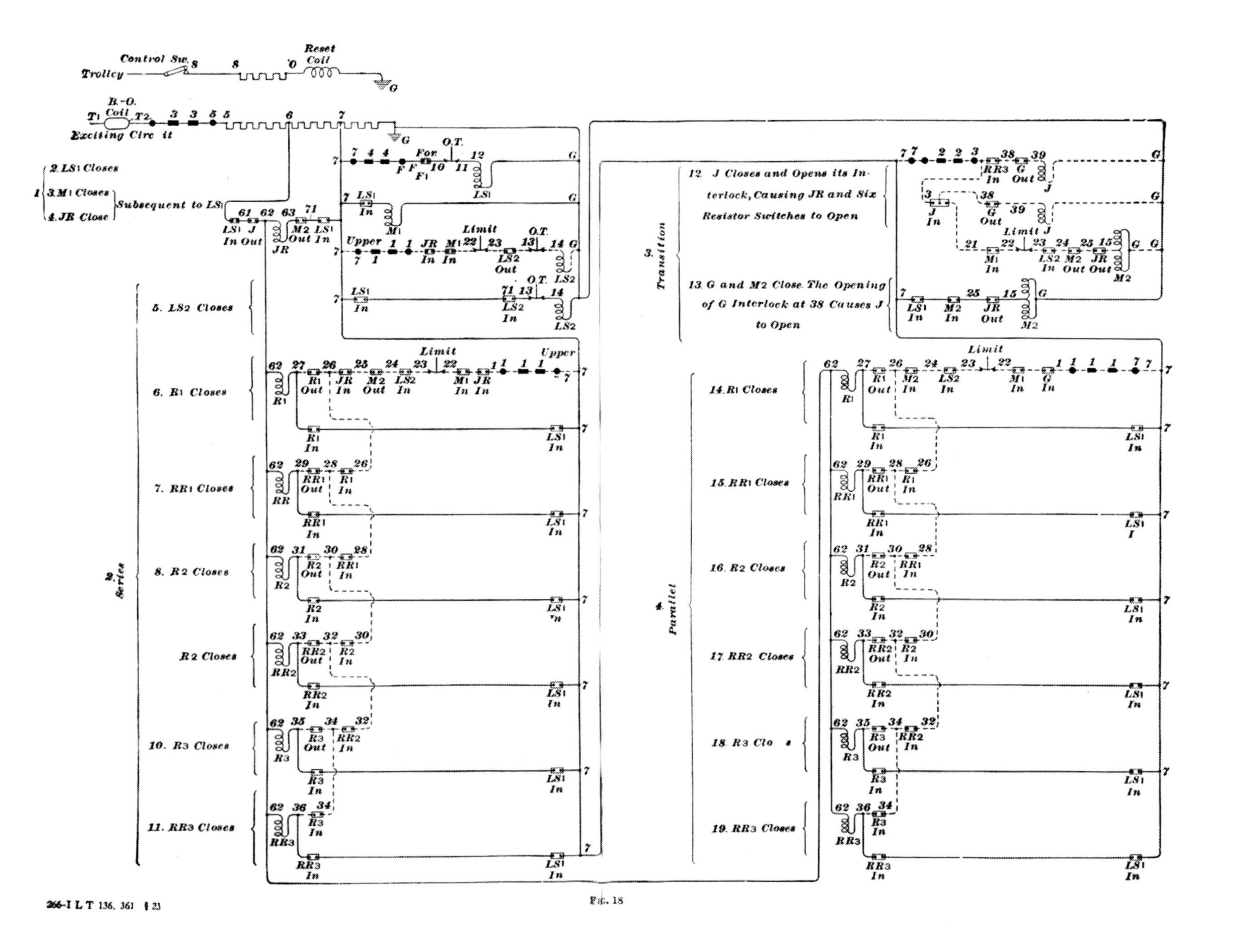

Trolley
Control Sw.
Reset Coil
B.-O. Coil
Exciting Circ it
1
2. LS1 Closes
3. M1 Closes
4. JR Close
Subsequent to LS1
2. Series
5. LS2 Closes
6. R1 Closes
7. RR1 Closes
8. R2 Closes
R2 Closes
10. R3 Closes
11. RR3 Closes
3. Transition
12. J Closes and Opens its Interlock, Causing JR and Six Resistor Switches to Open
13. G and M2 Close. The Opening of G Interlock at 38 Causes J to Open
4. Parallel
14. R1 Closes
15. RR1 Closes
16. R2 Closes
17. RR2 Closes
18. R3 Clo s
19. RR3 Closes
Limit
Upper
O.T.
For.

Fig. 18

cannot close until the motor current falls enough to allow the limit switch to close. As soon as switch *LS2* closes, its own interlock transfers its magnet valve to another circuit containing the same overload trip contact, the other overload trip contact being in series with the coil of switch *LS1*. These two switches, therefore, open in case of overload, followed immediately by the opening of all the other switches because of the opening of the *LS1* interlock in the circuit of wires *6* and *7*. The reset switch enables the motorman to start operations again. In Fig. 18, temporary connections are represented by dotted lines, as in the first circuit of magnet *LS2*.

It is well to consider the motor circuit, Fig. 17, for the various steps. Note from the table of sequence of switches the switches that are active and consider that the small gaps by which the switches are indicated are closed. Then trace the motor circuit. For instance, on the second step of the progression with switches *LS1*, *LS2*, *M1*, and *JR* closed, the two pair of motors and six sections of resistors are connected in series. Each pair of motors consists of two motors in parallel.

On steps *3* to *8*, Fig. 17, sections of the main resistor are cut out of circuit. Fig. 18 shows the control circuits that first pick up these resistor magnets in circuit with the limit switch interlock and then transfer each in turn to a retaining circuit that is not controlled by the limit switch. In case the limit interlock opens, due to an unsafe current, it does not affect the switches that have been closed, but delays further closing of new switches until the current has been reduced to a safe value by the increased speed of the motors.

Switch *J*, Figs. 17 and 18, closes on the transition positions and its interlock opens, causing the switches controlled by the circuit containing wire *6* to open. As soon as switch *G* closes, its interlock opens the circuit of wire *38*, and switch *J* opens.

On step *9*, Fig. 17, the two pair of motors are in parallel and each pair has three sections of the resistor in series with it. On the succeeding steps, the resistor sections are cut out. Switch *JR* remains inactive on the parallel steps since the interlock of switch *M2* is in its *in* position.

On step *12*, the motors are in parallel with the resistor cut out.

If the controller handle is advanced to the second notch, the switches will close automatically until the motors are in series-running position. The active switches are then indicated by step *8* of the sequence of switches.

On reverse operation, the pair of motors are in series on notch *2* of the controller. The finger *R* is then active and the reverse will be thrown to its proper position for backward motion of the car.

SINGLE-PHASE SPEED CONTROL

GENERAL ARRANGEMENT OF PARTS

34. In the United States, the development of alternating-current traction has been based mostly on the single-phase system employing a series-wound commutating motor. A single-phase system of the Westinghouse type is here considered.

The car speed is controlled by connecting successively the several taps of an autotransformer or of the secondary coil of a two-coil transformer to the terminals of the motors. The autotransformer or the primary coil of the two-coil transformer is connected between the trolley and the rails. By this means the voltage impressed on the motors is varied and the car speed adjusted. In some equipments, a resistor is also used to aid in controlling the car speed.

The voltage between the trolley and rail may be, for different installations, from a few thousand to 11,000 volts; the maximum voltage impressed on the terminals of each motor is, however, only about 275 volts. When two motors are employed, they are usually connected in parallel. With four-motor equipments, the two motors of a group are usually connected in series and the two groups in parallel.

With a two-coil transformer, the motor circuit is insulated from the ground and the motors and auxiliary apparatus are not connected to the high-tension line.

The control-circuit and motor-circuit combinations are made by means of a master controller and unit switches in the same general manner as previously described. A storage battery is used to furnish current for the control circuit and the magnet valves are, therefore, designed for direct-current operation. A small motor-generator set is provided so that the storage battery can be charged as required.

PREVENTIVE-REACTANCE COILS

35. In an equipment where no resistor sections are used to control the speed, the motor circuit is connected to the transformer taps through reactors, called **preventive-reactance coils.** This method as applied to a two-coil transformer is indicated in Fig. 19. One end of the motor circuit is connected to the center of one coil, each end of which is joined to the center of other coils, and the four ends of these two coils to the transformer taps. As the unit switches close, the terminals of the two coils nearest the transformer are connected successively to the higher voltage taps, as indicated by the dotted connection of the lower coil. The coils prevent sections of the transformer from being short-circuited and the motor circuit from being opened during the progression of the switches. The voltage impressed on the motors is thus increased in steps.

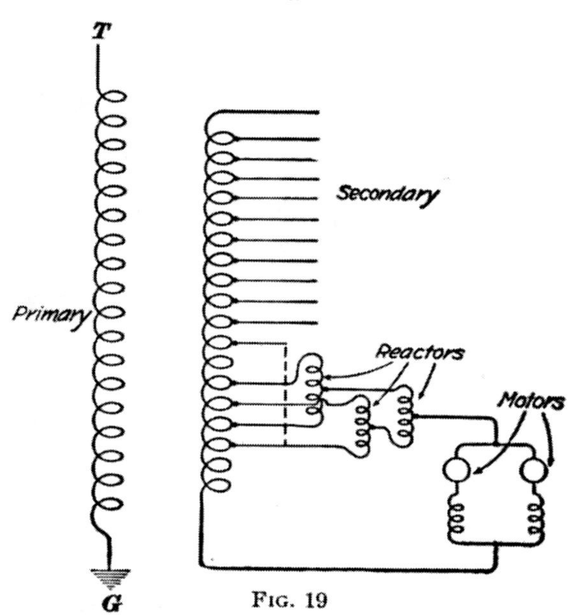

FIG. 19

ISBN #978-1-935327-99-8   1-935327-99-2
www.PeriscopeFilm.com

CPSIA information can be obtained at www.ICGtesting.com
Printed in the USA
BVOW06s0922300914

368886BV00008B/386/P